An Introdu[ction to]
Functional A[natomy]

DAVID SINCLAIR

MA (Oxford), MD (St Andrews)
DSc (Western Australia), FRCS (Ed)

Regius Professor of Anatomy
University of Aberdeen
Formerly Professor of Anatomy
University of Western Australia

FIFTH EDITION

BLACKWELL SCIENTIFIC PUBLICATIONS
OXFORD LONDON EDINBURGH MELBOURNE

ISBN 0 632 00227 1

First published 1957
Second edition 1961
Third edition 1966
Fourth edition 1970
Fifth edition 1975

Distributed in the United States of America by
J. B. Lippincott Company, Philadelphia
and in Canada by
J. B. Lippincott Company of Canada Ltd., Toronto.

Printed and bound in Great Britain by
Alden & Mowbray Ltd
at the Alden Press, Oxford

Contents

Contents

Part 2
The body as a whole

Part 3
Functional topography

Contents

Preface to the fifth edition

In this new edition the entire text has been revised and pruned in order to cut out unnecessary words and non-essential material. This has been done primarily in an attempt to prevent rising costs from rendering the price of the book unacceptable, but I hope that the resulting tightening up of the text has improved its clarity without reducing its readability or information content. The chapter on 'Size and Shape' has been deleted, but otherwise the scope of the book is as before. As always, it is intended for students of the para-medical specialties but it may also be found useful in elementary courses of anatomy or human biology for science students and others.

In preparing the new edition two illustrations have been deleted and seven new ones added; many of the others have been redrawn or relabelled. The sections on the mechanism of standing and walking have been completely rewritten, and in several other places extensive alterations have been made.

The first part of the book is concerned with the structure and activities of the various tissues and systems, and is introduced by a brief description of the general features of living cells. The second part deals in outline with some aspects of the functioning of the body as a whole. Such situations as the response to physical exercise involve the interaction of many systems, and may help the student to think of each system as part of a smoothly running whole. The third section comprises material selected from the mass of topographical information because of its functional importance. The attachments of muscles and the details of their nerve supply have been simplified, and only the main blood vessels of each region are mentioned. The objective throughout has been to present structure and function as an integrated pattern of study.

David Sinclair

Introduction

The word 'anatomy' literally means 'a cutting up', and it was in the fourteenth century that the dissection of executed criminals provided the foundations of our knowledge of what is now called **gross anatomy.** The subsequent invention of the microscope allowed anatomists to study the minute structure of the various **tissues** and **organs** from which the body is built up, and this branch of anatomical knowledge is known as **histology**.

An anatomical **system** comprises all the structures connected with a given general function, and the study of such systems is known as **systematic anatomy.** The **locomotor system** includes the bony skeleton, the joints and the muscles, and is concerned with movement. The **circulatory system** mediates the flow of blood round the body, nourishing the tissues and removing their waste products. The **respiratory system** comprises the air passages and lungs, and the **digestive system** deals with the intake and digestion of food. The **nervous system** co-ordinates the activities of the other systems and organizes their responses to changes in the environment. The **genitourinary system** is an association of the urinary organs concerned with producing, storing, and discharging urine, and the reproductive organs concerned with the procreation of children. The **endocrine system** groups together organs which release into the blood chemicals which produce effects on distant parts of the body.

Topographical anatomy deals with the form and relationships of the component parts of the body. It is commonly studied by examining the structures found in a given region, irrespective of the systems to which they belong. **Regional anatomy** is naturally very important for the surgeon. Though topographical anatomy is usually studied by dissection, a great deal can be learned by observing and feeling the surface of the living body.

Descriptive terms

Topographical description is based on the **anatomical position**, in which the body stands erect with the feet parallel and pointing directly forwards; the arms are by the sides and the palms face forwards. In this position structures which are further away from the ground are said to be 'superior' to structures which are closer to the ground; the converse of this word is 'inferior'. Thus the head is superior to the heart, and the heart inferior to the head. This relationship is still expressed in the same way even if the patient lies down, so bringing the head and the heart equidistant from the ground. We use a similar convention in map reading; one place remains north of another even though the map is twisted round.

Structures near the front of the body are 'anterior' or 'ventral' to structures near the back; the converse words are 'posterior' or 'dorsal'. The vertebral column always remains posterior to the stomach, even though the patient may lie on his face or on his back. The words 'lateral' and 'medial' refer respectively to structures further away from and nearer to the midline of the body. Something exactly in the midline is 'median'. 'Medial rotation' and 'lateral rotation' of the hip or shoulder joint are movements in which the anterior surface of the limb turns to face medially and more laterally respectively.

In the limbs, alternative terms are sometimes used. For example, in the forearm 'ulnar' is synonymous with 'medial', and 'radial' with 'lateral'. (The ulna is the medial bone of the forearm, and the radius the lateral.) The terms 'proximal' and 'distal' are often substituted for 'superior' and 'inferior'.

Notice that the words 'arm' and 'leg', which in everyday speech are used for the upper and lower limbs, are in anatomical usage restricted to the regions between the shoulder and elbow and the knee and ankle respectively.

Anatomical nomenclature

The anatomists of every country originally named structures without reference to the names employed in other countries. In 1895 an agreed international nomenclature—the Basle Nomina Anatomica—was adopted. Several countries found the Latin terms cumbersome and sometimes unsuitable, so that they subsequently began to stray from the original: the Birmingham Revised terminology of 1933 was an attempt to rectify flaws and anomalies in the B.N.A. The drift from Basle began again to make international understanding difficult, and in 1955 another internationally agreed terminology was adopted in Paris. This, the present system, is based on the B.N.A., and each country is free to make translations but not alterations of the basic Latin terms. The result of this sequence of events is that four different terminologies have been used in British dissecting rooms and operating theatres this century, and the student may hear speakers of different ages call the same structure different names. Successive editions of anatomical textbooks talk slightly different languages, and American and British books may differ over the name of a given structure.

Even though this confusion exists, the names of structures are often helpful. The name of a muscle usually gives a clue to its position, its shape, its task in the body or its construction. Sometimes the name of the supposed discoverer of a structure is commemorated, and other names derive from the sometimes rather far-fetched descriptions given by the ancients, who recognized resemblances to writing instruments, the beaks of birds, and other unlikely objects among the components of the human body. It is always important to give every structure its proper name, and these names are not to be remembered parrot-fashion, but to be thought about and made use of in the understanding of the subject. It is one of the chief virtues of anatomy as an educational subject that it demands complete precision in the use of language.

Part 1
Tissues and systems

1 · Cells and Tissues

Protoplasm

The living material of which the human body is composed is **protoplasm**, which is the basis of all forms of animal and plant life. If we examine a thin slice, or **section**, of any part of the body under the microscope, we can see that scattered thickly or sparsely through it are small circumscribed portions of protoplasm called **cells**. A cell is the smallest unit of protoplasm capable of an independent existence, and the cells are responsible for laying down and maintaining the entire structure of the body.

Between the cells lies a variable amount of intercellular substance, or **matrix**, and this may contain **fibres** of different kinds. The body also contains a great deal of fluid, some in the protoplasm, some in the intercellular spaces, and some circulating round the body in the blood and lymph or being secreted into and reabsorbed from the alimentary canal (p. 214).

Protoplasm has a consistency similar to that of raw white of egg. About 75 per cent of it is water, and it contains many inorganic salts essential to its proper functioning. But its most characteristic constituents are large organic **macromolecules** constructed by linking together smaller molecules in a repetitive way. Three types of macromolecules are of particular importance, the polysaccharides, the proteins, and the nucleic acids.

Perhaps the most important **polysaccharide** is **glycogen** (p. 210), which is formed from sugar molecules, and serves as a store of readily available energy. **Mucopolysaccharides**, compounded from sugars and proteins, are constituents of the intercellular substance in such tissues as cartilage (p. 25).

Proteins are substances of very high molecular weight formed by the linkage together of aminoacids (p. 209). **Structural** proteins are responsible for the characteristic structure of cells and tissues. A much larger group, the **enzymes**, act as catalysts and ferments building up or breaking down other complicated molecules. Enzymes are essential because the cells can only continue their existence by assimilating food brought to them, incorporating it into their own protoplasm, and getting rid of the waste products formed by their own activities. All the steps necessary to these chemical operations of **cellular metabolism** are controlled by the enzymes, which are themselves manufactured by the cell. Still

3

other proteins act as chemical messengers, or **hormones**, which are released into the blood to produce effects on distant parts of the body (p. 18).

The last group of macromolecules is the smallest. The **nucleic acids** deoxyribonucleic acid (DNA), and the several varieties of ribonucleic acid (RNA) are formed from the union of substances called **nucleotides**, which in turn contain sugars, phosphates, and purine or pyrimidine bases. The nucleic acids are concerned with the manufacture of proteins, and also with storing the information which enables this manufacture to take place. They are thus essential, not only to the working of the cell, but also to the transmission of the 'know-how' of protein synthesis from one cell to its descendants; DNA constitutes the structural basis of heredity (p. 9).

In addition to the macromolecules, protoplasm contains **lipids.** These are somewhat simpler substances which, combined with protein, are essential to the formation of surface layers within and around the cells. It is in relation to such layers that vitally important chemical reactions take place.

Cell structure

The spherical **ovum**, which is discharged from the ovary of the female during the reproductive cycle (p. 194), is about 100 micrometres (1 μm $= 10^{-6}$ metres) in diameter, and just visible to the naked eye; most other cells are much smaller. The total number of cells in the body is astronomical. To take only one example, there are some 25^{12} erythrocytes (p. 153) in the blood of an average man.

Cells vary in shape as well as size. Some, like the ovum, are spherical, others roughly cubical or cylindrical. Some have thin processes which may stretch out for a considerable distance, and some are greatly flattened or elongated (Fig. 1). These variations in shape can often be correlated with the functions which the cells perform. But where cells are closely packed together, the pressures which they exert on each other result in a tendency for them to assume a 14-sided shape. A section through a solid body having fourteen sides usually has a six-sided outline, and this is the commonest shape seen in sections of tissues where cells are in close contact.

Each individual cell is surrounded by a boundary wall called the **plasma membrane,** only about 0·01 μm thick, and made up of three layers of lipids and proteins. It is of enormous importance as a barrier between the protoplasm and the material outside. The membrane is selectively permeable, and allows only certain substances to pass in and out of the cell; if it is torn, protoplasm flows out through the rent, and if the damage is great enough the cell will die. In cells such as those which line the small intestine (p. 177), the plasma membrane becomes thrown into tiny folds covering finger-like processes of the protoplasm; these **microvilli** increase the area of membrane through which absorption can take place.

Most cells have a well-defined accumulation of denser and more viscous protoplasm called the **nucleus** (Fig. 2); the nuclear protoplasm is called **karyo-plasm**, and is separated from the **cytoplasm** which fills the rest of the cell by a **nuclear membrane** similar to the plasma membrane. The nuclear membrane is perforated at intervals, and this facilitates the passage of the bulky macro-molecules from the nucleus to the cytoplasm and vice versa.

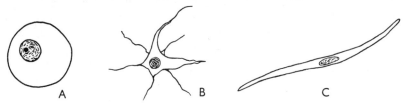

Fig. 1. Variations in the form of cells. A: ovum; B: nerve cell; C: muscle cell.

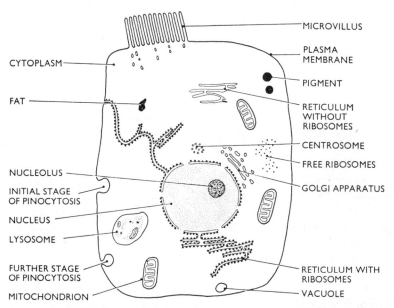

Fig. 2. Diagram showing some of the constituents of a cell.

The nucleus directs the activities of the cell. Some cells have more than one nucleus, and a few have many. If the nucleus is destroyed, the cell usually dies in a very short time. On the other hand, erythrocytes lose their nuclei in the process of specialization, and yet live for three or four months in the blood stream. But the loss of the nucleus means, as we shall see, that they have lost the power to reproduce themselves or to repair injuries to their cytoplasm or plasma membranes.

The karyoplasm of the nucleus contains a meshwork of very fine threads of protein called **chromatin**. Most cells not undergoing division also exhibit a denser rounded mass called a **nucleolus**, which lies eccentrically in the nucleus.

Both the chromatin strands and the nucleolus are composed of **nucleoproteins**—nucleic acids combined with proteins. The chromatin has a very high content of DNA, and the nucleus contains RNA, which it manufactures.

The nuclear chromatin shows irregular thickenings, but there is no clear indication of the fact that it consists of units called **chromosomes**; when the cell begins to divide, however, these chromosomes become readily distinguishable as 46 rod-like pieces of chromatin. Strung out along them are the functional units called **genes**, which store the 'information' necessary to the working of the cell. The DNA of the chromatin is a long thin molecule in the form of a double helix; this is a ladder-like arrangement twisted into a spiral, the rungs of the ladder being the purine or pyrimidine bases. The sequence of the bases forms a 'code' which controls the operations carried out by the cell by dictating the nature of the proteins it forms. Three consecutive rungs of the ladder are adequate to specify the manufacture of a given aminoacid, and a sequence of such triplets, constituting a gene, specifies the manufacture of a given protein by linking together a prearranged sequence of aminoacids.

The manufacture of protein is not carried out by the DNA directly; the molecules of DNA act as a sort of template for long thin molecules of RNA which line up alongside them, and the information carried in the genes is 'imprinted' upon the RNA molecules. These then leave the DNA and carry the instructions to the protein-manufacturing apparatus in the cytoplasm, and for this reason the particular kind of RNA involved in this transaction is known as **messenger RNA.**

The cytoplasm is just as complicated as the nucleus. It is permeated by a network of fluid-containing membranes and tubules similar in structure to the plasma and nuclear membranes and apparently connected with both. Some of this **reticulum** is associated with small bodies called **ribosomes**, which are rich in RNA. More ribosomes lie free in the cytoplasm. The ribosomes are the site of protein synthesis, the place to which the messenger RNA brings its coded instructions from the nucleus. The process of assembly of aminoacids into protein is not yet fully understood, but it certainly involves another type of RNA, which is called **transfer RNA** since it effects the lining up of appropriate aminoacids in the desired sequence, and the ultimate transfer of information from the messenger RNA to the ribosomal 'factory'.

The cytoplasm also contains small bodies called **mitochondria** (Fig. 2), which are specially numerous in cells actively engaged in chemical operations—for example, liver cells may contain hundreds of them. They are as a rule less than 0.5 μm in diameter, and can be $1-2$ μm in length. Each mitochondrion has a membranous sheath which is folded inwards to form shelf-like protrusions within its central cavity, and on these shelves are located enzymes which carry out vital chemical operations, particularly those concerned with the production of adenosine triphosphate (p. 58), the chief source of energy used by the cell. Mitochondria are therefore sometimes called the 'power-houses' of the cell.

The **Golgi apparatus** is a system of tiny tubes and sacs close to the nucleus. It seems to act as a temporary store for protein formed by the ribosomes, and it may also be concerned in the formation of polysaccharides.

The **centrosome** is a complex structure which plays an important part in the process of cell division (p. 9). It is found close to the centre of the cell, and contains two smaller masses known as **centrioles**. These are hollow cylinders composed of nine tubules, each of which is made up of three smaller tubules. This construction is very similar to that of a **cilium**. Cilia are tubular outgrowths, covered by plasma membrane, from the surface of the cell, and can move backwards and forwards actively, so sweeping away material accumulated on the surface of the cell. This cleansing action is very important in such situations as the respiratory tract (p. 159). There is evidence to suggest that in ciliated cells the centrioles multiply, migrate to the surface, and become converted into cilia.

Many cells contain small membranous sacs filled with digestive enzymes; these **lysosomes** 'digest' particles which may have been taken in by the cell through its plasma membrane.

This list of actively functioning bodies within the cytoplasm is not complete, but it serves to indicate that various functions are allocated to specialized anatomical regions of the cell. The cytoplasm may also contain various **inclusions**, composed of material which the cell has either absorbed or produced. For example, foodstuffs may be stored. Storage protein is distributed diffusely through the cytoplasm, but carbohydrate is stored as identifiable granules of glycogen. Fat is stored in special cells; a droplet of fat in the cytoplasm is added to until it comes to occupy most of the volume of the cell, the nucleus and the small amount of remaining cytoplasm being pushed to one side.

In cells which produce a secretion, **secretion granules** may be seen in the cytoplasm prior to being released. Other cells produce **pigments**, such as melanin (p. 20), which remain in the cytoplasm. Finally, cells which act as scavengers may contain foreign material, for example bacteria, which they have engulfed.

Cytoplasm is responsible for most of the behaviour regarded as characteristic of living material, even though the control of these activities ultimately resides in the nucleus. In the first place, cytoplasm can contract, and so produce movement. Not all cells move about, but some, such as certain leucocytes (p. 155), are intensely active. A process, or **pseudopodium**, is bulged out from one side of the cell, and the rest of the cell is then pulled along after this process by contraction of the cytoplasm. This ability to contract is specially developed in the cytoplasm of the muscle cells, whose sole duty is to shorten, and so to produce or resist movement. But many other cells show a more localized type of movement of cytoplasm near the periphery; the action of cilia and the lashing of the tail of a spermatozoon (p. 193) are examples. Similar localized contractility underlies the mechanism all cells use to take into themselves material which comes into contact with the plasma membrane. The membrane is made to form a cup, the lip of which surrounds the material to be taken in; this cup becomes a small spherical **vacuole**, or bubble, in the cytoplasm (Fig. 2), and eventually the wall of the vacuole breaks down and disappears. When the vacuole encloses a foreign particle, for example, soot taken from the lung, or a broken-down erythrocyte taken from the blood stream, the process is called **phagocytosis**; when the cell takes in a droplet of fluid, it is referred to as **pinocytosis.**

Another general function of cytoplasm is its responsiveness to stimulation, and is termed irritability. If a mobile cell is pricked by a needle, a pseudopodium appears on the opposite side of the cell, which moves away from the site of the injury. The opposite of this reaction to a harmful stimulus is seen when the cytoplasm is set in motion by chemical stimuli indicating the presence of foodstuffs, and the cell moves towards the source of food. The property of irritability is greatly developed in the cells of the nervous system, and also in the scavenging cells which are attracted to the scene of a bacterial invasion by the chemicals produced in the course of the battle.

Cells may use the raw materials they take in to manufacture other substances, including protoplasm itself. If more protoplasm is made than is needed for maintenance, the cell grows in size. This aspect of metabolism, concerned with building up complicated substances from simple materials, is called **anabolism**. But the cell also breaks down complicated substances, usually by oxidation or by hydrolysis, in order to produce the energy needed for other activities. This process is called **catabolism**, and results in the production of carbon dioxide and waste materials, both of which have to be got rid of as they poison the working of the cellular enzymes. Catabolism results in a diminution in the amount of material in the cell, and the cells maintain their size and activity by keeping a balance between anabolism and catabolism.

The products of cellular metabolism may thus be either stored in the cell, or extruded through the plasma membrane. If the materials extruded are useful to the body, the process is called **secretion**; if they are useless or harmful, the process is called **excretion**. Both excretion and secretion may occur either on to a surface, so that the product is lost to the body unless it is reabsorbed elsewhere, or into the tissue fluid or blood stream, which may carry the product to other parts of the body. The processes of secretion and excretion extend the influence of cells beyond the limits of their own cytoplasm. Some cells, such as the fibroblasts of connective tissue (p. 20), can manufacture intercellular material; others (p. 198) can control the activities of distant groups of cells; others still (p. 175) produce material which protects delicate surfaces from damage.

Finally, cells can reproduce themselves. The amount of material a cell contains, and thus the amount of chemical activity going on in it, increases as the cube of its diameter, but the surface of the cell, through which all the raw materials have to be taken in, only increases as the square of the diameter. It follows that when a certain size is exceeded, the efficiency of the cell falls, and some chemical stimulus of which we still know very little sets in motion the process of cell division.

Cell division

The usual method by which a cell divides into two daughter cells is called **mitosis.** In the first stage of mitosis the nuclear chromatin sorts itself out into well defined separate chromosomes. The nucleolus disintegrates and ultimately disappears. The two centrioles separate from each other and move towards the opposite sides of the cell, their movement being accompanied by the formation of very fine fibrils in the cytoplasm; these **astral rays** ultimately form a **spindle** joining the two centrioles. The nuclear membrane disappears, and the chromosomes move towards the centre of the spindle, where it can be seen that they have all divided longitudinally into two daughter chromosomes, each of which becomes attached to one of the astral rays. Eventually the daughter chromosomes are dragged by the astral rays to opposite sides of the cell, where they form the chromosomal basis of two separate nuclei, and ultimately submerge their identity in typical chromatin networks. Each new nucleus then forms a new nucleolus, and becomes surrounded by a nuclear membrane. Outside, in the cytoplasm, the astral rays disappear, and each centriole divides into two to form a new centrosome. Lastly, the whole cell constricts to form a 'waist', the cytoplasm becomes divided into two, and each new cell acquires a complete plasma membrane (Fig. 3). The various contents of the cytoplasm are more or less evenly distributed between the two daughter cells, which at first are somewhat smaller than their parent was originally; the deficiency is soon made up, and the offspring become indistinguishable from the parent.

The crux of the whole operation is the splitting of the nuclear material in such a way that the genetic content of the daughter nuclei is identical with that of the parent. It is accomplished by the duplication of the DNA molecules of the nucleus before division. The two intertwined chains of the DNA 'ladder' split apart by breaking their attachments at the 'rungs' (p. 6), and each separate chain then builds on itself a complementary chain which exactly reduplicates the original partner. In this way a double set of chromosomal material is formed prior to mitosis, and the genetic sequences which hold the 'instructions' passed on to the daughter cells are identical with those of the mother cell.

The process of replication of chromosomal material every time the cell divides is usually faultless in early life, but in old age (p. 224) disturbances tend to appear, so that some cells may have more, or less, than their normal complement, and may also contain chromosomes of abnormal appearance. This also happens in the cells of tumours.

The genetic material determines our hereditary endowment, since the instructions we receive in the fertilized ovum from which we grow are transmitted to every cell in the body, among them the sex cells which are to unite to form another individual. But if the mature male and female sex cells contained the same amount of nuclear material as the ordinary cells of the body, the new cell resulting from their fusion would have twice as much chromatin. To obviate this there is a special kind of 'reduction division' or **meiosis** in the course of development of both male and female sex cells. This extremely complex process results in each of the mature sex cells coming to contain half the number of chromosomes possessed by the cells of the rest of the body. At fertilization the original number is restored when the nuclear content of the spermatozoon is added to that of the ovum.

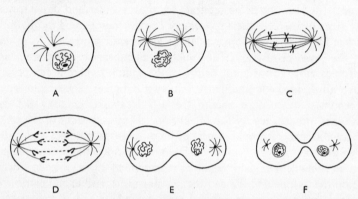

Fig. 3. Scheme showing mitosis: only four chromosomes are shown for the sake of clarity. A: normal parent cell with a nucleus showing threads of chromatin; the centrioles are beginning to move apart and the astral rays are forming. B: the chromatin has broken up into chromosomes which are not yet distinctly separated; the nucleolus has disappeared and the nuclear membrane has disintegrated. C: the chromosomes have partially split into two and have moved towards the centre of the spindle. The daughter chromosomes now move towards the two centrioles (D), and ultimately form two new chromatin networks (E), round which nuclear membranes form as the cytoplasm constricts (F) and finally divides.

Each cell of the normal female body (except the mature sex cells) contain 46 chromosomes, made up of 44 **autosomes** and 2 similar **sex chromosomes**, which are called X chromosomes; this arrangement can be written 44 XX. One of the X chromosomes may be visible as a darkly staining spot of material, about $0.5-1$ μm in diameter, usually close to the periphery of the nucleus. The other is not so tightly coiled on itself, and cannot be detected in the resting cell.

Normal male cells also have 44 autosomes and 2 sex chromosomes, but the two sex chromosomes differ markedly from each other. One is similar to the female X chromosome, but the other is much smaller, and is called the Y chromosome; the formula for a male is thus 44 XY. The Y chromosome contains the genes which result in maleness, and in its absence the embryo will develop as a female.

After meiosis, every ovum contains 22 autosomes and one X chromosome; there is no other possibility. However, spermatozoa can be of two kinds, for during meiosis the X chromosome is pulled into one daughter cell and the Y chromosome into another. Half the spermatozoa thus have the formula 22 X and the other half have the formula 22 Y. Should a 22 X spermatozoon meet and fertilize the ovum (22 X), the result will be a cell containing 44 autosomes and 2 X chromosomes, and will thus develop into a female; if a 22 Y spermatozoon should be successful, the result will be a 44 XY individual—a male.

Differentiation

Fertilization of the ovum not only restores the full complement of chromosomal material, but also triggers off a series of mitotic divisions. For a time all the resulting cells have equal potentialities. Thus, the daughter and even the granddaughter cells of the fertilized ovum are each capable of forming a whole and complete normal individual; they are said to be **totipotent**.

But as the cluster of cells formed by successive divisions increases in size, so specialization begins to appear. Certain cells adapt themselves to the task of lining the alimentary canal, others dip in from the surface of the cluster to form the nervous system, still others form blocks of tissue from which the muscles will develop, and so on; these are **multipotent** cells, and though they have wide potentialities, they cannot form a whole normal body.

In the adult body most cells are still more specialized; a cell which secretes acid into the stomach looks quite unlike, and is totally unsuited to perform the functions of, an absorptive cell lining the small intestine, even though both are derived from the same multipotent ancestors. A cell specialized to perform a given task is said to be **differentiated**, and this results in at least an impairment, and occasionally a complete abolition, of the power to reproduce. But differentiated cells are constantly being lost, either through normal ageing and death, or through local injuries; if normal function is to be preserved, these losses must be made good. In many tissues, therefore, there are **stem cells** which are not fully differentiated and can divide to form new cohorts of differentiated cells.

Some differentiated cells have a very transitory existence; for example, cells lining the small intestine die and are replaced within 48 hours. The stem cells in these regions are thus very active. In contrast, nerve cells are never replaced when they die, and each such death means a permanent structural and functional loss. It follows that the size of the nervous system is gradually reduced with increasing age, since in adult life something like 10,000 nerve cells die every day.

Since every cell division after fertilization (apart from the meiosis of the sex cells) involves the exact replication of the genetic material, it follows that every cell in the body contains all the information necessary to the operation of every other cell; the cells of the skin have stored within them the instructions for producing saliva, and the cells of the liver contain the instructions for determining the colour of the eyes. This information is there, but is not used. What is it that stops a nerve cell from secreting bile, or a kidney cell from forming bone? The present theory is that in a specialized cell many of the genes, though still present, are 'suppressed' by some unknown process, and that different cells suppress different groupings of genes, so allowing the remaining active genes to control the activities and structure of the cell. Under certain circumstances, such as injury, some of the suppressed genes may become active again, so altering the behaviour of the cell. If this theory is correct, differentiated cells do not make use of all their capacities, and if we knew how to activate some of the suppressed genes we might be able to improve the quality of wound healing, to allow amputated limbs to regenerate, and to improve the performance of diseased cells. At present, however, this is a dream for the future.

Tissues

The three basic components of the body—cells, intercellular substance, and fluid—combine in varying proportions and arrangements to form **tissues**, which are specialized to perform certain functions.

In some tissues, such as muscle, the prominent feature is the cell content; in others, such as cartilage, the intercellular matrix is of great importance. In tissues such as tendon, the fibres produced by the cells have the main functional significance, and in blood the high fluid content plays a vital part.

Despite the great number of tissues in the body, there are really only four basic types—**epithelia, connective tissue, muscle,** and **nervous tissue.** These are combined in various ways, and with various local specializations, to form functional units which may either be diffusely scattered throughout the body—for example, the conducting tubes of the vascular and lymphatic systems penetrate almost everywhere—or localized to form circumscribed **organs** (viscera).

Epithelia form sheets of cells which cover the external or internal surfaces of the body; they are also the active cellular constituents of secreting glands. Connective tissue may be either general or special; the second heading embraces such materials as cartilage, bone, bone marrow, blood, and lymphoid tissue. Muscle is subdivided into three main groups, in all of which the cells are specialized to contract. Finally, nervous tissue consists of cells which are specialized to respond to irritation and to conduct messages.

When a tissue or organ is not used for some time, the specialized cells which it contains becomes smaller, and may be reduced in number. This process is well seen in paralysed muscles (p. 92); if the muscle is not kept artificially exercised it 'wastes' and becomes **atrophied**. Conversely, if a normal muscle is actively exercised over a period of time, the individual cells in it become larger so that the whole muscle swells—a process called **hypertrophy**. The largest muscle cells do not enlarge further, but the smaller cells are brought up to the same size as the largest ones.

Injury to any tissue at once stimulates the connective tissue cells in the vicinity to action. The injured region becomes filled with blood clot which is then invaded by fibroblasts (p. 20) which manufacture a meshwork of fibrous material. The fibres in this meshwork later contract to form a tough **scar** firmly uniting the edges of the wound. Healing by **scar tissue** of this kind is the rule throughout the body. When, however, there has been a loss of specialized cells, an attempt may be made by their remaining representatives to proliferate. Tissues vary greatly in their ability to do this. Nervous tissue cannot replace itself at all, but if a third or even more of the liver is removed at operation, the specialized cells can return to their normal numbers within a short time.

If it is not possible to make up the normal number of cells, the remainder may hypertrophy to effect a functional repair. Most tissues and organs have a wide safety margin, for only a relatively small proportion of the cells in them are ever working at any given time. Although kidney cells do not replace themselves, one kidney may safely be removed, for the other, by hypertrophying and by working its cells a little harder, can perform the functions usually carried out by both.

2 · Epithelia and glands

Epithelia

The sheet of cells covering any exposed surface of the body, whether external or internal, is called an epithelium. It consists of cells which are in intimate contact with each other, and contains just enough intercellular matrix to cement the cells together. A **simple** epithelium is only one cell thick; a **stratified** epithelium has several layers of cells. Immediately deep to the epithelium in either case there is usually a **basement membrane**, a sheet of non-cellular fibrous material and matrix which strengthens and supports the epithelial cells which are attached to it.

Three types of simple epithelium are recognized, according to the shape of the cells. In simple **squamous** or **pavement** epithelium the cells are flattened plates with irregular margins which fit into each other like the pieces of a jigsaw puzzle (Fig. 4A). This kind of epithelium lines the ends of the smallest air passages in the lungs, the blood vessels and lymphatics (p. 133), and the body cavities, such as the pleura, pericardium and peritoneum (pp. 140, 165, 181). The nucleus is usually central, and makes the flat cell bulge slightly; in surface view the general outline of the cells is very roughly hexagonal (p. 4).

Cuboidal epithelium differs from pavement epithelium only in the fact that the cells are about as thick as they are wide, so that in vertical section they appear square. It is found as the secreting cells of many kinds of glands (p. 16) and as the lining of their ducts.

The third type of simple epithelium is **columnar** epithelium, in which the cells are taller than they are wide; it is found in such places as the lining of the alimentary tract. The cells are packed tightly together in a honeycomb pattern (Fig. 4B), just like the pavement cells, and surface view again shows hexagons. Columnar cells are frequently highly specialized; for example, they may be ciliated (p. 159), or contain pigment, or develop into sensory receptors, or exhibit microvilli (p. 4).

Stratified epithelium is categorized according to the shape of the cells in the most superficial of its layers. **Stratified cuboidal** epithelium occurs in the lining of the urinary tract from the kidney down to the upper part of the urethra (p. 191), where it is often called **transitional epithelium** (Fig. 4C). There is no

14

apparent basement membrane, and when the cavity which the epithelium lines is empty there are about six layers of cells; when it is fully stretched, the layers diminish to one or two. The cells appear to slip over each other, lubricated by a slimy intercellular material.

By far the most extensive stratified epithelium in the body is **stratified squamous** epithelium, in which several layers of cells become successively more and more flattened towards the surface. This type of epithelium is found wherever there is hard usage; if forms the superficial covering of the skin, and lines the mouth and upper part of the alimentary canal. It is also found in the urethra (p. 191), the anus (p. 184) and the vagina (p. 194). The cells in the more superficial layers are formed by the division of stem cells in the deepest layer, the formation of new cells forcing the older ones further towards the surface. As

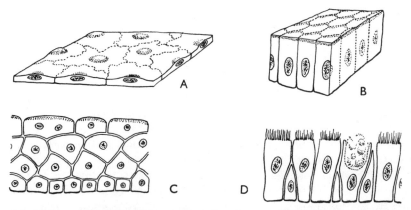

Fig. 4. Types of epithelium. A: pavement; B: columnar; C: transitional; D: ciliated columnar (among the ciliated cells lies a goblet-shaped cell producing mucus).

they rise, they lose their nuclei and accumulate in their cytoplasm granules of a horny protein called **keratin**. In some situations the whole cell becomes converted into a flake of keratin, and one must distinguish between **keratinized** epithelium, such as that found in the palms of the hands and the soles of the feet, and **nonkeratinized** epithelium, such as occurs in the mouth and oesophagus. The outer layer of dead and dying cells has a protective function, for keratin withstands friction, is waterproof, and resists bacterial invasion (p. 201).

Pseudostratified epithelium appears to exhibit layering of cells, but actually consists of cells of varying heights, all of which are attached to the basement membrane, though some do not reach the exposed surface. This kind of epithelium is found in the upper part of the respiratory tract; the exposed cells are columnar and carry cilia, which play an important part in cleaning the surface (p. 159).

Some surfaces in the body are covered by layers which are not true epithelia. The most important of these are the lining membranes of joints (p. 40) and bursae (p. 42), which are really specialized connective tissue surfaces.

The surface area of a sheet of epithelium through which absorption takes place is often increased by various devices; for example, the lining of the small intestine is thrown into folds, and the epithelium covers innumerable finger-like processes or **villi**, each containing blood vessels and lymphatics in a connective tissue base. The same principle is used to increase the absorptive surface area of an ordinary bath towel. These villi, which are about 0·5 mm long, and can be seen with a hand lens, must be contrasted with microvilli (p. 4), which are processes of the cytoplasm of individual cells, and are only distinctly visible with the aid of an electron microscope. Nevertheless, both have the same purpose, to increase absorptive capacity.

Glands

A **gland** is a circumscribed collection of epithelial, or rarely of other, cells producing a secretion (p. 8).

Exocrine glands

The exocrine glands deliver a fluid or semifluid secretion, either directly or by means of tubes known as **ducts**, on to the surface of an epithelium. The cells of an exocrine gland can be arranged in various ways. In the simplest form, every cell of the epithelium may contribute, but in others only certain parts of it may be specialized to secrete.

In all internal epithelia the surface is kept moist, in order to protect the delicate surfaces of the cells. Small rounded **goblet cells**, scattered about at intervals in the epithelium, produce a moist sticky material called **mucus**, containing mucopolysaccharides. Their efforts may be supplemented by larger aggregations of cells known as **mucus glands**, which form secretory patches among the surrounding epithelial cells. Internal epithelia are thus often called **mucous membranes**, or **mucosae**. Other cells in a mucous membrane produce a more watery, or **serous** secretion, which may have actions other than mere protection; the alkaline secretion of the duodenal epithelium (p. 178) neutralizes the acidity of the material fed into the duodenum by the stomach, and the serous secretion of the small intestine contains digestive enzymes.

Specialized patches of secretory cells can become recessed from the surface in a number of ways, diving into the underlying connective tissue (Fig. 5). This allows the total mass of secreting cells to be greatly increased.

If the secreting cells are carried some distance away from the surface where their secretion is needed, the portion of the ingrowth close to the surface becomes a duct, which may or may not add its complement of secretion to that of the main body of cells, though its chief function is that of conducting the secretion to the appropriate place. There may be one or several ducts. If they do not branch, the gland is a **simple** one; if they do, the gland is **compound**. If the active cells at the end of the ducts lie in the walls of minute structures like testtubes the gland is **tubular**, but if the blind ends of these testtubes are blown out into tiny sacs or **alveoli**, the gland is **alveolar**. Most glands are classified as a mixture of the two, since usually the terminal tubules of the duct system are involved in the secretory process along with the alveoli, or else both tubular and alveolar units are present (Fig. 5D).

The sweat glands (p. 204) are simple tubular glands; the lacrimal gland (p. 125) which produces the tears is a compound tubular gland. The parotid salivary gland (p. 171) is an example of a compound alveolar gland, and the breast (p. 205), the pancreas (p. 179) and many other glands are compound tubuloalveolar glands.

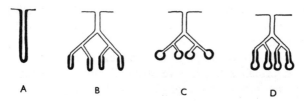

A   B   C   D

Fig. 5. Types of exocrine glands. A: simple tubular; B: compound tubular; C: compound alveolar; D: compound tubuloalveolar.

Finally, the liver (p. 178) is unique among exocrine glands in having its secretory cells arranged in branching cords forming a network or **reticulum**: it is therefore called a **reticulate** gland.

Only the smallest glands can be accommodated actually in the surface where the secretion is needed; the larger the gland, the further it must remove itself from the scene of action, and the longer becomes its duct. This implies that large glands must have serous secretions, since the passage of thick sticky material like mucus along a lengthy duct would lead to a risk of blockage.

The method of secretion of the individual cells varies with the consistency of the material produced. Thick secretion such as sebum (p. 204) may accumulate within the cell until it bursts and dies; the process is called **holocrine** secretion. In **apocrine** secretion only part of the cell disintegrates and is shed with its contents; this happens in the breast. The nucleus of the cell is left intact and the cell does not necessarily die. **Epicrine** secretion is easier still; the thin watery secretion passes through the intact plasma membrane without damaging the cell in any way. The majority of the exocrine glands secrete in this manner.

The secretion of exocrine glands naturally depends on the amount of blood they receive, since all their raw materials arrive via the blood stream. The volume of blood supplied per unit time is controlled by the autonomic nervous system (p. 86), so that secretion is indirectly under nervous control. In certain glands, such as the sweat glands, autonomic nerve fibres end around the secreting cells, and stimulate them directly. Other exocrine glands, for example the breast, are stimulated to produce secretion by hormones (p. 4).

Endocrine glands

Endocrine glands pour their secretion, not on to a surface or into a duct, but into the blood stream. Most of them consist of epithelial cells which have migrated from their original position, but have lost their original connection with the surface. A few, such as the interstitial cells of the gonads (p. 200) are not epithelial cells at all. Taken together, the endocrine glands form a chemical system for integrating the activities of the tissues throughout the body.

The endocrine glands are naturally provided with a very rich supply of blood. Because of the necessity for intimate contact with the blood, their cells tend to be arranged in branching cords or plates separated by large vascular channels; this reticulate arrangement is similar to that of the cells in the liver (p. 178). The **hormones** which the glands produce circulate round the body and produce effects on distant 'target' organs.

All endocrine glands have a tremendous reserve of tissue, and their satisfactory performance can survive serious losses; for example, about three-quarters of the suprarenal cortex (p. 200) or of the anterior lobe of the pituitary gland (p. 198) may be removed without functional disturbances resulting.

The simplest endocrine glands, such as the **interstitial cells** of the gonads (p. 193), consist of scattered individual cells. The **islets of Langerhans** (p. 200) are little isolated clumps of endocrine cells in the pancreas, which is an exocrine digestive gland. Finally, there are large and well organized glands such as the **thyroid** gland (p. 199). Each **suprarenal** gland (p. 199) consists of two distinct endocrine glands, one packed inside the other.

3 · Connective tissue

Connective tissue is the relatively generalized material which surrounds and permeates the more highly differentiated tissues. It binds together and supports weaker and more delicate tissues and organs, and is therefore sometimes called **supportive** tissue. Cartilage and bone are supportive tissues of a special kind, and will be considered in separate chapters.

The outstanding characteristic of connective tissue of all kinds is the preponderance in it of intercellular matrix, which is of very varying consistency. Connective tissue proper consists of a meshwork of fibres among which several different kinds of cells lie bathed in **tissue fluid**. In some types of connective tissue the cells are scanty, and in others they are plentiful.

There are three kinds of fibres. The commonest are the white **collagen fibres** which are composed of the structural protein collagen, and are very resistant to being stretched. Each fibre, which may measure up to 30 μm in breadth, is composed of bundles of **collagen fibrils**, which are less than 0·5 μm in breadth. A fibril may leave its parent fibre and run to join another, but individual fibrils never branch. Collagen fibres are found in bulk in such structures as ligaments and tendons (p. 23), which are constantly being subjected to tension, but they occur throughout the body. Boiling them produces gelatin, and tannic acid converts them into tough insoluble material; this is the basis for the tanning of leather.

The collagen fibres in connective tissue are not normally at full stretch, so that they appear wavy and relaxed under the microscope. Pulling such fibres straightens them out, and eventually breaks them; when they do break, the fibres fray like those of a rope.

The second kind of connective tissue fibre is yellow, and is made of **elastin,** a protein with elastic properties. Unlike collagen fibres, **elastic fibres** never occur in bundles, but are always single; they branch frequently and form a network. They are under tension even at rest, and so appear tight and straight in a section; if one is broken, the ends recoil and curl up. Elastic fibres are larger (about 1–4 μm in diameter) than collagen fibrils, but much smaller than the average collagen fibre.

Elastic fibres occur in large numbers in the ligamenta flava of the vertebrae (p. 249), which for this reason look yellow. The elastic fibres of the skin allow it to adapt itself to the varying positions of the body which it clothes; during life these fibres progressively lose their elasticity and the flaccid skin of old age is their memorial. Elastic fibres are of the great functional importance in the lungs (p. 165) and in the larger arteries (p. 133).

Reticular fibres are considerably smaller than collagen fibres (less than 1 μm in diameter), and branch to form a fine network. However, they resemble collagen fibres in being composed of inelastic fibrils, and there is very little difference between collagen and **reticulin**, the protein of which the fibres are composed. It is believed that reticular fibres can become converted into collagen fibres in a healing wound.

The matrix in which these three kinds of fibres are dispersed contains varying quantities of water and of mucopolysaccharides (p. 3). The tissue fluid which permeates this material is derived from the plasma (p. 152) of the blood, and contains proteins, salts, metabolic products and gases (carbon dioxide and oxygen).

Many different kinds of cells are found in connective tissue. One of the most numerous is the **fibroblast**, a thin, flat, branching, irregular cell which forms the collagen fibres, the intercellular matrix, and possibly also the reticular and elastic fibres. The present view is that the fibres 'crystallize out' of the matrix in the vicinity of the cell. The collagen fibres are laid down along the lines of strain or tension in the connective tissue, so that they are in the best position to resist whatever forces normally act on it. Thus, in the dermis of the skin they tend to lie in parallel bundles, usually longitudinal in the limbs and circumferential in the trunk. Incisions made along the lines of these bundles heal with much less scar tissue than those made at right angles to them. During healing of a skin wound, the fibroblasts forming the scar tissue line up in such a way that the fibres they produce will unite the raw surfaces (p. 23).

The next most abundant cells in connective tissue are the **histiocytes**, which are scavengers specialized in the direction of phagocytosis (p. 8). Some of them are mobile, and wander about actively picking up any strange material, such as particles of India ink introduced by tattooing. Other histiocytes wait in fixed positions to snap up whatever comes past them. Even these, however, can move about if need be. The connective tissue histiocytes form part of the **macrophage system** of cells whose duty it is to attack and destroy invading micro-organisms. Such material, once taken in, is dealt with by the numerous lysosomes (p. 7) in the cytoplasm, but inert foreign material may merely be deposited again in the vicinity.

Fat cells are conspicuous. They resemble fibroblasts to begin with, but fat accumulates in their cytoplasm in a progressively enlarging vacuole (p. 8) which comes to occupy most of the cell. The fat can be discharged again when needed, by the reversal of the mechanism by which it entered the cell.

Other kinds of cells are found in connective tissue. The **pigment cells** of the skin and the iris of the eye manufacture granules of **melanin**, a black pigment which remains in their cytoplasm and protects the deeper tissues against the damaging effects of sunlight. **Leucocytes** from the blood (p. 155) may wander into connective tissue after escaping from the capillaries. A few **plasma cells** normally live in connective tissue, but they are not prominent, except when there is some invasion of the tissue by bacteria. These cells contain many rib'osomes, and manufacture antibodies which are part of the chemical defence system of the body (p. 149). There are also a few **mast cells**, which are fairly large and contain large granules; they produce **heparin** (which hinders coagulation of the blood), and two other important substances, histamine and serotonin.

Finally, the small and undistinguished looking **mesenchymal cells** are stem cells, which, if stimulated to divide, can replace all the other types of connective tissue cells. They provide many of the extra fibroblasts which lay down scar tissue during healing and many of the connective tissue macrophages which deal with infection.

Loose connective tissue

Three main varieties of connective tissue have a loose arrangement of fibres and a considerable fluid content. In **reticular** connective tissue reticular fibres are associated with fibroblasts and macrophages in a fine meshwork. Such tissue occurs in the bone marrow, the liver, the various lymphoid organs (p. 148), and in relation to the basement membranes of the epithelia of the alimentary and respiratory systems.

Areolar tissue is found all over the body, for it is the main packing material between other tissues, and its loose texture allows movement to occur between them, while its elastic fibre content ensures a return to the original relationships once the movement is over. For example, the deeper layer of the skin merges into a layer of areolar tissue called the **superficial fascia**, in the interstices of which lie many cells containing fat. The skin can thus be moved with ease over the underlying structures in most regions of the body. Here, as elsewhere, areolar tissue provides pathways for the vessels and nerves, and its mobility helps to preserve them from injuries which they might suffer if they ran in rigid fixed positions. In the pads of the fingers, the toes, and the heel, the superficial fascia contains many dense fibrous partitions which divide it up into a kind of honeycomb arrangement enclosing small collections of fat. A similar arrangement occurs in the portion of the buttock which takes the weight on sitting down. These loculations are probably protective devices in areas subjected to friction or pressure. In other regions, such as the face, the neck and the scrotum, muscle fibres develop in the superficial fascia.

Areolar tissue contains more different cell types than any other kind of connective tissue, and the spaces in its texture are filled with tissue fluid; the main fibres present are collagen fibres.

Adipose tissue, in which fat cells predominate, is also a packing material, and may help to cushion important viscera against shocks applied to the body from outside. The fat of the superficial fascia forms an insulating layer which retains the body heat, and so protects the body against external cooling. The deposition of fat varies with the state of nutrition of the body, and such tissues as the superficial fascia may be classed as adipose tissue when fat is plentiful, or areolar tissue when it is not. There is relatively more subcutaneous fat in the female, and its distribution is one of the secondary sexual characteristics. Thus, at puberty much fat is deposited round the mammary gland (p. 205), as well as in the buttocks and hips. Later in life, fat is deposited in the abdominal wall, the upper part being favoured in males, and the lower part in females. In both male and female there is very little fat under the skin of the nose and ears.

Fatty deposits are found in many deeper situations, for example, a packing of fat surrounds the kidney and helps to maintain it in position. In such situations the fat is found in localized collections called **lobules,** which may be separated by areolar tissue. There is no fat in the lungs or in the nervous system.

A special kind of **brown fat** is found in the human embryo and newly born infant, but not in the adult; it resembles the fat which is found in hibernating animals, and is concentrated in the region between and around the scapulae, in the neck, and in the posterior abdominal wall. It is thought to act as a store of easily available energy-providing material which can be catabolized to maintain body heat in infants, who cannot shiver when they are cold.

Dense connective tissue

In striking contrast to the fatty areolar tissue of the superficial fascia is the layer of connective tissue immediately deep to it, in which the fibres predominate. In some regions this **deep fascia** forms a thin glistening membrane of no great strength, but in the limbs it makes a tough dense sleeve which surrounds the muscles and dips between them to gain attachment to the bones. The partitions or **septa** formed so split the limb up into 'compartments', and to these septa many of the muscles are attached (Fig. 6). Other dense fascial partitions of this kind are found between the muscles of the back.

At the wrist and ankle the deep fascia becomes strengthened to form **retention bands** which strap the tendons down and prevent them from springing away from the bones. It can also form pulleys and slings to alter the pull of tendons or check their action. As the deep fascia passes over bony prominences it tends to adhere to them, and in this way the enveloping sheaths of the limbs are stabilized and supported. The deep fascia is not easily penetrated by fluid, and fluids such as pus therefore tend to take the line of least resistance and spread along the 'fascial planes'.

Deep fascia is made of interwoven sheets of collagen fibres arranged so as to resist stresses applied from several directions. It contains few cells (mostly fibroblasts). Other examples of this kind of dense tissue are the **capsules** which surround many internal organs, the **periosteum** which envelops bone, and the

dura mater which surrounds the nervous system (p. 99).

In other dense connective tissues the collagen fibres are arranged in parallel bundles designed to resist stresses applied mainly in one direction. A typical example is the **tendon** of a muscle (p. 54). Such tissue has a greater tensile strength than bone of the same diameter; a tendon 5 mm thick can support a weight of nearly 500 kg before breaking. The tendons of some muscles are spread out flat to form **aponeuroses** (p. 54), and these are similarly constructed, though the fibres do tend to run in slightly different directions. **Ligaments** are bands of similar tissue which have no connexion with muscle and are developed where great strength is needed to hold structures together. All such structures contain no elastic fibres and so cannot be stretched beyond their original length without rupturing (p. 48). It also follows that they can only be effective as protective and restraining agents when they are pulled tight (p. 46).

Fig. 6. Diagram of a cross-section through the thigh. Under the skin lies the fatty layer of superficial fascia, and deep to this is a complete enveloping layer of deep fascia, which is greatly thickened laterally to give attachment to muscles. Passing in from this layer are three fascial septa which are all attached centrally to the femur, and thus divide the thigh into three muscular compartments.

In contrast are the unusual structures called **elastic ligaments**, in which there is a parallel arrangement of elastic fibres. The ligamenta flava have already been mentioned (p. 19), and another example is provided by the vocal cords (p. 161). Such ligaments can maintain tension under varying degrees of stretch, and can recover their original length when the pull is relaxed.

Some structures receive the name of ligament although they are not strictly ligaments at all; the inguinal ligament is simply the tendinous edge of the external oblique muscle of the abdomen (p. 280), and the round ligament of the uterus (p. 282) is the functionless remnant of an embryological structure.

Regeneration of connective tissue

Following injury, fibroblasts, which are normally sluggish cells, migrate to the site of the damage. They recover the ability to divide, and form new matrix and collagen fibres. Some new elastic fibres may also be produced, but not in such numbers as to restore normality. At the same time mesenchymal stem cells divide to produce the necessary specialized cell types. In this way areolar tissue can be roughly repaired, but the result is usually a dense fibrous scar, for it seems to be easier to make new collagen fibres than to make other fibres or cell types. Scar tissue of this kind is very liable to stretch during its formation, but afterwards it tends to contract, so pulling the edges of the wound together and minimizing the size of the defect. Fat cells do not divide; all new fat cells derive from mesenchymal cells. A tendon or ligament repairs itself in much the same way; fibroblasts and blood vessels grow in to the damaged areas and new fibres are laid down along the lines of stress. These fibres are at first thin, but later thicken up by the addition of new fibrils. Elastic ligaments are generally repaired by collagenous tissue rather than by new elastic fibres.

4 · Cartilage

Cartilage is a supportive connective tissue in which the intercellular matrix is prominent and more solid than in ordinary connective tissue; it is principally composed of **chondromucoid**, a protein contain large amounts of the mucopolysaccharide **chondroitin sulphate.** Within the matrix there is a fine groundwork of collagen fibrils, which sometimes combine into collagen fibres. If there are few fibres, the matrix looks structureless and glassy, and the tissue is called **hyaline** cartilage (Fig. 7); if there are many, it is much more opaque, and is called **fibrocartilage**. A third type of cartilage contains not collagen but elastic fibres, and this uncommon tissue is called **elastic** cartilage.

The specialized cartilage cells or **chondrocytes** are to be found lying in cavities (**lacunae**) in the matrix, and the cartilage is surrounded by a dense connective tissue covering called the **perichondrium.** The outer layer of this is mainly collagen fibres, and the inner layer contains stem cells called **chondroblasts**, which during growth divide and give rise to the chondrocytes. The chondrocytes in turn surround themselves with matrix, so that newly formed cells at the surface of the growing material gradually sink in towards the centre as successive layers of material are formed external to them. This is a process called **appositional growth**, and pushes the perichondrium further outwards. The chondrocytes in the centre of the growing mass continue to divide for a time and produce more matrix; this is called **interstitial growth** (Fig. 8). The chondrocytes near the surface of the mass are flattened, but those in the centre are fat and rounded, and usually occur in family groups resulting from cell division.

Cartilage in general has little or no blood supply. Its metabolic activity is low, and enough nourishment can diffuse through the matrix (which itself has a high water content) to supply the relatively small needs of the chondrocytes. Hyaline and elastic cartilage have no nerve supply, but fibrocartilage in joints is usually well innervated, and plays a part in the sensory mechanism of the joint (p. 41).

Elastic cartilage is restricted to the mobile part of the nose, external ear, and certain parts of the larynx (p. 161). Its elastic qualities are demonstrated when the nose is pulled or squashed and bounces back to its normal shape immediately.

Fibrocartilage is more widely distributed, and forms such important structures as the intervertebral discs (p. 245), and the discs and menisci found in many synovial joints (p. 41). It usually merges gradually into either fibrous tissue or hyaline cartilage. With increasing age there is an increase in the amount of collagenous material in all forms of cartilage, so that hyaline cartilage may be converted into fibrocartilage and fibrocartilage into a mass of fibrous tissue, with a few 'islands' of cartilage cells here and there.

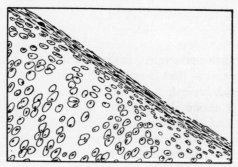

Fig. 7. Hyaline cartilage. The flattened cells on the surface form the perichondrium, and the rounded cartilage cells lie in irregular clumps embedded in translucent matrix.

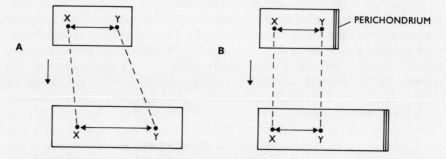

Fig. 8. Interstitial (A) and appositional (B) growth. In A the distance between the cells X and Y increases as growth takes place, because more matrix has been formed between them. In B the distance between the cells remains the same, the increase in size being wholly due to new material laid down by the perichondrium to the right of cell Y.

Hyaline cartilage is the commonest form of cartilage. It forms the precursor of the whole bony skeleton except the clavicle and the bones of the vault of the skull (p. 31). At birth quite a substantial proportion of the skeleton is still made of hyaline cartilage, though the shafts of the long bones of the limbs are largely composed of bone. The conversion of cartilage to bone continues in restricted regions of the body until middle or old age, but is mostly complete by the age of 21 years. After this time hyaline cartilage persists chiefly in the respiratory tract (p. 162), in the cartilages which join the ends of the ribs to each other and to the sternum (p. 268), and in the synovial and secondary cartilaginous joints (p. 39).

The perichondrium covering adult cartilage contains cells capable of forming new chondroblasts, and in young people a limited degree of regeneration of injured cartilage is thus possible. In older people damage to cartilage is usually repaired by fibrous scar tissue, some of which may eventually be replaced by bone.

In summary, cartilage is light but strong, resilient and smooth. It needs no blood vessels, and can survive prolonged pressure without damage. It can grow much more rapidly than bone, and is thus useful in the stage of fetal development to provide a scaffolding for the immense proliferation of soft tissues; later on, cartilaginous epiphyseal plates (p. 33) play a vital part in the growth of the bony skeleton.

5 · Bone

Functions and description

The skeleton forms a rigid supporting framework on which the softer tissues are built up, and resists the action of gravity and other forces applied to the body. The lower limb would be of little use in supporting the weight of the body if its stiffening core of bones were removed, and a boneless upper limb would not be much good in a fight. The soft and vulnerable brain is enclosed in the strong bony box of the skull, and the ribs are a more mobile protection for the viscera in the chest and the upper part of the abdomen.

A few of the 206 bones in the skeleton are fused with their neighbours to form composite bones, but the majority are connected together by joints which permit movement between them. The resulting system of bony levers, actuated by the muscles, produces movements of the various parts of the body relative to each other. This in turn liberates the body to move about in space, for it can be pulled along by the grasping hands, or propelled by the thrust of the lower limbs on the ground.

Bones are difficult to classify. **Long bones**, such as the femur (Fig. 171), have a shaft and two ends, which are expanded and thickened to provide large bearing surfaces for the joints in which they take part. **Miniature long bones** are similar, but much smaller. Examples are the phalanges of the fingers and toes (Fig. 135). **Short bones** have no recognizable shaft and are usually about as broad as they are long; they are found in the wrist and the ankle (Figs. 135, 172). **Flat bones** are never really flat; they are always slightly bent or curved, like the ribs and the bones which form the vault of the skull. Finally (in desperation), we have a class of **irregular bones,** to which belong all those which cannot be included in the other classes. The vertebrae are irregular, and so are the bones of the face.

The surface of a bone is not uniformly smooth. Where muscles are attached there is usually a roughened area which may be either depressed below the general surface or elevated above it. A depression is called a **fossa** or a **pit**, and an elevation receives different names according to its shape and size. The most imposing bumps are called **trochanters** (Fig. 172); the **tuberosities** (Fig. 173), and the **tubercles** (Fig. 131) are smaller. A **process** is any kind of projection, and

28

a **spine** is a sharp pointed process. A **facet** is a smooth articular surface, and a **foramen** is a hole. A **condyle** is a rounded swelling covered with articular cartilage, and an **epicondyle** is a projection in relation to an articular eminence. Other names will be encountered in the descriptions of the individual bones.

The skull and vertebral column form the **axial skeleton,** and the bones of the limbs the **appendicular skeleton.** In addition a number of **sesamoid** bones occur in the substance of tendons and muscles in different parts of the body. They are usually very small, and some of them are not invariably present, but the **patella** (p. 347) is a constant sesamoid of some size. The **hyoid** bone, which forms part of the larynx, is usually considered along with the axial skeleton.

Structure of bones

With the exception of the enamel of the teeth, bone is the most durable tissue in the body, because it contains about 60 per cent by weight of inorganic salts. Chief among these is calcium phosphate, but there are several others, including calcium and magnesium carbonates. These salts are deposited in a matrix which surrounds a framework of collagen fibres, and this framework is tough but pliable, as can be demonstrated by dissolving out the inorganic material with acid. If for any reason the calcium content of the bones is reduced, their rigidity is lessened, and they begin to bend under the stress of resisting gravity.

The architecture of individual bones is of importance. A surface layer of dense **compact bone** surrounds a central core of **spongy bone** which has a much looser texture. This arrangement is necessary because a skeleton completely made of solid bone would weight so much that enormous muscles would be needed to move it about.

Compact bone is not solid, for it is penetrated by innumerable fine canals which communicate with each other and are on the average about 50 μm in diameter. These **Haversian canals** convey into the interior tiny arteries and nerves derived from the vessels and nerves of the deeper layer of the **periosteum**—the membrane which covers the surface of bone in the same way that perichondrium covers cartilage. The outer layer of the periosteum is fibrous and tough, and is 'nailed' down to the bone by fibres which penetrate the compact bone at intervals and become embedded in it.

Spongy bone is formed by the honeycomb-like criss-crossing of thin irregular plates of bone called **trabeculae**. The trabeculae are placed so as to strengthen the bone against the external forces which are usually applied to it, and in this way a characteristic pattern develops for each bone of the skeleton (Fig. 9).

In the short and irregular bones, and in the ends of the long bones, the layer of compact bone is thin, but in the shafts of the long bones it is very thick, and the spongy bone in the middle is completely excavated away, so that the shaft forms a hollow cylinder.

In the interstices of the spongy bone lies the **red marrow**, a very cellular tissue, richly supplied with blood, which manufactures most of the cells which circulate in the blood (p. 154). In the embryo red marrow is found in all the bones, but in the adult it is confined to such places as the sternum (p. 268), the vertebrae, and the ends of the long bones. A sample of red marrow can be obtained for examination by the operation of **sternal puncture**, in which a hole is punched in the sternum and a little marrow sucked out. In the adult the shafts of the long bones are occupied by **yellow marrow** which has largely become transformed into fat. It may be looked on as red marrow which is having a rest because of the reduced demand for blood cells, and is at the same time acting as a fat depot. However, if the need arises, as for instance after severe and repeated haemorrhages, the yellow marrow may take over its former functions and become red again.

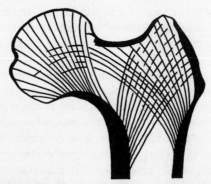

Fig. 9. Diagram showing the pattern of the bony trabeculae in the upper end of the femur. Note the variations in the thickness of the layer of compact bone.

Bone therefore acts as a container for a very active issue. But it would be a mistake to think of the container itself as quiet and inactive. The soft tissues of the rest of the body require for their efficient functioning a fairly constant concentration of calcium and phosphorus in the blood stream. Now the amount of calcium and phosphorus taken in the diet varies from day to day, and the skeleton acts as a vast reservoir of calcium phosphate which can be drawn upon to make adjustments in the blood levels of these substances. The relationship between the amount of calcium and phosphorus in the blood and the amount in the bones is partly controlled by **parathormone** liberated into the blood stream by the **parathyroid glands** (p. 199).

Ossification and growth of bone

The process of bone formation, or **ossification**, begins in the clavicle and the mandible about the fifth or sixth week of intra-uterine life; a little later changes can be detected in the long bones of the limbs. After this stage, therefore, an X-ray can disclose the existence of pregnancy.

Some bones—those of the face and the vault of the skull, the mandible and the clavicle—are said to be ossified 'in membrane', because bone formation occurs directly in the primitive connective tissue of the embryo. All the other bones ossify 'in cartilage', the connective tissue being first converted into cartilage, which is subsequently destroyed and gradually replaced by bone. The process of ossification is exactly similar in the two instances, except that the cartilage bones go through an extra stage.

Certain developmental defects may single out the membrane bones without affecting the cartilage bones and vice versa. Thus, in the condition of achondroplasia the cartilage bones are much shorter than usual, while the membrane bones are normal. Circus dwarfs are usually of this type.

The point at which ossification begins in a given bone is known as the **primary centre** of ossification. In the case of cartilage bones the cartilage cells at the primary centre become swollen; calcium salts are deposited in the matrix, and eventually the cells die. At the same time the perichondrium develops into an outer fibrous layer and an inner cellular one. From the inner layer blood vessels grow in to the region of the dying cartilage cells and bring to the site **osteoblasts** and **osteoclasts**. The osteoclasts are large cells containing several nuclei, and they erode and remove the calcified cartilage. In the spaces so produced the osteoblasts lay down collagen fibres round which inorganic materials are deposited to form bone. Some of the osteoblasts become imprisoned in the material they produce, and settle down to live there as **osteocytes**. There is evidence that in emergency these osteocytes can once again become active osteoblasts and lay down new bone.

In a short bone the lump of hyaline cartilage which is its precursor is continually growing all over its surface by the multiplication of the chondroblasts in the perichondrium. At the same time the cartilage in the heart of the mass is being replaced by bone. A race develops between these two processes, and eventually the ossification catches up with the growth of the cartilage, and the whole cartilaginous mass becomes replaced by bone with the exception of a thin rim left on the surface wherever the bone takes part in a joint. When this has happened growth of the bone stops (Fig. 10).

The growth of a long bone is more complicated. The primary centre occurs in the shaft, and bone is formed in the interior in the same way as in a short bone. But the chondroblasts in the deeper layer of the perichondrium become converted into osteoblasts which remain there and begin to form bone on the surface, so adding to the girth of the bone by surface accretion. Since the perichondrium is now surrounding bone and not cartilage, it changes its name to **periosteum**, of which the active cells form the **osteogenic layer**. When the shaft of the bone reaches a certain diameter, the osteoclasts inside the bone scoop out a cavity which is not built in by the osteoblasts. This is the **marrow cavity** (Fig. 11), and as bone formation extends along the shaft in both directions, so the marrow cavity extends. The bone grows stouter because of the deposition of new bone on the outside of the shaft, but the osteoclasts in the walls of the marrow cavity continually remove bone from the inside, so that the thickness of the wall of the cylinder remains proportionate to its total diameter.

Ossification spreads up and down the shaft of the bone, and as it does so the growth of the cartilage at the ends of the bone keeps pace with the increase in

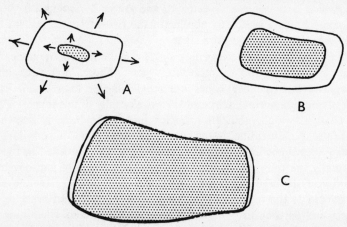

Fig. 10. Ossification of a short bone. A: the bony mass in the centre grows at the same time as the cartilaginous precursor of the bone expands. B: ossification has proceeded faster than growth of the cartilage. C: ossification is complete. A thin rind of cartilage is left at each end, where the bone takes part in the adjacent joints.

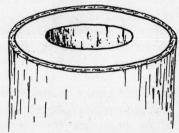

Fig. 11. Diagrammatic cross-section through the shaft of a long bone, showing the central marrow cavity and the closely enveloping periosteum.

size of the shaft. At a time depending on the individual bone, **secondary centres** of ossification appear in the cartilaginous ends. Some of these are present before birth, but others may not appear until several years later (Fig. 12).

Radiographs are useful in assessing the progress of children towards maturity. By a given chronological age we expect certain primary and secondary centres to have appeared and to have reached a certain size and shape (Fig. 13). Although there are wide variations in the timing of individual centres, it is possible, after examining a radiograph of a child's hand, a region

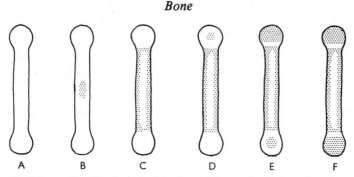

Fig. 12. Scheme of ossification of a long bone (changes in size not shown). A: the cartilaginous model. B: the primary centre appears in the shaft. C: the whole shaft has ossified from the primary centre, and the marrow cavity has been scooped out. D: a secondary centre appears at one end. E: this centre has formed an epiphysis separated from the bone of the shaft by an epiphyseal plate of cartilage; another secondary centre appears at the other end of the bone. F: the bone has an epiphysis at each end, and growth continues at the epiphyseal plates till these are obliterated.

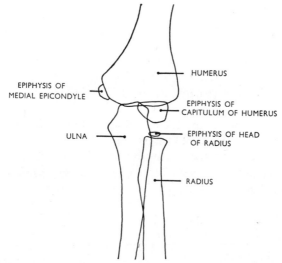

Fig. 13. Tracing of a radiograph of the elbow of a child of 7, showing three epiphyses. (Compare the adult elbow shown in Fig. 132.)

which contains many centres, to determine how he stands in relation to the average development of children at different ages, and so to allot to him a **radiological** or **skeletal age.**

Events in the secondary centres are exactly similar to those which took place in the primary centres, and eventually a mass of bone called an **epiphysis** is formed, separated from the shaft of the bone by a thin portion of still growing cartilage called the **epiphyseal plate;** in a radiograph the plate shows up as an **epiphyseal line** (Fig. 12). The growing cartilage cells in the plate increase its thickness at the same time as ossification from the shaft is converting the cartilage into bone. The two opposing processes keep exact pace with each other, and so the cartilage remains approximately the same thickness until the bone attains its adult length. This happens for most of the long bones at about 16–18 years of age in the female and 18–21 in the male (the clavicle is the last of the limb bones to stop growing, at about the age of 25 years). Growth stops when the ossification process catches up with the multiplication of the cartilage cells and converts the whole of the epiphyseal plate into bone. In a few people the epiphyseal plates are not obliterated and may remain open throughout life, as they do in the rat. Further growth does not take place at such unfused epiphyses unless they receive some abnormal stimulation, for example, from the activity of a tumour of the pituitary gland.

Interstitial growth of bone does not occur. If two metal pins are driven into the shaft of a bone in a baby, they will still be exactly the same distance apart in the adult; all growth in length takes place at the epiphyseal plates. The force exerted by the growth process in this situation is considerable; a pressure of about 900 pounds is required to suppress growth at a lower limb epiphysis.

In a given bone the secondary centres which appear last are the first to unite with the shaft. One end of a long bone thus grows more than the other, and is sometimes misleadingly spoken of as the 'growing end'. Any injury to an epiphyseal plate is always serious, for growth in length of the bone may be permanently impaired, but the resultant shortening compared to the normal bone of the opposite side is naturally worse if it is the 'growing end' which is damaged.

Secondary centres of ossification may occur also in some of the prominences and processes of individual bones. Where the position of such an epiphysis corresponds with the attachment of a muscle it is sometimes spoken of as a **traction epiphysis**, as opposed to the **pressure epiphyses** which lie at the ends of the long bones and take part in joints. However, not all epiphyses fit neatly into such a classification.

In some instances several secondary centres may unite together. For example, at the upper end of the humerus (p. 292) three centres occur, one for the head of the bone, one for the greater tubercle, and one for the lesser tubercle (Fig. 14). All these subsequently fuse together to form a single epiphysis separated from the shaft by an epiphyseal plate at which growth in length of the bone takes place.

As a bone grows, it acquires markings and prominences (p. 28). Most of these are genetically determined, and appear even if the bone is grown in tissue culture, but their fine detail is partly due to the pressure of nerves, the pull of muscles, etc. Thus, the stronger muscles of an active male create much more definite and obvious surface features than are seen in the corresponding female bone. A bone preserves its shape during growth to a remarkable degree; an adult humerus is almost exactly the same shape as the humerus of a child. This means that as growth proceeds, a considerable amount of material has to be removed from the outer surface of the growing bone (Fig. 15), as otherwise the ends of the bone would become clumsy, thickened and distorted. This **remodelling** is carried out by osteoclasts.

The growth of the skull and the vertebral column is even more complicated than that of the limb bones. At birth the vault of the skull is still partly

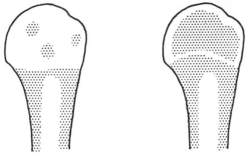

Fig. 14. Ossification of the upper end of the humerus. The three secondary centres fuse together to form a single epiphysis.

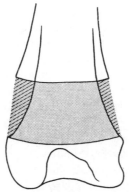

Fig. 15. Remodelling of bone. The diagram shows the amount of new bone added to the lower end of the femur between the ages of 16 and 20 years (stippling) and the amount of bone which has in consequence to be removed if the shape of the bone is to be maintained constant (cross-hatching).

membranous (p. 256), and this allows the bones to slide over each other during the passage of the skull down the birth canal. Growth takes place at the sutures (p. 39) between the bones, and surface deposition adds to their thickness while internal erosion by osteoclasts increases the size of the cavity for the brain. Similar growth occurs in the bones of the face, where large spaces, filled with air instead of marrow, and communicating with the nose, constitute the **air sinuses of the skull** (p. 258).

The replacement of the **primary dentition**, in which the teeth are suitable in size for the jaws of a young child, by the **permanent dentition**, in which teeth suitable for an adult jaw must temporarily be accommodated in a jaw which is still growing, is an awkward problem. A great deal of the complicated growth pattern of the face and the base of the skull is determined by the growth and eruption of teeth and the continuing need to maintain a satisfactory bite.

The normal growth of bone depends on many factors, both local and general. Disuse of a limb because of paralysis results in diminished growth of the bones concerned, both in length and in thickness. A local injury, such as a badly set fracture, may distort bone growth by leading to alterations in the mechanics of the limb. And any general illness, if sufficiently severe, may affect normal bony development. Even lying in bed for a long time may interfere with growth.

The diet must contain an adequate supply of the mineral salts which are necessary to give bone its rigidity. Food is not always rich in calcium, and for this reason bread is sometimes 'fortified' by the addition of a little chalk. In pregnancy, when the bones of the baby are being formed, there is a considerable drain on the calcium reserves of the mother, and this must be rectified by giving her extra calcium.

The diet must also contain a sufficiency of vitamin D, for without this the necessary calcium and phosphorus cannot be utilized properly, and the condition known as **rickets** occurs, in which the softened bones become distorted and bent. If the diet contains too little vitamin A the bones become thickened and clumsy, especially the ends of the long bones, the skull, and the vertebral column. This is because the osteoclasts fail to function properly. The resulting faults in the remodelling process (p. 35) may lead to foramina in bone failing to enlarge to correspond with the growth of nerve trunks running through them, so that the nerves are compressed and paralysed.

If too much parathormone (p. 199) is liberated, as may occur when the parathyroid glands are affected by tumours, the bone throughout the skeleton may become decalcified and softened. The **thyroid gland** (p. 199) controls the rate of growth of all the tissues, and failure of thyroid secretion is one cause of a child remaining a dwarf. Dwarfism may also be produced by a deficiency in the secretion of the anterior lobe of the **hypophysis**, or **pituitary gland** (p. 198); conversely, an excess of this secretion may make the patient into a giant. The secretions of the **gonads** (**testes** or **ovaries**) influence the time of closure of the epiphyseal plates; in consequence castration may lead to another form of giantism.

Fractures

Normal healthy bone can only be broken by considerable violence. The violence may be direct, as when a fracture is sustained at the site of a blow, or indirect, when force applied to one part of the body causes a bone some distance away to be snapped by leverage. It is even possible for a bone to break as a result of the unexpected contraction of the patient's own muscles, as, for example, when a sudden tremendous effort in made to prevent overbalancing.

In young people the bone may crack in a spiral fashion, part of it remaining intact. This is sometimes called a 'greenstick' fracture from its resemblance to the way in which a green twig tends to break incompletely. In old people the bones are much more brittle, and readily snap across. A similar fragility of the bones may occur after prolonged rest in bed (p. 230).

In a typical greenstick fracture the fragments are not displaced relative to each other, but in most other fractures there is some degree of displacement and consequent deformity. The main factors which determine this are the direction of the violence applied and the pull of the muscles attached to the fragments. For this reason fractures of the limb bones at a given anatomical site tend to result in predictable directions of displacement. For example, in a fracture of the humerus just below its middle, the upper fragment is pulled outwards by the deltoid muscle (p. 312) and the lower fragment is pulled upwards and inwards, so over-riding the upper fragment and causing the arm to become shortened. Sometimes the fragments may be rotated relative to each other by muscle action, or actually pulled apart (distraction). If the bony fragments are allowed to unite in the unnatural relationship they have assumed, there will be permanent disability. Accordingly, treatment is designed to reduce the deformity by bringing the fragments into normal alignment. Often this can be done by manipulation, or by making the limb assume a compensating posture during healing (in the example given, the arm can be maintained at right angles to the body in order to bring the lower fragment into line with the upper one and hold it there). Occasionally, however, such measures are ineffective, and it is necessary to operate and secure the fragments together by an internal splint.

A fracture in which any displacement at all occurs results in a rigid jagged structure being forced among the soft tissues, to which considerable damage may be done. After every fracture it is necessary to test the integrity of the peripheral nerves which might have been injured, and it is most important to know the relationships of nerves to individual bones.

Even though no large vessel may have been torn, a fracture is always accompanied by considerable bleeding, forming a collection of blood (a **haematoma**) between the broken ends. Within a few days this blood clot is invaded by cells from the osteogenic layer of the torn periosteum. These give rise to material like cartilage, which gradually becomes impregnated with calcium salts. Osteoblasts produce in this material a partly ossified mass called **provisional callus** which surrounds the broken ends and binds them together. The fragments may have been separated by the violence which caused the fracture and the periosteum may have been torn. In consequence the provisional callus may be very much bulkier than is necessary to restore the strength of the bone, and it is now reconstructed by osteoblasts and osteoclasts to form a more streamlined structure called the **definitive callus**. If the alignment of the fragments is not good, there may also be extensive remodelling of the broken ends themselves. Gradually, over a period of months, the definitive callus becomes permeated by Haversian canals, and a new internal architecture is formed to correspond with the new stresses which the reconstructed bone has to bear.

The process of repair is thus essentially similar to the process of normal ossification, and is under the control of similar dietary factors. An adequate intake of vitamin C is essential to secure good bony union, and the blood supply to the region of the break is also vital. Fragments of bone deprived of their blood supply may die and be extruded as **sequestra** without seriously interrupting the process of repair. However, when the general circulation is not very good the bone may remain alive, but have an insufficient margin of activity to allow it to undertake such an energetic process as repair. Accordingly, the fracture may remain ununited, and this is a serious risk in the aged.

6 · Joints

Description and classification

The place where one component of the skeleton meets another is known as a joint. Joints can be temporary or permanent, movable or immovable. Temporary joints are found during the period of growth, and later become obliterated. For example, the epiphysis of a long bone is united to the bone of the shaft by hyaline cartilage in a **primary cartilaginous joint**, which disappears when the epiphysis fuses with the shaft. At such a joint no movement is permitted; its function is to allow growth and not to permit locomotion.

There are three main types of permanent joints. In the first the bones are united by fibrous tissue to form **fibrous joints**. If the attachment is so tight and intricate that no movement can take place the joint is called a **suture**. Sutures occur only in the skull, and are also a device for permitting growth. They are not always strictly 'permanent', for some of them are invaded by bone in later life. If there is enough fibrous tissue to permit a certain amount of stretching or twisting movement, the joint is called a **syndesmosis.** An example occurs between the lower ends of the tibia and fibula (p. 354); the movements permitted are extremely small, but are nevertheless important to the mechanics of the ankle.

The second type of permanent joint is a **secondary cartilaginous joint**: the joints between the bodies of adjacent vertebrae are examples. The surfaces of the bones taking part are covered with hyaline cartilage, and the cartilaginous surfaces are firmly united by a plate of fibrocartilage or fibrous tissue. The centre of this plate either becomes softened, as in the intervertebral discs (p. 245), or develops a small cavity, as in the pubic symphysis (p. 278).

The third type of permanent joint, the **synovial joint** (Fig. 16), permits free movement. The surfaces which move on each other are covered by a thin layer of intensely smooth material. In most synovial joints this material is hyaline cartilage, but in the temporomandibular joint (p. 261) it may be fibrocartilage or even fibrous tissue. In all cases the **articular cartilage** is devoid of nerve fibres or blood vessels, and is not covered by perichondrium. The cartilage is generally thickest in the large weight-bearing joints, where, in young people, it may measure as much as 0·5 cm in depth. Its thickness varies from point to point

39

within a given joint, and tends to decrease with advancing age, which also brings a decrease in the resiliency of the material.

Stretching from one bone to the other and forming a sleeve round the articular surfaces is the **joint capsule**, made of collagen fibres running criss-cross in different directions. Here and there definite bands of more or less parallel fibres are developed in directions which enable them to resist the most important stresses to which the joint is subjected. These bands are **intrinsic ligaments**, and must be distinguished from **extrinsic ligaments**, which are similar bands uniting the bones at some distance from the capsule.

Ligaments vary considerably in their shape and attachments. Some are flat and others are cord-like; some merely connect the bones participating in the joint, and others range further afield. Many pass over two or more joints, and are functionally associated with several movements. Ligaments are usually attached primarily to bone, but some are connected to other structures, such as the cartilaginous discs which are found in certain joints (p. 41), and most of them also afford an attachment for muscles or tendons.

Whatever they are attached to, the union is generally very strong, and an injury involving the forcible stretching of a ligament can lead to a fracture of the bone rather than to rupture of the ligament itself.

Lining the joint capsule is a thin vascular **synovial membrane** which is attached all round the margins of the articular cartilages, and therefore covers all those bony surfaces within the attachments of the capsule which are not covered by articular cartilage. It also covers any tendons or ligaments which may lie within the joint cavity (p. 41).

The synovial membrane varies in thickness; in areas exposed to pressure it may almost disappear. In the quiet backwaters of the joint, however, it develops folds and fringes which bulge into the joint but never intervene between the articular surfaces. In these regions there is often some fat between the membrane and the joint capsule, and it has been suggested that this is a factor conducing to efficient lubrication.

The synovial membrane produces the **synovial fluid**, which filters out from the blood stream through the walls of the vessels in the synovial membrane and thence into the joint cavity. It is a clear or slightly yellow viscid fluid which lubricates the moving surfaces, nourishes the articular cartilages, and conveys into the joint macrophages which remove any debris resulting from the working of the surfaces. The amount of fluid is very small—the knee joint has the largest joint cavity in the body, but it normally contains only about 0·5 ml of fluid. If, however, the joint is damaged, as in a **sprain**, there may be a tremendous leakage of fluid into the joint, stretching and distending the already torn capsule. In such circumstances the knee may come to accommodate anything up to about 300 ml. If the joint cavity becomes infected, the synovial membrane acts as a filter in the reverse direction, readily absorbing poisons from the fluid in the joint, and so causing a marked general disturbance.

Sometimes additional structures are found in a synovial joint. The tendon of the long head of the biceps muscle passes inside the capsule of the shoulder joint (p. 312), and the hip and knee joints (pp. 351, 352) contain ligaments. In all three cases the invading structures are excluded from the joint cavity by an investment of synovial membrane from which they derive their nourishment. Another kind of intra-articular structure is the fibrocartilaginous **labrum** which deepens the socket for the shoulder joint (p. 298); a similar labrum is found in the hip joint.

In some joints there occur fibrocartilaginous **discs** or **menisci**. A disc is a roughly circular structure interposed between the articular surfaces, and attached by its circumference to the joint capsule. If it is complete it will divide the joint cavity into two separate cavities. If it has a hole in the middle, the two

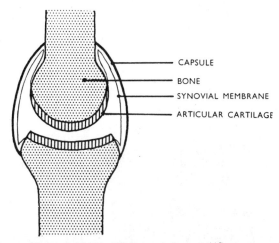

CAPSULE

BONE

SYNOVIAL MEMBRANE

ARTICULAR CARTILAGE

Fig. 16. Diagram of a synovial joint. The joint cavity is greatly exaggerated; in fact the cartilaginous surfaces would be in contact, separated only by a thin film of synovial fluid.

cavities communicate with each other through the hole (Fig. 17). Menisci are semilunar in shape, and can therefore only form an incomplete partition in the joint. Two menisci are found in the knee joint (p. 352), but no other joint has more than one. The function of discs and menisci is still a matter for argument. They must act as shock absorbers, for they have a considerable resilience under compression. They may increase the stability of the joint by increasing the congruity of the differing profiles of the articular surfaces which are presented to them during movement. They are often richly innervated, and may act as sensory receptors feeding information to the nervous system about joint movements. Other suggestions include protection of the margins of the joint, and a more effective distribution of stresses by increasing the area of contact between the bones. One of the most plausible theories is that discs and menisci allow different movements to take place in the two compartments into which the joint is divided; for example, in one compartment a gliding movement could take place, and in the other a hinge movement. Thus in the temporomandibular joint (p. 261) the mandible has to move bodily in relation to the base of the skull and also to swing open and shut; the disc in this joint is a complete one, and is structurally complex. Finally, many believe that the main

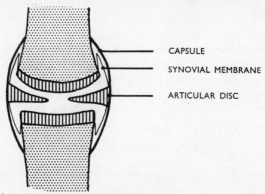

CAPSULE

SYNOVIAL MEMBRANE

ARTICULAR DISC

Fig. 17. Section through a synovial joint containing an articular disc. The disc is incomplete centrally, allowing the two parts of the joint cavity to communicate with each other.

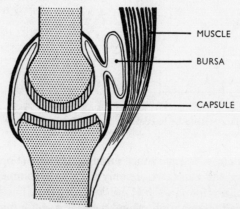

MUSCLE

BURSA

CAPSULE

Fig. 18. A bursa communicating with the joint cavity. The walls of the bursa are formed by the synovial membrane plus a thinned-out layer representing the capsule; in practice it is impossible to distinguish the two layers.

function of the discs is to assist in the lubrication of the joint, for their movement during joint activity may help to distribute the synovial fluid between the moving surfaces (p. 46).

In some joints the synovial membrane pushes outwards through a gap in the capsule to form a little recess or pocket lying outside the capsule (Fig. 18). Such recesses are known as **bursae,** and they develop between the capsule and a muscle which plays over it; they are thought to be devices for reducing friction, for instead of the muscle rubbing on the capsule, one smooth surface of the bursa glides on the other, lubricated by synovial fluid. Other bursae have no direct communication with the cavity of any joint, and develop in the connective tissue spaces. They are smooth walled cavities containing viscid lubricant fluid, and irritation or inflammation within them may lead to a considerable increase in the amount of the fluid. Well known examples of this condition are 'housemaid's knee' and 'student's elbow'.

Some elongated bursae are wrapped round tendons for a considerable distance, particularly in the wrist and ankle regions, to form **synovial sheaths** (Fig. 19). These too may become inflamed, thus giving rise to a 'tenosynovitis'.

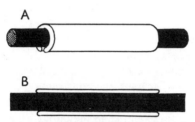

Fig. 19. Diagram of a synovial tendon sheath. A: surface view. B: section. A long bursa wraps itself round the tendon as a rubber hot-water-bottle might be wrapped round the leg of a table. The inner layer becomes adherent to the tendon and moves with it on the outer layer, which is attached to the surrounding tissues and remains stationary.

Movements permitted at synovial joints

Synovial joints can be classified according to the movements which they permit, and these are determined by the shape of the articulating surfaces. The simplest cases are those in which movement is permitted in only one plane. In **hinge joints**, such as are found in the terminal parts of the fingers, the angle between the bones can be made greater or smaller, but no other kind of movement is normally possible. Bending is known as **flexion**, straightening is called **extension**.

Pivot, or **trochoid** joints resemble the hinges of a gate, and like them allow only the movement of **rotation.** A good example is found between the first and second cervical vertebrae (p. 252), where much of the rotation of the head takes place. When the body is in the anatomical position the pivots of all the trochoid joints lie vertically.

If the articular surfaces are more or less flat, as in **plane** joints, **gliding** movements will be allowed in a single plane. Examples of this kind of joint are found between the short bones of the wrist or ankle.

In a **condylar** joint one bone articulates with another by two distinct and often completely separate articular condyles (p. 29). Perhaps the best example is the knee joint (p. 352), but other condylar joints occur in the vertebral column.

They behave very much as hinge joints, permitting flexion and extension, but in addition they allow other movements, as when one condyle rotates on its own axis and the other is forced to glide in an arc centred on the point round which rotation takes place.

Saddle-shaped, or **sellar** joints, also permit movement in more than one plane. The opposing surfaces are shaped like two saddles, one placed crosswise and upside down on top of the other; the profile of the articular surface may thus be concave or convex according to the plane of section. Rocking movements are possible in two main planes, and to a certain extent in other intermediate planes, while a small amount of rotation can also occur. The best example is the carpometacarpal joint of the thumb, and in view of the immense importance of movements of the thumb the mechanism of sellar joints deserves close attention. Many other joints in the body have this basic form and this type of mobility (p. 45), though the curvatures of the joint surfaces are not so marked as in the saddle joint of the thumb.

The type of joint which permits the greatest variety of movement is the **spheroidal**, or **ball and socket** joint, the construction of which is clear from the name. Good examples are the hip and shoulder joints. The socket is not always formed by a single bone; for example, in the talocalcaneonavicular joint (p. 356) two bones and a ligament receive the rounded head of the talus. In a ball and socket joint movements take place round any axis drawn through the middle of the ball, but for descriptive purposes several main directions of movement are recognized. In the shoulder, for example, flexion and extension are considered to take place round a horizontal axis drawn in the plane of the body of the scapula. When the hands are clasped behind the back the shoulder joint is extended, and when the nose is rubbed with the elbow the joint is flexed. Movements away from the midline of the body are called **abduction,** and movements towards it are called **adduction**. (In the hand and foot it is convenient to fix other arbitrary 'midlines' from which to assess abduction and adduction; in the hand the line passes through the middle finger, and in the foot it passes through the second toe. This apparently complicated and confusing procedure makes the actions of the small muscles on the digits (pp. 321, 370) much easier to describe and remember). Rotation in ball and socket joints is defined as movement round an axis passing through the shaft of the long bone concerned, as in pivot joints, and **circumduction** is the name given to a combination of all these movements, as when the toe is used to trace out a circle on the ground.

Ellipsoid joints are modified ball and socket joints in which both ball and socket are not spherical, but ellipsoidal. Such a joint permits flexion and extension, abduction and adduction. It will even allow circumduction, but rotation is impossible. The most important example is the radiocarpal joint (p. 302).

The foregoing account gives the generally accepted classification of joints by the movements permitted. However, all the joint surfaces in the body can be reduced, in essentials, to two main types. An **ovoid** surface presents a convex profile, no matter where it is sectioned, and fits into another ovoid surface which is concave. A **sellar** surface has already been defined; it may be incomplete, but there is always a direction of maximum convexity and one of maximum concavity, and it fits into another similarly shaped surface.

The two kinds of movement which may take place between such surfaces have been termed **spinning** and **sliding**. Spinning occurs when the moving bone rotates about an axis perpendicular to the fixed bone at the area of contact, and sliding refers to all other movements between the bones. Spinning may be a movement on its own, sometimes called **adjunct rotation,** or it may happen during a sliding movement, when it is called **conjunct rotation.** In fact, it is probably true that pure sliding movements, without some degree of conjunct rotation, are relatively uncommon.

In ovoid joints such as the shoulder joint, it is possible to have a considerable amount of adjunct rotation. What is commonly called 'rotation' of the shoulder—i.e. movement in the long axis of the humerus—really involves sliding of the joint surfaces. If a true spin takes place at the shoulder, the result in terms of gross movement is flexion and extension, not 'rotation'. This is because the articular surface of the head of the humerus is more or less at right angles to the shaft of the bone (Fig. 131). Conjunct rotation at ovoid joints is a result of the directions of pull of the muscles grouped together to produce the primary movement, and the rotation can be eliminated by muscular action.

In sellar joints, adjunct rotation is less free, and such conjunct rotation as occurs is the result of the shape of the articulating surfaces. It may be considerable, as in the carpometacarpal joint of the thumb, or minimal, as in the elbow joint.

Sliding movements may be classified into **swings**, in which the angle between the two bones alters, or **translation** movements, in which it remains more or less constant, as when the mandible is pulled straight forwards on the skull. A combination of translation and swinging movements occurs in the knee joint (p. 352), where the femoral condyles move backwards and forwards on the tibia as the joint is flexed and extended. The translation component may serve the purpose of bringing different areas of the tibial articular surface under pressure during different phases of movement, and so of minimizing wear and tear at any particular point. Translation of the lateral condyle of the knee joint necessarily accompanies adjunct rotation of the medical condyle (p. 44).

Very few movements involve only one joint. Many muscles and ligaments take part in the mechanics of two or more joints, and movements at one of them may automatically produce an effect on another. Many joints work together to influence the normal postures of the body (p. 226), and an injury—often apparently insignificant—to one of them can produce far-reaching alterations in the behaviour of the others. Finally, certain pairs of joints are unavoidably coupled together. Thus rotation of the radius in the movements of pronation and supination (p. 300) must take place at both the superior and inferior radio-ulnar joints or it cannot take place at all.

Stability of joints

Many factors co-operate to maintain the strength of synovial joints. There is first the shape of the articulating surfaces. A deep socket almost completely enveloping a closely fitting ball, as in the hip joint, is clearly a strong mechanical plan. If the bony surfaces are well adapted to each other, as in the sacroiliac joint (p. 279), **dislocation** of the joint surfaces is rare, but if they are not, as in the shoulder joint, dislocation is common. The degree of congruity varies as the bones move, and in the middle of the range of movement there may be very considerable disparity between the profiles of the two articular surfaces. Usually the surfaces become much more fully congruous at the extremes of the range, and the joint is then said to have reached a **'close-packed'** position, in which it is better adapted to weight bearing or stress. At first sight it would seem desirable to have the bones fitting together well in all positions of the joint, but the relatively poor 'fit' in most positions may be a device for ensuring proper lubrication of the articular surfaces; if the fit were perfect, synovial fluid would not be able to form an adequate film in the proper places. Probably for this reason joints are very seldom maintained in the close-packed position for long. Even in the upright posture, in which the joints of the lower limb are close-packed and take a great deal of strain, there are frequent small movements, which serve to bring different articular areas in contact and to allow synovial fluid to creep between them. Synovial joints must not be thought of as places at which a passive posture can be maintained; they are devices to allow active movement, which is necessary for their wellbeing. If a joint is maintained in a fixed position for any length of time, as by a surgical plaster, its tissues will begin to degenerate.

The strength and direction of the capsular ligaments are of importance in maintaining joint stability. Since ligaments are made of inelastic collagen fibres they can only be tight at the end of the movement they are adapted to resist, and many of the ligaments in a joint become fully tense only when the joint is in the close-packed position. It may at first sight seem curious that the ligaments should become taut when the bony surfaces are most closely fitting, but in many joints a conjunct rotation takes place towards the end of a movement, and this rotation not only brings the bones into the close-packed position, but also twists the ligaments and 'winds them up' tight. The braking effect of one ligament after another becoming twisted tight helps to bring the joint smoothly to a halt (p. 48).

Although some ligaments are only effective at the end of the range of movement, others maintain stability throughout. In hinge joints, such as the elbow, strong ligaments are developed at each side, and remain tight throughout the entire range of flexion and extension; they hold the bones together and prevent any sideways movement which might endanger the joint.

But the main influence on the stability of most joints is the integrity of the surrounding muscles. Unlike the inelastic ligaments, muscles can be tight throughout the whole of the movement. For example, when a joint is flexed, the muscles which extend it do not suddenly give way, but gradually 'pay out', so retaining continuous control of the movement and preventing any dangerous overaction of the flexors. Some muscles, indeed, are specially arranged for just this duty, and are known as 'extensile ligaments'. Muscles of this kind are found particularly in relation to joints in which great freedom of movement is possible, such as the shoulder joint (p. 298), where the shape of the bones and the slack and roomy capsule are factors which would favour dislocation. The muscles prevent this by holding the head of the humerus firmly against the shallow socket of the scapula. Dislocation of a joint primarily protected by muscles need not tear ligaments or break bones, as it commonly does when the joint stability depends largely on the shape of the bones and the presence of strong ligaments.

Muscles which act more or less along the line of the moving bone hold the bony surfaces together and in rapid movements prevent them from being pulled apart by centrifugal force. Such muscles are sometimes termed **shunt muscles**, in contrast to the **spurt muscles**, which act at right angles to the moving bone. Thus, at the elbow joint the brachioadialis (p. 313) acts as a shunt muscle, while the brachialis and biceps act as spurt muscles (p. 312).

The joints of the lower limb, which have to bear the weight of the body, have specially well developed bony, ligamentous, and muscular mechanisms which come into play as the strain is taken (p. 371). These were formerly described as 'locking mechanisms', and in the upright position the joints were said to be 'locked'. This is a bad way of looking at it, since it implies a rigid and unmodifiable response to strain.

It follows from what has been said that each joint is most protected as it approaches the extremes of the movements of which it is capable, for then the ligaments and muscles resisting that particular movement are both tight. In contrast, each joint has its **'position of rest'**, when the ligaments are all slack and the surrounding muscles are under the least strain; this is the position the joint takes up when it is not in use. The position of rest is naturally about the midpoint of the various movements undertaken by the joint. It is not always easy to determine, and thus the zero point for pairs of opposing movements is also difficult to obtain. In consequence many people take as the zero point the point mid-way between the extremes of the range of movement, and some confusion thus exists between different books about the extent of a given movement permitted at a given joint.

In any given joint a 'choice' has to be made between stability and mobility. The factors which make for stability—good fit of bones, strong tight capsule, closely applied strong muscles—militate against freedom of movement. On the contrary, a joint with a weak, lax capsule, marked incongruity of the bony surfaces, and controlled by distant muscles, may have a wide freedom of movement, but is unlikely to be strong. It is instructive to compare the strong hip joint (p. 351) with the shoulder joint (p. 298), which has great mobility.

Limitation of movement

Movement in any direction at a joint is ultimately brought to a stop by a variety of factors, some of which, in specific instances, are more important than others. For example, extension of the elbow is limited by the collision of bony prominences, but flexion is limited by the development of the muscles on the front of the arm; the more powerful and strong the biceps, the more it gets in the way of flexion, the very movement which the muscle itself produces. Ligaments are also important, the moving bone being stopped in the same way that a dog is brought up short at the end of his chain. Under ordinary conditions this mechanism is a secondary one, for movement is usually restrained by the muscles before the ligaments become fully tightened. This may be considered a defence mechanism to prevent damage to the capsule of the joint. Because ligaments are inextensible, they are liable to be ruptured by a *sudden* maximal strain, but they are well able to cope with *continuous* stresses such as occur in the foot during standing (p. 372).

The importance of the muscles may be seen when they are paralysed. The capsule becomes stretched, and the range of passive movement is greatly increased, leading to the formation of a 'flail joint'. Similarly, under an anaesthetic, more movement is possible than when the muscles are alert. Conversely, if the muscles are abnormally tense, as in a spastic paralysis (p. 98), the range of movement is very much reduced.

A distinction should be made between **active** and **passive** movements. The satisfactory performance of active movements demands the integrity of both the sensory and motor sides of the central nervous system, as well as intact muscles and a healthy joint. The mental attitude of an injured patient is also important, for he may be unwilling to try to move the joint through its whole range for fear of pain. Passive movements occur in response to an externally applied force, such as gravity or the manipulations of the examiner, and their range is nearly always greater than the range of active movement. Often a joint will allow passive movements which cannot be produced by active muscular contraction; for example, the joint surfaces can often be pulled apart or **distracted**, and rotation can be produced at joints (such as the interphalangeal joints) where no rotator muscles exist.

The range of passive movement permitted decreases with age. The ability of very young children to entwine their feet behind their necks while carrying on casual conversation does not outlast the first five or six years unless special steps are taken to preserve it by constant practice, and most acrobatic tricks involving an abnormal range of joint movements must be begun in childhood, while the capsule and the muscles are capable of being stretched. It is, however, possible for an adult to acquire an increased range of joint movement by training, though it is more difficult than for a child.

A dislocation of a joint usually results when its protective mechanisms are taken unawares—when the muscles are caught off guard, or when some trick of leverage is brought to bear. The capsule is always either stretched or torn, and the muscles attached to the bones may pull them into a still more unnatural position. There is thus a considerable risk of damage to the surrounding soft parts. Once the capsule has been stretched in this way it becomes easier for a subsequent dislocation to take place, and recurrence is not uncommon.

It may become necessary, because of disease, to fix a joint in one position by excising the joint surfaces, so that the bones grow together. This position must be very carefully chosen so as to allow the maximum function of the limb, and a compromise is often necessary. For example, the position in which the hip joint is best fixed is one of about 30° of flexion. This allows the patient to stand up and, by increasing the curvature of his lumbar spine, to achieve a virtually upright posture. At the same time it allows him to sit down without too much discomfort.

7 · Muscle

Muscles are composed of cells in which the capacity of protoplasm to contract has become highly developed and has been given direction. Most muscle cells are long and thin, and therefore receive the name of **fibres** (Fig. 20); they react to appropriate stimulation by shortening along their long axis.

Visceral muscle

The simplest variety of muscle fibre is a thin spindle-shaped cell with a central oval nucleus, and a length of about 0·5 mm. Such fibres are woven together in a thin connective tissue matrix to form sheets of **visceral muscle**. Muscle of this kind occurs in the walls of the alimentary canal, the urinary passages and the blood vessels; it is found in the capsules of various glands and also in the eye, where it regulates the size of the pupil and the focussing of the lens. In the skin, visceral muscle controls the activity of the sweat glands and is responsible for making the hairs stand on end. The wall of the uterus, which expels the baby at birth, is made of visceral muscle.

In the walls of hollow organs the sheets of visceral muscle are disposed in layers. In the alimentary canal, for example, there is an outer longitudinal layer, in which the long axis of the fibres is directed along the length of the gut tube, and an inner circular layer whose fibres run at right angles to those of the first group. In certain places, such as the stomach, a third layer of obliquely disposed fibres is added. These muscle sheets propel along the contents of the alimentary canal, and produce a constriction of the gut wall which travels down the canal in much the same sort of way as a twitch can be transmitted along a rope if its free end is shaken. Just ahead of the constriction wave is a wave of relaxation, and these successive waves of alternate contraction and relaxation are called **peristalsis**. The anatomical basis for peristalsis lies in the fact that both the longitudinal and the circular muscle coats are really spirals, the first being extremely loosely wound and the second being very tightly wound. If the muscle layers of the gut are stimulated at a given point by stretching, a contraction wave travels from this point in both spirals. The result is to produce a local constriction which travels slowly along the gut via the tightly wound 'circular' coat. At the same time the more longitudinal fibres pull the gut tube towards the site of the stimulus, so relaxing the tension in its walls and allowing it to bulge.

At intervals along the alimentary canal the circular coat becomes specially thickened to form **sphincter** muscles. When a sphincter contracts, the walls of the gut are drawn together as if a string had been tied round it, so preventing regurgitation when pressure rises in the more distal part of the canal.

In the vascular system similar sphincter muscles are developed in the arterioles (p. 134), and can shut off the stream of blood passing through these small vessels. The muscular arteries (p. 134) have a muscle coat which runs mainly circularly, though there are a few longitudinal fibres.

Visceral muscle contracts and relaxes relatively slowly; it is capable of sustained contraction for a long time without fatigue. The whole muscle sheet must contract together, for there is no mechanism to control the activity of a single fibre or even a small group of fibres. Visceral muscle can contract even though the nerve fibres which supply it have been destroyed. These contractions are often rhythmic; the alimentary canal or the uterus will go on contracting and relaxing regularly in a nutrient solution long after they have been removed from an animal's body. In the intact animal the contraction of visceral muscle is mediated by the autonomic system of nerves (p. 86) which is not under the control of the will, but functions automatically, though it is influenced by emotion. For this reason visceral muscle is often called **involuntary muscle**, but this is not a very satisfactory name, since many 'voluntary muscles', such as the diaphragm (p. 271) are also not completely subject to the will. Visceral muscle is also sometimes referred to as **smooth muscle**, or **unstriped muscle**, and all four terms are used interchangeably.

In some situations visceral muscle is influenced by circulating hormones. For example, the muscle of the bronchioles (p. 163) can be relaxed by adrenaline secreted by the suprarenal medulla (p. 88), although this same hormone will cause contraction of the muscle in the walls of the blood vessels in the skin. The reason for such differences is apparently to be found in receptor mechanisms which react to the presence of different chemicals, but the nature of these mechanisms is not understood.

In contrast to cardiac and somatic muscle, visceral muscle can multiply and regenerate satisfactorily. Thus, during pregnancy, the wall of the uterus thickens enormously because new fibres are added at the same time as the existing fibres enlarge. Again, when new blood vessels are formed during the repair of an injury, new visceral muscle develops in their walls.

Cardiac muscle

The **cardiac muscle** which forms the wall of the heart is unlike that found anywhere else in the body. The cells are thicker and shorter than those of visceral muscle, but like them have central oval nuclei. Cardiac muscle also resembles visceral muscle in having an inherent rhythmicity which is manifest in the absence of a nerve supply, though in the living body the rhythm is modified and controlled by the autonomic system of nerves. In other features, however, there are marked differences. The cytoplasm of cardiac muscle shows a series of longitudinal striations produced by **myofibrils** embedded in a clear jelly-like material called **sarcoplasm**. Each myofibril is transversely striated by alternate light and dark bands which are 'in step' with the bands in adjacent myofibrils, so that the whole muscle fibre appears to be divided into light and

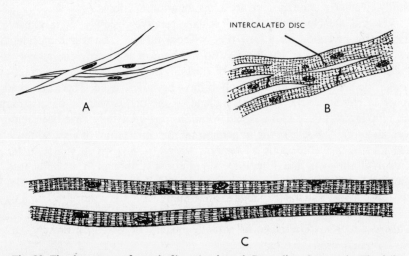

Fig. 20. The three types of muscle fibre. A: visceral. B: cardiac. C: somatic. The full length of the somatic fibres is not shown. Notice that their nuclei, which lie to one side of the fibre, may appear to be in the middle of it when they happen to be on the side nearest the viewer.

dark slices. The light and dark bands are called **isotropic** and **anisotropic discs** respectively (Fig. 20). When the fibre is contracting the appearance changes, because of changes in the arrangement of the proteins of which the myofibril is composed (p. 56).

Cardiac muscle cells are irregular in shape and often show branching. There are often several nuclei to each fibre, and the fibres adhere to each other to form a continuous network. The boundaries between cells are recognizable as irregular dark lines known as **intercalated discs** (Fig. 20).

When the heart beats, the contraction spreads from cell to cell throughout the sheet of muscle, the frequency of contraction being determined by those cells which have the most rapid inherent contractile rhythm. The 'fastest' cells lie in a specialized region called the sinuatrial node (p. 144), but if this is destroyed, other regions can take over the task of 'driving' the heart.

Somatic muscle

The muscles which effect movements demanded by our will are composed of **somatic, skeletal, voluntary**, or **striped** muscle fibres, all four terms being synonymous. Somatic muscle fibres are long thin cylindrical cells which in extreme cases may reach a length of over 30 cm and a breadth of 50 μm, but usually do not branch. Each fibre contains many nuclei, sometimes several thousand, and is intimately surrounded by a sheath, corresponding to the cell membrane, called the **sarcolemma.** The nuclei of somatic muscle fibres, unlike those of visceral and cardiac fibres, lie to one side of the fibre, just under the sarcolemma. As in cardiac muscle, there are longitudinal and transverse striations, which are due to a similar underlying structure. They are extremely prominent, and are responsible for the name 'striped muscle' (Fig. 20).

Two types of somatic muscle fibre can be recognized. The **red** fibres have more sarcoplasm and less regular cross-striations than the **white** ones, and their nuclei may be in the middle of the fibre. The red colour is due to the presence of a pigment called **myoglobin**, which resembles haemoglobin (p. 153), and like it has the ability to combine with and subsequently to liberate oxygen. Red fibres are mixed with white ones in all muscles, and there are many fibres with intermediate characteristics. However, the red fibres are present in greater numbers in muscles which maintain the posture of the body. The ability of these **postural muscles** to maintain a prolonged steady contraction depends in part on their ability to draw an extra supply of oxygen from the myoglobin within the cells.

Outside the sarcolemmal sheath of the individual fibre lies a loose layer of connective tissue called the **endomysium**, and the fibres are collected together into bundles or **fasciculi** by a stronger wrapping of connective tissue called the **perimysium**. The fasciculi are in turn bound together into a **somatic muscle** by a sheet of collagenous material called the **epimysium** (Fig. 21). In some muscles this sheet is inconspicuous, but in others it is dense and strong, and affords attachment to some of the muscle fibres.

Attachments of somatic muscle

If the contraction of somatic muscle is to be useful the two ends of the muscle must be attached to other structures which can thus be moved relative to one another. In the simplest case the muscle has two attachments, one at each end, but there may be a multiplicity of attachments to many different structures. Sometimes one end of a muscle may be attached to structures quite widely separated from each other; the muscle is then said to have more than one **head.** A familiar example is the **biceps** muscle of the arm, which has at its upper end two attachments, one from the region of the shoulder joint, and one from the coracoid process of the scapula (p. 312). A muscle with three heads is said to be **tricipital**, and the **quadriceps femoris** (p. 361) has four.

Usually somatic muscles are attached to bone, but they may be attached to skin, to fascia, to other muscles, and so on. Two somatic muscles, the **orbicularis oris** and the **sphincter ani externus**, run round the upper and lower ends respectively of the alimentary canal in a continuous circle, just as the sphincters of viceral muscle do in its intermediate portions. If a muscle is attached to bone the attachment is always indirect, through the medium of collagenous tissue. Each end of every muscle fibre is replaced by a collagen fibre which is attached to the sarcolemma. The collagen fibres then blend with the outer collagenous layer of the periosteum, which is itself anchored to the bone by perforating fibres (p. 29).

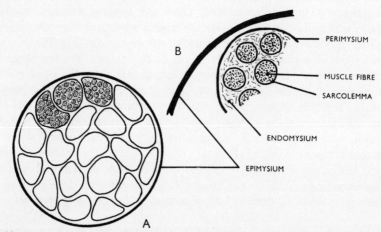

Fig. 21. Diagrammatic cross-section of a voluntary muscle. A shows a number of fasciculi within the epimysium; only three fasciculi are shown in detail. B is an enlarged view of one fasciculus, showing individual muscle fibres surrounded by endomysium and bound together by perimysium.

Often the collagen fibres at the end of a muscle are very short, and cannot be seen by the naked eye; the muscle is then said to have a **fleshy** attachment. But sometimes they are long and conspicuous, and the transition from muscle to fibrous tissue may begin some distance from the point of attachment; if so, the attachment is said to be **tendinous**. Tendons are composed of parallel bundles of collagen fibres with fibroblasts scattered about between them; the whole structure is extremely stong. Sometimes a tendon is a thin round cord, sometimes it is a broad and flattened sheet called an **aponeurosis.**

Tendons may occur in the middle of a muscle, which then has two fleshy **bellies**, one at either end; such a muscle is said to be **digastric**. Tendinous patches may develop in muscles where they are exposed to friction, and sometimes these patches may form an extensive smooth glistening surface to the muscle. The two main muscles of the calf (p. 364) have tendinous surfaces of this kind which allow one to move over the other with a minimum of friction. Where a tendon is subject to constant friction because it runs through a tight tunnel, as at the wrist and ankle, it is often surrounded by a synovial sheath (p. 43).

Arrangement of fasciculi

The transition from muscle fibre to collagen only rarely begins at the same point in all the fibres composing the muscle, and for this reason the tendon appears to send prolongations up into the fleshy belly of the muscle. Their existence demonstrates that not all the muscle fibres extend throughout the whole length of the muscle belly.

Muscles in which the fasciculi run more or less parallel to each other from one attachment to the other are called **strap** or **fusiform** muscles. **Unipennate** muscles develop a tendon on one side of the muscle belly, and the fasciculi are set into the tendon obliquely. If the tendon runs into the centre of the muscle, fasciculi may be set obliquely into it from two sides; this is the **bipennate** arrangement. Finally, in a **multipennate** muscle the tendon splits up into several portions, each serving as an attachment for fasciculi approaching it from two or more directions (Fig. 22).

When a muscle fibre contracts it becomes roughly half its original length. It follows that the longer the fibres in a given muscle, the greater the range of movement which that muscle can produce. But a short fibre can pull just as hard as a long one, and so the strength of the muscle depends on the total number of fibres it contains, whether these are long or short. A given volume of muscle can contain either many short fibres or few long fibres, and each muscle therefore has to strike a balance between range of movement and strength. Muscles accommodate fibres which are shorter than the length of the muscle belly by adopting the pennate arrangement, in which the fasciculi are set obliquely into the tendon. This obliquity has the disadvantage that the force exerted in the line of the tendon is not so great as if the fasciculi had a straight pull.

A strap or fusiform muscle has the greatest possible range of movement for its size. A long tendon allows a muscle to act at a distance, without its fleshy belly getting in the way; this is desirable in such places as the wrist and hand. The longer the tendon, however, the shorter is the muscle belly, and this reduces the range of movement. If, as in the long muscles acting on the fingers and thumb, a considerable range of movement is required, it follows that the muscle bellies must be fusiform, or at most very obliquely pennate. In places where strength is the main consideration and neither bulk nor restricted range of movement is a disadvantage muscles may adopt the multipennate arrangement.

Contraction of somatic muscle

Each myofibril consists of longitudinally arranged interlocking filaments of the proteins **actin** and **myosin**. The thin actin filaments lie in the light isotropic bands, but their ends extend into the dark anisotropic bands (p. 52), where they insinuate themselves between the thicker filaments of myosin which occupy this

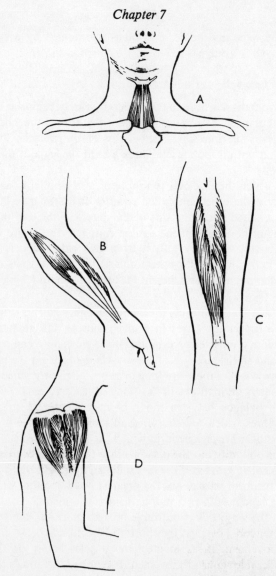

Fig. 22. Types of voluntary muscle. A: strap muscles, the **sternohyoids**. B: a fusiform muscle (the **flexor carpi radialis**) and a unipennate one (the **flexor pollicis longus**). C: the bipennate **rectus femoris**. D: the multipennate **deltoid** muscle.

situation. During contraction, the isotropic bands disappear, because the actin filaments slide further between the myosin filaments, causing a telescoping of the muscle fibril. A single fibre is not capable of a graduated response; it either contracts to its fullest extent or does not contract at all.

Despite this, the muscles which the fibres go to form can readily produce a graduated contraction because it is possible to bring into play a varying number of fibres at any one time. In somatic muscle, in contrast to cardiac and visceral muscle, the nervous system exercises its control over small groups of fibres instead of over the muscle as a whole (p. 72). If the strength of the contraction has to be increased, more of these groups are activated by the cells of the central nervous system, and if the muscle is required to relax, the active fibres cease contracting group by group. The muscle fibre forming a group, together with the single nerve cell which drives them, are known as a **motor unit**. In muscles such as those of the thigh, in which fine gradations of movement are not required, each nerve cell may control several hundred muscle fibres. Where fine adjustments of tension are a necessity, as in the muscles of the eyeball, the motor units are small, and a nerve cell may supply as few as half a dozen muscle fibres. In a given muscle the motor units usually interlock with each other, so that one nerve cell innervates muscle fibres which are scattered at random throughout a sizeable volume of the muscle (up to 10 mm in diameter in the muscles of the lower limb) instead of lying side by side in a closely packed fasciculus.

It is thought that muscles which contract for long periods have a sufficient reserves to permit the working of shifts; one assembly contracts for a time and is then replaced by fresh ones which have been resting; there is, however, no proof of this. If a strong electric current is applied to the nerve supplying a muscle, it is possible to cause nearly all the fibres in it to contract simultaneously. Under normal circumstances, however, it is very seldom that anything like all the fibres in a muscle are thrown into contraction at once. In sustained contraction of a muscle, the 'all-or-none' responses of the individual motor units are smoothed out by two circumstances. Firstly, the nerve cells driving the motor units do not become excited synchronously; at any given time some motor units are in action, some are relaxing, and some are resting between contractions. Secondly, the pull of the muscle fibres is transmitted to their attachments through connective tissue which is not quite rigid, so that a certain amount of irregular activity can be masked in this way.

Muscle with an intact nerve supply is always firm and slightly resistant to the touch; it is said to possess **tone**. In contrast, muscles whose nerve supply has been cut are soft and flaccid. The tone of a muscle thus depends on the integrity of its nerve supply, but other factors are also important, e.g. the state of nutrition of the body, the presence of disease, etc. The existence of normal tone in a muscle means that it is always in a state of readiness, and there is no slack to take up when the time comes for it to contract. It is 'standing at ease', rather than 'standing easy'.

Work and energy

In an **isotonic** contraction the muscle shortens but the tension it develops is kept constant. Muscles may also be in action without shortening, merely to prevent their attachments being pulled further away from each other. The muscles of the calf, for instance, do not shorten when we are standing upright, yet they are in constant activity to prevent the force of gravity from making us fall forwards (p. 371). This kind of contraction, where the length is constant but the tension increases, is called **isometric.** Isometric contractions are powerful, for the maximum tension a muscle can develop is greater if the muscle remains stretched than if it is allowed to contract. At the beginning of any movement the muscles are in a virtually isometric condition, and this confers an advantage which fits in well with the need to overcome inertia. Finally, muscles may be active while they are actually lengthening, as when they 'pay out' during the action of their opponents to maintain control over the movement and protect the joint. A good example is the gradual relaxation of the elbow flexors during the movements of setting down a cup of tea held in the hand. The active lengthening of a muscle is called **excentric** action, in contrast to the **concentric** action which occurs when the muscle shortens.

In all three types of activity the muscle fibres perform work, and the source of the energy which enables them to do this lies in an extremely complicated series of chemical reactions inside the cell. It will simplify matters if we consider only the main processes. The raw material is a substance called **adenosine triphosphate;** when a contraction of the fibre takes place this is broken down by a special enzyme to **adenosine diphosphate** and phosphoric acid. This process liberates a great deal of energy, which is used for the contraction; the vital raw material is, however, destroyed, and has to be built up again in readiness for the next contraction. The energy for this purpose is provided by the conversion of glycogen (p. 3) to **lactic acid**. This is poisonous to the cell, and is got rid of by converting it into carbon dioxide and water, which are liberated into the blood stream (p. 158).

While oxygen is essential for the recovery process, the process which yields the energy for contraction needs no oxygen, and it is therefore possible for muscles to undertake spurts of work in the absence of sufficient oxygen to complete the series of chemical reactions. They 'borrow' energy against the security of their future oxygen supply. As a result lactic and other organic acids accumulate in the muscle fibres, and have to be disposed of in the period following the cessation of the work. Until the final conversion is complete the body is said to be in **oxygen debt**, and excess oxygen is taken in until the debt is paid. The situation is familiar to everyone who has ever run for a bus; during the sprint there may be no panting at all, but after it the necessary extra oxygen has to be obtained by an increase in the rate and depth of respiration.

Activity in a muscle requires more blood to be supplied to it than is provided in the resting state, and during exercise (p. 231) the muscle capillaries dilate, while those in other parts of the body are shut down. The heart rate increases, and this and other changes increase the flow of blood through the muscle. Nevertheless, when the muscle is actively contracting, the capillaries between the fibres are compressed and closed, so depriving the active part of the muscle of its oxygen supply just when it needs it most. This is probably the reason why the mechanism of muscle metabolism is so complicated, and why red muscle fibres (p. 53) are found in postural muscles.

Muscular work gives rise to heat, which is transferred to the blood stream and circulated round the body. Sawing wood on a cold day thus warms up the whole body. If the body is cold and there is no exercise of this kind to perform the central nervous system causes opposing groups of muscles to contract spasmodically, so producing **shivering**; this warms the body up by burning the chemical fuel stored in the muscles.

Active muscle fibres produce tiny changes in electrical potential, and an electrode placed on the skin over the surface of a working muscle will detect **action potentials** inside the muscle. If the muscle to be investigated lies deeply it is necessary to insert an insulated needle electrode into its substance. The technique is called **electromyography**, and is often of great value in determining whether or not a given muscle has been paralysed after a nerve injury.

Actions of somatic muscles

All muscles have at least one attachment which remains relatively stationary while the muscle is acting; this attachment is called the **origin** of the muscle, and the attachment which moves is called its **insertion**. A spring arranged to close a door could be said to have its origin from the door frame and to be inserted into the door. No muscle in the body has such a clear cut arrangement as this, and the terms 'origin' and 'insertion' are only relative ones. For instance, the muscles that pass from the chest wall to the arm are customarily employed in moving the arm about; they are said to originate from the trunk and to be inserted into the arm. Nevertheless, when the upper limb is used for climbing these same muscles are used to pull the trunk up on the fixed arms, and are now acting from a fixed 'insertion' on a movable 'origin'. This reversibility of muscle action is perhaps most important in the lower limb in connexion with the act of walking (p. 373).

The action of a muscle depends on its position relative to the joint which it crosses, and upon the types of movement permitted at that joint. At a hinge joint, for example, all the muscles which cross the joint on one aspect are flexors, and all those which cross it on the other are extensors; there is no other possibility. At a ball and socket joint, however, a given muscle may be capable of exerting a number of different actions. For instance, the deltoid muscle (p. 312) has several actions on the shoulder joint. The part of the muscle which passes obliquely downwards and laterally in front of the joint (Fig. 148) flexes and medially rotates it (i.e. turns the front of the arm inwards). The fasciculi which lie behind the joint (Fig. 146), on the other hand, are extensors and lateral rotators. The central portion of the muscle, which covers the joint on its lateral side, is the main abductor of the arm. Many other muscles resemble the deltoid in having several different, and sometimes mutually opposing, actions on a single joint.

In such cases the parts of the muscle having contradictory actions are naturally separately controlled by the central nervous system, though they may be supplied by a single peripheral nerve.

A specified movement at a joint is very seldom entrusted to a single muscle; it is usual for several muscles to combine together as a group, though obviously some may be more important than others. This means that if one muscle should be paralysed, the remaining contributors to the movement will usually be able to compensate, at least partly, for the missing one. Further assistance may be obtained from other muscles which normally take no part in the affected movement, provided their attachments allow them to be recruited under special conditions. For example, flexion at a joint may still be possible, even though all the usual flexors are paralysed, because one of the abductors becomes capable of flexing if the joint is held in a certain position. Such 'trick movements' are often used by patients with muscle paralysis.

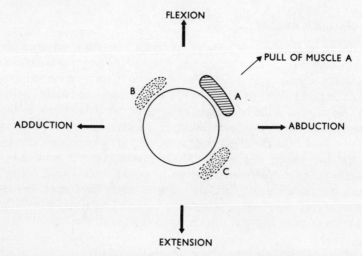

Fig. 23. Cross-section through the ball of a ball and socket joint to illustrate group action of muscles. Contraction of the muscle A produces a combination of flexion and abduction, but if it is combined with muscle B the result is pure flexion; combination with muscle C gives pure abduction.

If a muscle has several actions which are not under separate nervous control, it will naturally produce them all when it contracts. A muscle which passes over a ball and socket joint in such a position that it produces both flexion and abduction (Fig. 23) cannot contract so as to give rise to abduction without flexion. However, unwanted actions of this kind can be cancelled out by the pull of the other members of the group with which the muscle is operating. Thus, if the required movement is one of pure abduction, the unwanted flexion is prevented by simultaneously calling into play one or more extensor muscles; if these also contribute to abduction, so much the better. If, on the other hand, the movement required is pure flexion, the abduction is cancelled by combining the muscle with a group of adductors and flexors. One and the same muscle therefore may contribute to different results according to the company it keeps. Similar considerations apply when a muscle crosses two or more joints. If only one of these joints is required to move, then the pull of the muscle on the others is counteracted by the appropriate contraction of other muscles. For example, when it is desired to flex a finger, the extensors of the wrist contract to prevent the finger flexors from also flexing the wrist.

The members of such a group of muscles are coordinated by the nervous system, and make contributions of different kinds to the performance of the movement. Several terms are used to describe these contributions. A muscle is said to act as a **prime mover** when its contraction is directed towards the primary purpose of producing the movement. When it acts in a subsidiary

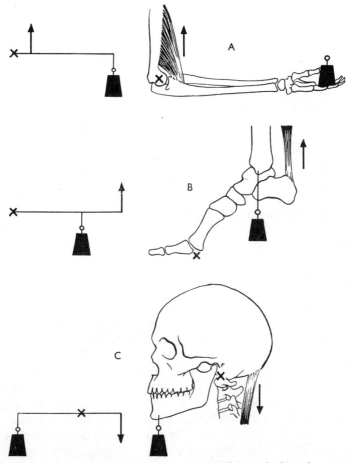

Fig. 24. The lever action of muscles. In each drawing the fulcrum is shown by a cross. A: the **brachialis**. B: the **calf muscle**. C: the **semispinalis capitis**.

capacity it is called a **synergist**. Synergists not only prevent unwanted movements inherent in the attachments of the prime movers, but may also help by fixing other joints, so that the prime movers can obtain a firm purchase. In such cases they are called **fixators**. Thus, contraction of the prime movers passing from the freely movable scapula to the humerus (p. 310) is accompanied by the stabilizing contraction of the muscles tethering the scapula to the trunk (p. 307). The opponents of a movement are called **antagonists,** and the gradual active relaxation of the antagonists protects the underlying joints and acts as a continuous control of the activity of the prime movers (p. 63).

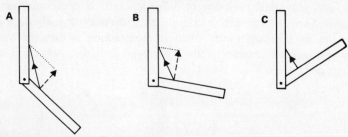

Fig. 25. Effect of position on the power exerted by a muscle. The pull of a muscle (solid line) can be represented by a force (dotted line) in the line of the moving bone, and another (dashed line) at right angles to it. In A the force bending the joint is only about two-thirds of the force exerted by the muscle; in B the proportion has increased, and in C the whole of the pull is now bending the joint.

In the limbs the muscles, bones and joints form a system of levers, the muscles supplying the force and the joints acting as the fulcra. The commonest type of lever in the body is the one in which the force is applied between the fulcrum and the weight, but the other varieties are also found (Fig. 24). In every case the **mechanical advantage** of the muscle depends on its distance from the fulcrum relative to the distance of the resistance which the muscle has to overcome. The mechanical desirability of inserting the muscle well away from the fulcrum is offset by the desirability of having it close to the joint so that it does not impede the movement. An insertion close to the fulcrum allows speed of movement rather than power; an insertion some distance away carries the reverse implication. The power exerted by a muscle varies from a minimum when its line of pull is nearly parallel to the long axis of the bone to which it is inserted to a maximum when the line of pull is at right angles to the bone (Fig. 25).

It is not always easy to determine what part a given muscle plays in the movements occurring at a given joint. Several methods may be used. The first is to consider the relationships of its fasciculi to the centre around which movement takes place. It is sometimes helpful to imagine a sheet of elastic passing between its various attachments. Secondly, it is often possible to feel the muscle contracting during the performance of a given movement, and this can be supplemented by electromyography, which has recently afforded much information in regard to standing and walking. Thirdly, the muscle may be electrically stimulated. This method is open to error because the body never singles out muscles in this artificial way, but it does at least tell us what movements the muscle is capable of producing on its own. The fourth method, that of observing the results of paralysis of the muscle, is open to a similar objection because the loss of one member of a group of muscles at once establishes an abnormal situation, with other muscles trying to compensate for the loss. It does not, therefore, tell us what the muscle normally does in the intact body, but information on what happens when a muscle is paralysed is perhaps just as important.

Not all somatic muscles produce movements at joints when they contract. For example, the sphincter muscles, two of which have already been mentioned (p. 54), have either no attachment to bone or at most an attachment which merely serves as an anchorage. Thus, the **orbicularis oculi**, which protects the eye by sweeping round the orbit and eyelids (p. 262), is attached to bone at only one point in its course.

Other muscles only produce movements of soft tissues. The muscles which move the soft palate (p. 161) and many of the muscles of facial expression (p. 262) are of this type. Still others have a supportive function, such as those forming the pelvic diaphragm (p. 282), which takes much of the weight of the pelvic viscera.

Even muscles which cross a joint may serve primarily as extensile ligaments (p. 47), though they can contribute to movements if required. The fact that nearly all movements involve groups of muscles reinforces this protective system, since it follows that there is an increase in muscular tension on several aspects of the joint during contraction.

Some somatic muscles are important by virtue of their mere presence. The gluteus maximus, for example, is a very efficient soft pad which covers many of the bony points underlying the buttock. If, in a bed-ridden patient, it is paralysed and atrophies, these bony points rub against the skin, and this may lead to damage to the skin and the production of **bed sores.**

Types of movement

In most movements the antagonists are in a state of active contraction throughout, 'paying out' to control the activity of the prime movers and so helping to make the movement a smooth and steady one. But when maximum effort is required, as when a sprinter starts a race, the antagonists may be almost completely relaxed in order to reduce any opposing force to a minimum. In movements such as striking a tennis ball, not only are the antagonists relaxed, but the prime movers also tend to relax once the stroke has been made, the 'follow-through' being due to the momentum acquired by the limb. In both cases the antagonists contract towards the end of the movement, to prevent damage to the joints, ligaments, and bones (p. 48).

When great steadiness and accuracy are required over a limited range, as in writing, the prime movers and the antagonists contract together, and the resultant movement is due to fluctuations in the tone of the opposing groups. Such movements gain considerably in control, since the net force exerted can be made very small.

Methods of assessing muscular function

The strength of a muscle can be measured directly by the resistance it can overcome; both isometric and isotonic apparatus may be used for the purpose. A familiar example of an isotonic apparatus is the 'try your strength' machine in which the subject tries to pull together two iron bars by using the flexor muscles of his fingers. In an isometric apparatus, strain gauges measure the force exerted on a bar which does not move.

Electromyography (p. 59) gives a rough idea of the amount of muscle substance in action at a given time, and can also define which muscles are active in a given movement. It is of value in determining which muscles are paralysed after an injury. Clinically it is often possible to feel the muscle contracting by palpating it with the finger, and in assessing the degree of impairment of function in a given muscle a standard classification is often used:

    0 = No contraction
    1 = Flicker or trace of contraction
    2 = Active movement with gravity eliminated
    3 = Active movement against gravity
    4 = Active movement against gravity and resistance
    5 = Normal power

A system of this kind is very useful when following up a patient over a period of months to estimate the gradual recovery from paralysis.

Treatment of paralysed muscles

A muscle which is completely paralysed must be kept in good condition while its nerve supply is in the process of recovery. This is best done by exercising it. Electrical stimuli will cause the muscle to contract in the absence of its nerve supply, and passive movements of its attachments will stretch the muscle over its full range. Massage can increase the local blood supply and preserve the nutrition of the muscle.

Once power has started to return to the muscle it can exercise itself, and this is done in a series of stages. The first exercises are **assisted**; that is to say, the therapist's muscles are used to help the required movements. At this stage it may be valuable to perform the movements in a bath of water, so as to diminish the pull of gravity.

The next stage in the rehabilitation of a muscle is to allow it to perform movements unassisted; the muscle is encouraged to move throughout its full range, at first with gravity and later against it. Finally, as the muscle improves, the stage of **free** exercise is superseded by the stage of **resisted** exercise, in which the patient has to perform the movements against the resistance of the therapist's muscles or against a mechanical force such as a spring or a weight.

The patient may have been deprived of the use of that particular muscle for perhaps two or three years, and its functions may have been partially taken over by other muscles. He must therefore learn all over again to combine the action of the paralysed muscle with the healthy ones, and this is far from easy. Frequently a muscle which has recovered a fair amount of strength may still be so 'clumsy' as to be of little use. This is particularly so in older patients; a young child makes the necessary adjustment in its stride, but an adult may have considerable trouble in welcoming back to full use the missing member of the neuromuscular group concerned. The main factor here is the patient himself; patients will perform prodigies if they are determined to do so, but if not, they may let a promising degree of recovery slip back into uselessness.

Co-ordinated movements do not spring full-fledged from the co-operation of nerve and muscle; they have to be learned. Just as the pianist who nonchalantly copes with the difficulties of Ravel had to start with one finger melodies, so the child has to learn the use of his muscles, and particularly the use of his hands and legs. When he is born he can neither sit nor stand, and the use of his hands is limited to grasping. Individual movements of the fingers develop only slowly. Learning to feed himself is difficult, and the proper use of forks and knives is often not achieved until the age of six or seven. As is to be expected, he finds the management of his centre of gravity difficult, and falls frequently. Balancing feats like riding a bicycle come only after prolonged practice.

Motor co-ordination thus comes only with experience, and once it is lost it has to be learned all over again. When dealing with a child with paralysis it is important to realize the stage of co-ordination which at his age he may be expected to have achieved, and when dealing with a paralysed adult it is equally important to realize that he has to begin again like a child.

8 · The nervous system

Evolution and plan of construction

A unicellular animal has no need for an internal communication system, but as the size and complexity of the body increases **nerve cells** become adapted to act as message bearers, and send out thin processes into the surrounding tissues. In very small animals these **nerve fibres** connect the surface of the body with the muscle cells of the mouth and locomotor apparatus, so that direct messages may pass between these organs. This very simple arrangement necessarily limits the reactions of the animal. In more highly developed creatures the nerve cells group themselves together into small clusters or **ganglia** from which their processes radiate to different parts of the body, including other ganglia. In the ganglia the processes of one cell may enter into communication with those of its neighbours, and so messages can be transferred from one pathway to another in a manner reminiscent of a telephone exchange. A plan of this kind allows the animal a considerably increased range of behaviour. But it is not until we reach the vertebrates that the full potentialities of a communication system based on nerve cells are realized. Most of the nerve cells retreat to the middle of the body and become entangled in an immense network of their own fibres: this is the **central nervous system**. The forepart of this system is the **brain**, and lies within the skull: in higher vertebrates it becomes greatly enlarged and complicated. The hinder part remains relatively simple and subordinate to the brain; it is called the **spinal cord,** and is housed in the vertebral column.

The brain may be compared to an aircraft control room: messages are received by it from all available sources, and on the basis of the summated evidence of these messages a course of action is planned. This action is broken down into its components, and appropriate messages are then sent out to various individual destinations in order to produce the desired effect. The spinal cord more nearly resembles an automatic telephone exchange: when certain messages are fed into it the response it initiates is automatically determined.

The central nervous system is built up on a supporting framework of connective tissue cells known as **neuroglia**; these cells take no part in the transmission of messages. In a slice of brain or spinal cord parts look white and other parts grey. The **grey matter** is tenanted by the nerve cells, and large

66

collections of grey matter are therefore called ganglia even though they may be submerged within the substance of the brain. Smaller collections go under the name of **nuclei**. The **white matter** consists largely of nerve fibres.

As the nerve fibres leave the central nervous system to run to their destinations elsewhere they become grouped together into bundles called **nerves**. The nerves which are attached to the spinal cord are known as **spinal nerves**, and those which spring from the brain are the **cranial nerves**: together they form the **peripheral nervous system**. Under this heading we distinguish the **autonomic system,** which includes those cells and fibres which supply smooth muscle, cardiac muscle, and the cells of certain glands, and the **cerebrospinal system,** which supplies the whole of the rest of the body.

Anatomy of the neurone

The basic unit of the nervous system is the **neurone**, which is the name given to a nerve cell together with all its processes. It is perhaps easiest to appreciate just how specialized these cells have become by imagining one enlarged. A big nerve cell in the spinal cord measures perhaps 75 to 100 μm in diameter, and may send a process down to supply a muscle in the foot; this process will be something like 10 μm thick and a metre long. If the nerve cell were the size of a golf ball, the process would resemble a thick piece of string a quarter of a mile in length.

The simplest type of neurone has only two processes, an **afferent** one which brings messages *to* the cell and is called the **dendrite**, and an **efferent** one which takes messages *from* the cell and is called the **axon.** This kind of neurone, which is called **bipolar** because the processes leave the cell body at opposite poles, is plentiful in the embryo, but persists in the adult only in relation to the special senses (Fig. 26). Because they are concerned with bringing information to the brain from sense organs, the bipolar neurones are classed among the **afferent** or **sensory** neurones. The more usual type of sensory neurone in the adult is **unipolar** (Fig. 26); the afferent and efferent processes are continuous and bypass the cell body, being connected to it by a common root, so that the cell lies alongside the sensory pathway like a booster unit on an electric cable. Such neurones are found in the majority of peripheral sensory ganglia.

But the commonest variety of neurone has several processes, and is therefore **multipolar** (Fig. 26). One of these is the axon; all the rest are dendrites. Multipolar cells are found within the central nervous system and their dendrites do not leave it. Their axons, too, may run their whole course inside the brain or spinal cord, ending in relation to other nerve cells. Such a neurone is termed a **connector** or **intercalated** neurone. On the other hand, some multipolar cells have axons which leave the central nervous system and run in the peripheral nerves to supply the fibres of muscles: these are **motor** neurones.

An axon or dendrite which ventures outside the shelter of the grey matter to run to some other part of the nervous system may acquire a coating of glistening fatty material called **myelin**. It is the presence of large numbers of such **myelinated** fibres which gives the white matter its characteristic appearance.

In the central nervous system the myelin sheaths are derived from **oligodendroglial cells**, which wrap their plasma membranes like a bandage round the fibres. If a myelinated fibre leaves the central nervous system, the job of forming and maintaining the myelin sheath is taken over by **Schwann cells**, which behave in a similar fashion, the whole cell being wrapped spirally round the fibre (Fig. 27). Outside the Schwann cells there lies a basement membrane surrounded in turn by delicate connective tissue, the **endoneurium** (p. 91).

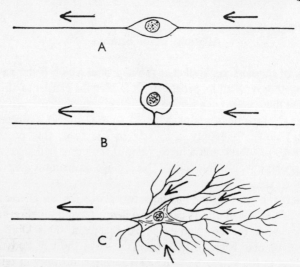

Fig. 26. Types of neurone. The arrows indicate the direction along which messages pass. A: bipolar. B: unipolar. C: multipolar.

The thickness of the myelin sheath varies with the thickness of the **axis cylinder** (a non-committal term applied to both dendrites and axons). If this is thick, its coating of myelin will also tend to be thick; if it is thin, its myelin sheath will be thin. At intervals along the course of the fibre the myelin sheath is interrupted by indentations where one Schwann cell ends and the next one takes over. These are the **nodes of Ranvier**, and at each node the Schwann cells interlock with each other. Their plasma membranes, however, remain quite separate, and the basement membrane outside them thus comes into effective contact with the axis cylinder. At this point, therefore, the axis cylinder is vulnerable to the action of chemicals and ions lying outside the fibre. This fact has an important bearing on the mode of transmission of messages, and it also explains how local anaesthetics can attack the axis cylinder. Nodes occur at intervals which vary with the thickness of the fibre; in a 10 μm fibre they are about 2 mm apart.

As a myelinated fibre approaches its peripheral destination, its myelin coating becomes finer and finer and may eventually disappear, though both sensory and motor fibres still retain the protection of a continuous sequence of Schwann cells until just before their terminals.

Both in the central and in the peripheral nervous system the myelinated fibres are in the minority; in peripheral nerves they are outnumbered by **unmyelinated** fibres by almost four to one. The unmyelinated fibres in the central nervous system do not appear to have any special protection, but those

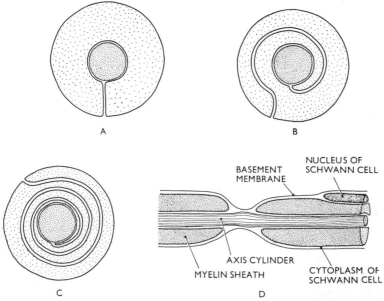

Fig. 27. Myelination. A, B, and C show the way in which the plasma membrane of a Schwann cell is wrapped round the axis cylinder. D is a longitudinal section of a myelinated fibre showing a node of Ranvier; the interlocking of the Schwann cells is not shown, for the sake of clarity.

in the peripheral nervous system are, like the myelinated fibres, ensheathed by Schwann cells, although in a different way (Fig. 28). The same Schwann cell provides cover for several fibres, some of which lie in shallow grooves in the surface of the cell, while others penetrate more deeply into the shelter of the cytoplasm, though they always remain outside the plasma membrane. Unmyelinated fibres may leave one Schwann cell sequence to join another, and, like the myelinated fibres, remain protected in this way until very close to their terminations.

The neurone theory

The axis cylinder is merely a very long and thin filament of the cytoplasm of the cell. The neurone theory states that the cytoplasm of one neurone is never directly continuous with the cytoplasm of another: there is always a distinct gap between the axon of one neurone and the dendrite of another. These gaps are known as **synapses**, and are only about 0·02 μm wide. Attempts have been made to show that in the skin and other places there is protoplasmic continuity between different neurones, but the neurone theory remains unshaken.

Intercalated neurones

The axon of an intercalated neurone may run for a considerable distance within the central nervous system. It commonly divides into a number of branches, each of which eventually breaks up into a large number of fine twigs bearing tiny swellings. Some of these become entangled with the dendrites of other neurones, or twine round their cell bodies; others terminate in little **terminal buttons**, which lie among the dendrites or close to the cell body. A single motor cell in the spinal cord may have anything up to 30,000 such synaptic buttons in relation to it.

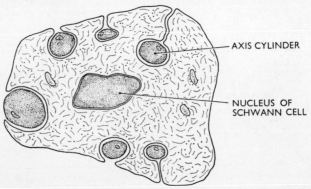

AXIS CYLINDER

NUCLEUS OF
SCHWANN CELL

Fig. 28. Six unmyelinated fibres running partly or wholly in the shelter of the cytoplasm of a Schwann cell.

The axon is usually finer than the dendrites, which branch to form an enormously complicated thicket of twigs (Fig. 26), into which the terminals of many axons may find their way. The branching of both axons and dendrites allows any one cell to communicate with many others, and to receive messages from a great many more. In this way a message may be relayed to many different parts of the central nervous system, and this process of diffusion and redirection to allow of a variety of responses is the chief function of intercalated neurones.

Sensory neurones

The cell bodies of the sensory neurones lie outside the central nervous system, in the ganglia which occur on the course of each nerve which has a sensory component. The cells are more or less spherical, and are often embedded in a protective covering of smaller cells. The axon runs from the ganglion into the central nervous system, where its branches end in the usual manner round the dendrites of a number of intercalated neurones. The single dendrite of the sensory cell is more specialized, and because it conducts impulses in the same way as an axon and not in the manner of a dendrite in the central nervous system, it is sometimes referred to as a 'sensory axon'. It stretches from the ganglion to a termination which may lie in the skin, in muscle, in the gut, or in any structure which is sensitive to stimulation. The dendrite may branch, but does so infrequently until close to its termination, where many small twigs come off and spread out in the structure innervated; the end of each twig becomes modified into a specialized **endorgan**.

There are two main kinds of sensory endorgan. The first is an open network of very fine beaded fibres which pervades the skin all over the body and embraces the roots of the hairs in a complex basketwork. Similar networks are also found in the deeper structures. This is the simplest, and from its ubiquity, the most important form of sensory innervation. The second kind of endorgan is formed when the dendritic terminals coil up on themselves or form a small spray occupying a limited area. This punctate type of ending may contain an elaborate spiralling of the nerve fibre within a connective tissue capsule; in other forms the little terminal knobs are devoid of protection.

The **lamellated corpuscle** occurs in such places as the mesentery of the gut (p. 181) and the deeper layers of the skin, and is sensitive to vibration. It is composed of a laminated cellular capsule within which ramify the fine nerve fibres, the whole looking rather like an onion when sectioned. Other endings with a characteristic structure are the **tactile corpuscles** of the finger pads (Fig. 29), which probably respond to light touch.

In voluntary muscles are found the **muscle spindles** (Fig. 29), which consist of open coils of fine nerve fibres wrapped round two or three 'intrafusal' muscle fibres. The muscle spindle sends a message to the central nervous system when the muscle is stretched, and it is therefore classified as a **stretch receptor**. The intrafusal fibres have a special motor supply which is thought to 'set' the spindle to the desired level of sensitivity. Other stretch receptors with variable structure are located in tendons close to their junction with muscle. There are many other varieties of endorgan in such places as the periosteum, the capsules of joints, and the skin. The terminals which are found in the organs of special sense—the eye, the ear, the nose, the tongue—will be considered separately.

Sensory endings are sometimes classified according to the type of information the brain obtains as a result of their activity. **Proprioceptors** provide information resulting from stimuli arising within the body itself, in contrast to **exteroceptors,** which record changes in the outside world.

Motor neurones

The motor cells lie in the grey matter of the central nervous system. They are usually multipolar, and some of them are called **pyramidal cells** because of their shape. Their axons travel out into the peripheral nervous system to reach voluntary muscles. There are many such axons in the **motor nerve** which enters each muscle, but this nerve is by no means exclusively motor, for anything up to half its fibres are sensory ones coming from endorgans in the muscle.

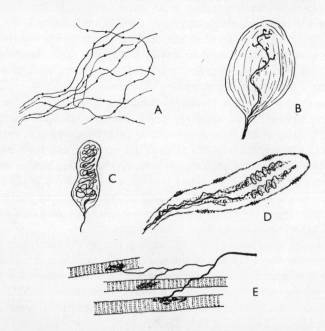

Fig. 29. Varieties of nerve ending. A: diffuse sensory network of beaded fibres. B: a lamellated sensory corpuscle. C: a tactile corpuscle. D: a muscle spindle. The two muscle fibres are enveloped in coils of nerve fibres and the whole is surrounded by a delicate capsule of connective tissue. E: a group of three motor end-plates.

Once inside the muscle, each axon breaks up into branches, each of which seeks out an individual muscle fibre and ends upon it in a complex structure called a **motor end-plate**. At the end-plate the basement membrane of the nerve fibre becomes continuous with the sarcolemma of the muscle fibre, and within this shelter the axis cylinder breaks up into club-shaped coiled terminals which lie among several small nuclei (Fig. 29). Each muscle fibre receives one end-plate, but each axon, by virtue of its branching, controls several muscle fibres. The apparatus consisting of a single motor neurone and all the muscle fibres it innervates is called a **motor unit** (p. 57).

Approximately one third of all the motor fibres reaching the muscle are destined for the specialized intrafusal fibres of the muscle spindles. Taking this in conjunction with the large number of sensory fibres coming from the spindles, it is easy to deduce the immense importance of the spindle system, which continuously monitors the tension in a muscle. If this falls too low, the central nervous system recruits more motor units: if it rises, some of the active ones are rested. The mechanism is an excellent example of 'negative feedback' (p. 111).

In visceral muscle the arrangement is different; the axons wander about and finally end in simple terminal knobs on the surface of the sheet of muscle fibres. So complex is the network formed by the interlacement of different axons that the territory controlled by an individual axon cannot be determined. Under the microscope the network resembles the contents of the darning basket after the kitten has got at it—each strand retains its identity and does not fuse with its neighbours, but it may be impossible to sort out its individual course among the general tangle. The motor neurones which innervate visceral muscle belong to the automatic system, and will be considered again later (p. 86). In the meantime we must note that, just as a 'motor nerve' contains many sensory fibres, so a 'sensory nerve' coming from the skin contains many autonomic fibres running to supply such structures as the cutaneous blood vessels and the sweat glands with motor innervation.

The transmission of messages

The nerve impulse

Detectable changes take place in a nerve fibre transmitting a message. For example, if a special electrode is placed on the fibre and connected to a very sensitive amplifier we can record a series of waves of depolarization travelling along the surface of the fibre in one direction only (towards the cell in dendrites and away from it in axons). These waves are known as **nerve impulses**. The rate at which they travel depends on the size of the fibre; in the thick myelinated fibres coming from muscle spindles it may reach 120 metres per second, and in very small non-myelinated fibres it may be as low as a tenth of a metre per second. The electrical changes are due to a temporary leaking of ions through the surface of the axis cylinder; these rapidly return again, so that after one impulse has passed the fibre quickly returns to its normal condition. During the short time before normality is restored, the fibre is incapable of conveying another impulse, and is said to be **refractory**. The message sent up the nerve fibre can thus never be a continuous one, for a limit is set to the rapidity with which impulses can succeed each other.

Conduction in unmyelinated fibres is similar to conduction in a coaxial cable, and is known as 'cable spread'. In myelinated fibres, however, the electrical charges 'jump' from one node to the next, a mode of transmission known as 'saltatory'. So far as our apparatus allows us to judge, the impulses in a given fibre are all qualitatively identical.

In axons and sensory axons (p. 71) the **all-or-none law** operates. This means that impulses are not produced in a fibre unless the stimulus applied to it reaches a certain intensity known as the **threshold**; if this is exceeded, impulses will pass, but no increase beyond the threshold will cause any difference in their nature, though the rapidity with which they succeed each other will increase up to a maximum.

Conduction in the central nervous system, through the bodies of the neurones and their dendrites, does not obey the all-or-none law, and graded responses to graded stimulation are possible. This allows one message to sum with, or to cancel out, another arriving at the same cell, and this has a most important effect on the transmission of impulses from one neurone to the next one in sequence (p. 97).

Sensory transmission

Sensory endorgans act as transducers—structures which can alter the nature of the energy applied to them, as a microphone converts sound waves into electricity. Some endorgans in the skin can transduce many kinds of energy but others are more specific; the spindles in the voluntary muscles respond to stretching, the lamellated corpuscles to deformation by pressure. Whatever the type of stimulation, the response is in all cases the same—an impulse or series of impulses is sent up the parent nerve fibre.

Just as the nerve fibre has a refractory period, so the endorgan cannot respond to continuous stimulation at more than a certain limiting rate; it has its own refractory period, during which it gathers strength, so to speak, to fire off another impulse. Under continuous stimulation an endorgan will fire off rapidly to begin with, but after a time it ceases to do so; it is said to have **adapted**. If then the intensity of the stimulus is suddenly altered, a brief burst of impulses occurs, and this happens also when the stimulus is removed. Adaptation of the sensory receptors is one reason why we are normally unconscious of the pressure of our clothes unless we make a movement; another reason is that we cease to pay attention to the stream of impulses coming up from the cutaneous receptors. Adaptation can occur whatever the nature of the stimulus, but more readily when the stimulus is innocuous than when it is harmful and arouses pain. It is also possible experimentally to **fatigue** an endorgan so that it ceases to respond, no matter what stimuli are applied to it, until it has had time to recover: it is doubtful whether this often happens in the intact body.

The smallest stimulus that can be devised always arouses impulses in several different sensory fibres; this is because the territories supplied by the branches of different sensory fibres are superimposed on each other, so that any small area on the skin is supplied by at least two different fibres. This principle of **multiple innervation** means that in response to any stimulus the brain receives messages via many different pathways, some of which may conduct rapidly and others slowly. The stimulus is thus translated into a pattern of impulses dispersed in time and in space, and variations in the patterns received by the brain can be caused by variations in the number of active pathways and the frequency with which each is conveying impulses. There is no doubt that this is the sort of message the brain receives, but we still do not know how it is able to distinguish one form of stimulus from another. It is generally accepted that impulses in the nerves subserving each of the five senses can be distinguished by the fact that they travel to different parts of the brain (p. 106). But it has also been suggested that the same principle holds within each sense, so that one kind of cutaneous sensation can be distinguished from another because there are specific fibres for each which reach different destinations within the same general region of the brain. This has recently been questioned, and it is suggested instead that the brain recognizes characteristic variations in the impulse patterns which it receives. A crude analogy to the specific fibre theory is that of a control panel for a complex machine; if a light shows in a certain position the operator knows that something has gone wrong with the oil feed, while a light in another position tells him that a fuse has blown, and so on. A similarly crude analogy to the second theory is the operation of a television set, where different patterns of impulses generate different kinds of picture.

The synapse

When the axon of a sensory cell enters the central nervous system it divides into many branches and ends round different cells which take over the message and convey it to other parts of the nervous system. These cells are known as **second order** neurones, from their secondary position in the chain of neural transmission; their axons in turn may end round third order neurones and so on, depending on the complexity of the pathway to be traversed. Transmission from one neurone to another involves the message jumping the gap at the synapse, a process which involves the liberation of a chemical (see below). The passage of the impulse across the synapse takes approximately a thousandth of a second, and the impulses aroused in the second neurone are exactly similar to those in the first. The axons of many cells may converge on a single synapse, which therefore has its own protective refractory period, to prevent the second order cell being overloaded—just as an endorgan will not respond to every stimulus if these are offered to it too rapidly.

In the synapses of the motor pathways of the autonomic system (p. 111) the mechanism of transmission has been fairly well worked out. An impulse arriving at the terminal of the axon releases into the gap a minute quantity of **acetylcholine**, which stimulates the second order cell to fire. If the acetylcholine were allowed to stay in the region of the synapse after it had done its work it would continue to excite the second neurone, and chaos would rapidly result. For this reason the enzyme **cholinesterase** is concentrated round each synapse, where it breaks down the acetylcholine formed so that the synapse can be cleared to receive another impulse. The whole process of **chemical transmission**—formation and destruction—takes little more than a thousandth of a second, roughly the same time as synaptic transmission in the central nervous system.

Motor transmission

The impulses in motor fibres are qualitatively identical with those in sensory or intercalated fibres. The fibres which supply voluntary muscle pass straight to their objective without 'relaying' to another fibre, and when the impulses arrive at the motor end-plate acetylcholine is liberated. This stimulates the muscle fibre to contract, and is later removed by cholinesterase, just as in the synapses of the autonomic system. Neuromuscular transmission is attacked by various diseases and drugs. The best known drug is **curare**, which blocks the chemical reaction so as to cause a complete motor paralysis, and is used in surgery to obtain complete muscular relaxation. On the other hand, organic phosphorus compounds, which are used in large quantities as insecticides for crops, inhibit the action of cholinesterase, so that they cause the unrestricted accumulation of acetylcholine at the nerve endings, with fatal results. In the disease **myasthenia gravis** the muscles rapidly become unable to react to impulses reaching them, and require a long rest period to allow the acetylcholine/cholinesterase complex to return to normal again. In consequence a repeated movement—such as eating—gradually becomes weaker and is eventually paralysed for some minutes or even hours. Fortunately several drugs can counteract this condition by interfering with the chemical reactions involved.

The autonomic system (see p. 86) is divided into two distinct functional parts called the **sympathetic** and the **parasympathetic** systems. The nerve endings of the parasympathetic system produce acetylcholine just like the cerebrospinal endings, but those of the sympathetic system act by liberating **noradrenaline**. It follows that drugs which act on one part of the autonomic system will not do so on the other.

Reactions and reflexes

The central nervous system responds to the sensory data presented to it by issuing motor instructions designed to cope with the situation. There are two ways in which this is done, the first involving consciousness and the second being automatic. An example of the first type is the sudden application of the brakes we make when we round a corner and find an unexpected steamroller in our path. Such **reactions** require a conscious appreciation of the nature of the stimulus, and the brain must always take part. The **reaction time** to a given stimulus, such as a light or a sound, may be measured by asking the subject to complete an electrical circuit as soon as he is aware of the stimulus. Reaction times vary from person to person, from day to day, and with the intensity of the stimulus. On the whole, the quickest reactions, about a tenth of a second, are obtained in response to touch stimuli. The nervous pathways involved in a reaction contain many synapses. The message brought to the central nervous system is first relayed to the part of the brain which deals with the particular sense involved. From here it passes by intercalated neurones to the part of the brain which issues the motor instructions, and thence the motor pathways take it down to the spinal cord, where it finally synapses to pass out to the appropriate muscles of the hand. The reaction time is thus the sum of the times taken to traverse the sensory and motor pathways, to cross the synapses, and to evaluate the nature of the stimulus.

In contrast to a reaction, a **reflex** is an automatic response not necessarily involving consciousness. Familiar examples are the contraction of the pupil of the eye when a light is shone into it and the onset of sweating in a hot room. The simplest are the **local reflexes**. For example, when an irritant is applied to the skin, the area becomes flushed because impulses pass up one branch of a cutaneous nerve fibre and run back again to the skin down another. These impulses are called **antidromic** because they run in the reverse of the normal direction. When they reach their destination they liberate **histamine**, which in turn dilates the walls of the small blood vessels.

More important are the **long path** reflexes. Some of these, like those concerning the eye, are called **cranial reflexes**, because their pathways traverse the brain; others, the **spinal reflexes**, are mediated by the spinal cord.

Tendon reflexes

An important class of long path reflex is the **tendon reflex**, such as the knee jerk. When a sharp tap is given to the tendon of the quadriceps femoris muscle just below the patella, the muscle responds by contracting, so that the knee is suddenly extended, and the foot 'jerks' forwards. The sensory receptors in the tendon are stimulated, and send impulses up their sensory neurones to the spinal cord. The axons of these neurones end in relation to motor cells which send back impulses to the muscle. This return pathway to the starting point is called a **reflex arc**, and the particular arc mediating the knee jerk has only one synapse in it. The knee jerk is thus a **monosynaptic** reflex, and in this it may be unique, for it seems that all other tendon reflexes have reflex arcs consisting of at least three neurones, an intercalated neurone being interposed between the sensory and motor sides of the arc. This conclusion is arrived at by measuring the **reflex time** between the application of the stimulus and the resulting contraction; from this an idea of the number of synapses involved may be obtained. The afferent and efferent limbs of each reflex arc reach the spinal cord or brain at a specific level, and testing the tendon jerks (often called the **deep** reflexes) is thus a means of finding out whether a particular part of the central nervous system is affected by injury or disease.

If the spinal cord is cut off from the brain by the injury spinal reflexes persist unimpaired. Indeed, they are characteristically exaggerated after the first stage of 'spinal shock' has passed. This occurs because the **upper motor neurones** descending from the brain normally restrain the activity of the spinal cord cells which supply the specialized muscle fibres within each muscle spindle (p. 71). When this influence is removed, the tension of these fibres is increased, and this leads to an increase in the sensitivity of the apparatus. The result of tapping the patellar tendon may now be quite startling. The knee extends suddenly and powerfully, and may give a series of quivering jerks known as **clonus** before it subsides again.

When a muscle responds to a tap on its tendon by contracting, there is also a **reflex inhibition** of its antagonists. It is said that the afferent neurone is capable of producing at some of its terminals a **central excitatory state** which stimulates a second neurone, and at others a **central inhibitory state** which dampens down the existing activity of other neurones. To say this is merely to clothe our ignorance in words, but the inhibitory effect can readily be demonstrated.

Postural reflexes

Similar reflex arcs subserve the **stretch reflexes** which are a vital factor in the maintenance of **posture**. The centre of gravity is in such a position (p. 371) that the body tends to fall forwards at the ankle. This stretches the calf muscles, which contract reflexly to pull the body upright again. Any overcompensation in this direction stretches the muscles on the front of the joint, which in turn contract to restore equilibrium. The situation is basically similar, though more complicated, where several groups of muscles surround a ball and socket joint.

The degree of contraction of the postural muscles varies constantly with the varying postures of the body. To stretch out an arm is to alter the position of the centre of gravity, and this would cause a fall were it not for the stretch reflexes which this alteration produces in muscles at a distance. The muscles of the lower limb and back increase their activity in proportion to the degree of tension put upon them by the overbalancing body, and so prevent the imminent catastrophe. All such reflexes are subject to the co-ordinating control of the **cerebellum** (p. 227).

Superficial reflexes

When the sole of the foot of a normal adult is firmly stroked with a pencil, one of two things happens; either the great toe and one or more of the other toes flex towards the sole, or else an indeterminate movement occurs. In the first case the **plantar** reflex is said to be **flexor**; in the second, indefinite. In a newly born baby the response is different; the great toe, and perhaps some of the others also, turns up towards the front of the leg. A similar **extensor** or **Babinski** response occurs in adults when there has been damage to the motor pathway between brain and spinal cord. The reason for this is unknown, but the fact is very helpful in diagnosis.

The plantar reflex belongs to a group known as the **superficial** (in contrast to the deep) reflexes. Included in this group is the twitching of the umbilicus towards the site of a stroke made on the abdominal wall (this is the **abdominal** reflex).

Other types of reflex action

There are many other kinds of reflex, and they vary in the degree to which they are independent of conscious control. The most independent are those which are not accompanied by any conscious sensation, such as the reflexes which control the blood-pressure by varying the amount of constriction of the walls of the blood-vessels. In contrast to such automatic reflexes are those in which conscious effort can play a large part. Few people can without practice suppress the blink when something touches the cornea of the eye (the **corneal** reflex), but if the object is smooth and the subject is determined it can be done. When a crumb gets into the larynx the cough is usually uncontrollable, but some coughs due to irritation of the larynx can be consciously suppressed. A sneeze is a reflex action, yet it can be stopped. It is, in fact, difficult to draw any useful distinction between the more complicated reflexes and the simpler reactions.

The **conditioned** reflex is particularly associated with the name of Pavlov. When a dog is given his dinner, his mouth waters—the normal reflex result of food stimulating the inside of his mouth. If the giving of food is accompanied by the ringing of a bell, and this is repeated a sufficient number of times, the dog's mouth will water when the bell is rung, even though no dinner is given to him. He is said to have become **conditioned** to the bell as a stimulus signifying food. By such methods a great deal can be learned about the sensory capacities of different animals. For example, one can condition hens to pick corn from cards coloured blue but not from cards coloured red; one may deduce from this that hens can distinguish red from blue. Similarly, dogs can be made to salivate in response to a circle but not to a square; if they are then presented with a hexagon they may become anxious and disturbed, and by experiments of this kind it is possible to induce 'neurotic' behaviour. Conditioning is believed to be of considerable importance in human learning.

The peripheral nervous system

The spinal nerves

The spinal cord lies in the vertebral canal formed by the bodies and neural arches of the vertebral column (p. 245). It is shorter than the canal, and its lower end is at the level of the first lumbar vertebra. Connected to it are 31 pairs of spinal nerves, each containing many thousands of nerve fibres. The spinal nerves emerge in series through the **intervertebral foramina** between adjacent vertebrae. Most of the nerves are numbered to correspond to the number of the vertebra above them; there are 12 thoracic vertebrae, and 12 pairs of **thoracic** nerves, each pair distal to its corresponding vertebra. Similarly there are 5 pairs of **lumbar** nerves, 5 pairs of **sacral** nerves, and one pair of **coccygeal** nerves. However, in the cervical region there are 8 **cervical** nerves and only 7 cervical vertebrae. The extra pair of nerves come out above the first cervical vertebra, between it and the skull, so that the second pair are also superior to their corresponding vertebra, and so on down to the eighth pair, which lie inferior to the seventh cervical vertebra.

The intervertebral foramina increase in size from above downwards, but in the living body they are largely closed up by fibrous tissue. Each spinal nerve is attached to the spinal cord by two **roots**, which unite with each other at the intervertebral foramen. The **dorsal nerve root** is connected to the posterolateral aspect of the spinal cord by a series of rootlets arranged in a short vertical line. As it passes to join the **ventral nerve root** in the intervertebral foramen it develops a swelling called the **spinal ganglion**. The ventral root has no ganglion, and is joined to the anterolateral aspect of the spinal cord by rootlets which are arranged in an untidy bunch rather than in a straight line.

The length of these nerve roots varies considerably. They are attached to the spinal cord in series from above downwards, and we can think of the cord as being divided into spinal **segments** by imaginary planes drawn horizontally across it between the attachments of successive pairs of spinal nerves. For example, the portion of spinal cord to which the third cervical pair of nerves is attached by means of their nerve roots is called the third cervical segment, and so on. In the cervical region each spinal segment is very nearly opposite the point at which its spinal nerve emerges, but lower down, because of the discrepancy between the length of the cord and the length of the vertebral canal, the nerve roots have a progressively longer distance to travel in the canal before they reach the appropriate intervertebral foramen and can unite to form their nerve trunk. The fifth sacral segment lies opposite the first lumbar vertebra, and the fifth sacral nerve roots have to run all the way down to the lower end of the sacrum before uniting to form the fifth sacral nerve. The lower part of the vertebral canal is thus filled with a leash of nerve roots which are travelling more or less vertically downwards to their foramina, and these constitute the **cauda equina**, or 'horse's tail' (Fig. 30).

The dorsal roots contain only sensory fibres running into the spinal cord, and the ventral roots contain only motor fibres leaving it; this is known as the **Bell–Magendie law**. When the two roots meet at the intervertebral foramen, both types of fibre become intermingled, and the nerve which they form is therefore called the **mixed spinal nerve**.

Each spinal nerve runs only a very short course outside the vertebral canal before dividing into two main branches—the **dorsal ramus** and the **ventral ramus** (Fig. 31). Both rami eventually split up into medial and lateral branches. Those of the dorsal ramus supply the tissues of the back, and those of the ventral ramus travel to the tissues of the front and sides of the body. The ventral rami are the bigger and the more important; they alone grow out into the limbs. In the thorax the ventral rami are known as the **intercostal nerves,** because they run round the trunk between the ribs. In the neck and abdomen, with no ribs to guide them, the course of the ventral rami is less regular, and the arrangement becomes grossly distorted where the limbs arise from the body.

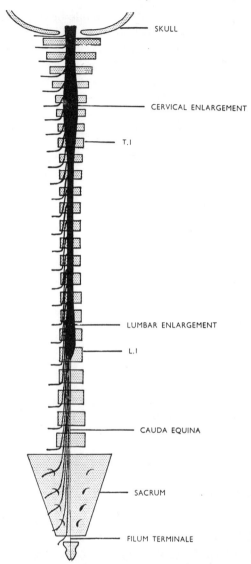

SKULL

CERVICAL ENLARGEMENT

T.1

LUMBAR ENLARGEMENT

L.1

CAUDA EQUINA

SACRUM

FILUM TERMINALE

Fig. 30. The relationship of the spinal cord to the vertebral column (highly diagrammatic) T.1. = 1st thoracic vertebra: L.1 = 1st lumbar vertebra.

Plexuses

The spinal nerves communicate with each other to form **plexuses**. The simplest form of plexus is the **loop plexus** which is to be found between all the dorsal rami. Each ramus is connected with the rami above and below by 'loops' consisting of fibres passing from one ramus to another (Fig. 32). Arrangements in the plexuses connecting the ventral rami tend to be more complicated, for the limbs introduce difficulties. Nevertheless the **cervical plexus** (p. 265) and the **lumbar plexus** (p. 285) are simple loop plexuses between the ventral rami of the first four cervical and the first four lumbar nerves respectively. The great **brachial** (p. 328) and **sacral** (p. 377) plexuses which serve the limbs are structurally more intricate, but their general principle is similar. On the spinal cord side of the communicating loops lie the rami which feed the plexus, and

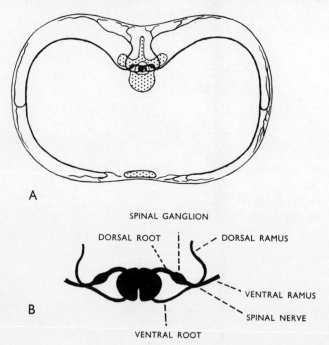

Fig. 31. Diagram of a spinal nerve. In A, a thoracic vertebra and the sternum are stippled, and a pair of spinal nerves and their branches are shown in black inside the outline of the body in horizontal section. B is an enlargement showing the mode of formation of the spinal nerve by its two roots.

from the peripheral side of the loops of communication arise the **branches** of the plexus. Branches which spring from the plexus formed by ventral rami are often given special names—an honour denied those of the dorsal plexuses.

The existence of a plexus allows a fibre emerging from the spinal cord in a given nerve root either to continue to the periphery in a branch passing outwards at the same level, or, by traversing the connecting loops, to enter another branch at a different level. Wandering fibres need not stop at the branch above or below, but may travel up or down for long distances (theoretically the whole length of the vertebral column) before they decide to resume their outward journey.

The branches of a plexus, like the rami which contribute to it, are mixed nerves, containing both motor and sensory fibres, and they subdivide into smaller 'motor' and 'sensory' branches which supply muscles and skin respectively: (it has already been pointed out that both 'motor' and 'sensory' nerves are really mixed). Each motor branch has a **motor distribution** to one or more muscles, and each sensory nerve has a **sensory distribution** to skin and/or deeper structures. The territory of every cutaneous nerve is overlapped at its edges by the territories of other nerves. The central undisputed area is the

Fig. 32. Diagram of a simple loop plexus. On the left are the rami constituting the 'roots' of the plexus, and on the right are its branches. Fibres entering the plexus by a given ramus may be distributed, via the connecting 'loops', into several different branches.

autonomous zone, while the peripheral areas of shared innervation are the **overlap zones** (Fig. 33). Overlap exists on a microscopic scale even within the autonomous zone, for it has been mentioned already that any one point on the skin is supplied by several different nerve *fibres*.) By blocking the cutaneous nerves with local anaesthetic, or by examining patients with nerve injuries, it is possible to obtain a fairly accurate idea of the average cutaneous distribution of a given nerve.

The overlap between the muscular distributions of different nerves is less obvious, but several muscles receive motor branches from two different named nerves. On the microscopic scale, the overlap and interlocking of motor units has already been mentioned (p. 57).

The general re-distribution of fibres in a plexus means that each branch of the plexus carries fibres derived from several different nerve roots, and, conversely, that the fibres in any one nerve root are distributed along several different pathways to the periphery. The group of muscles to which a single ventral nerve root sends fibres is called a **myotome,** and the individual muscles of the myotome may receive these fibres via several distinct named nerves. The muscles constituting a myotome may lie at some distance from each other, and need have no direct relation to the territory of skin supplied by the same spinal segment. Any given muscle is usually supplied by at least two spinal segments because its motor nerve contains fibres derived from several ventral roots.

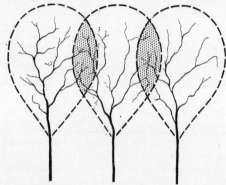

Fig. 33. Diagram showing the territories of three cutaneous nerves. The overlap zones are stippled.

The skin supplied by a given dorsal nerve root is called its **dermatome.** Adjacent dermatomes overlap each other much more extensively than do the cutaneous distributions of peripheral nerves, so that it is difficult to map out single dermatomes. However, it is possible to get an idea of their extent by mapping the loss of sensation in patients who have suffered damage to the spinal cord at a known level. If one patient has the cord cut across at the level of the fourth thoracic segment, and another at the level of the fifth, then the difference in the level of the sensory paralysis in the two patients roughly corresponds to the extent of the fourth thoracic dermatome. Alternatively, a known dorsal nerve root may be stimulated at operation in a conscious patient, and the area of pain produced may be mapped. A prolapsed intervertebral disc may press on the roots of the cauda equina, and the map of the patient's pain may be equated with the findings at operation.

The cranial nerves

The cranial nerves have little in common except the fact that they spring from the brain. There are said to be twelve pairs, though the **olfactory** nerves are not really a 'pair' since there are about fifteen or twenty of them on each side. Several cranial nerves (the first, second, third, fourth, sixth and eighth pairs) are wholly devoted to the sensory and motor supply of the organs of special sense, and are considered with these organs; the detailed distributions of the remainder are dealt with in the section on topography (p. 264).

The cranial nerves

Name	Category	Distribution	Remarks
1. Olfactory	Special sense	Mucous membrane of nose	15-20 small nerves on each side
2. Optic	Special sense	Retina of eye	Part of the brain; the fibres have thus no Schwann sheath
3. Oculo-motor	Motor	Muscles of eye	Parasympathetic supply to smooth muscle of eye
4. Trochlear	Motor	Superior oblique muscle of eye	
5. Trigeminal	Mixed	Sensory to face and underlying structures: motor to muscles of mastication	
6. Abducent	Motor	Lateral rectus muscle of eye	
7. Facial	Mainly motor	Muscles of facial expression	Parasympathetic to the salivary glands: taste from front of tongue
8. Stato-acoustic	Special sense	(a) Cochlear to organ of hearing (b) Vestibular to vestibular apparatus	Two completely distinct nerves
9. Glosso-pharyn-geal	Mainly sensory	Sensory to pharynx and tongue: motor to stylo-pharyngeus	Taste from back of tongue. A few para-sympathetic fibres to salivary glands
10. Vagus	Mainly motor	Larynx and pharynx: thorax and abdomen	Parasympathetic to thorax and abdomen
11. Accessory	Motor	Muscles of soft palate: sternocleidomastoid, trapezius	Two distinct nerves combined
12. Hypo-glossal	Motor	Muscles of tongue	

The cranial nerves do not form plexuses in the way that the spinal nerves do, though some of them communicate with their neighbours on leaving the skull; this is notably true of the **vagus** nerve. This nerve, as its name implies, wanders far afield, and supplies striated and visceral muscle in the neck, thorax and abdomen, as well as giving sensory branches to the viscera. In contrast to the wide distribution and catholic constitution of the vagus, the trochlear and the abducent nerves each supply nothing more than one somatic muscle (p. 123).

Such individual attention emphasizes the importance of movements of the eyeball.

The great sensory nerve of the head is the **trigeminal**, which has a large and well-developed **semilunar ganglion** lying just inside the skull, as well as a much smaller motor root which goes to supply the muscles of mastication. Other sensory ganglia are found in relation to the seventh, ninth, and tenth pairs of nerves, but there is no regular arrangement of motor and sensory roots such as there is among the spinal nerves.

In one instance fibres from a spinal nerve join forces with a cranial nerve: a twig from the first cervical nerve runs along with the **hypoglossal** (twelfth cranial) nerve, bound up in the same connective tissue sheath, and leaves it again later to supply certain muscles in the neck. A similar arrangement holds in the **accessory** nerve, which is a composite of two parts. The cranial part springs from cells in the brain; the spinal part is a curious motor nerve which contains fibres from the first five cervical segments of the spinal cord. Its roots come off from the lateral side of the cord, mid-way between the ventral and dorsal nerve roots. The spinal part travels up through the foramen magnum into the skull, and joins the cranial part in the same covering sheath. The two components soon part company again, and the spinal portion emerges from the skull to supply two muscles (p. 264) while the cranial portion joins the **vagus** nerve.

The autonomic system

The autonomic system controls most of the involuntary mechanisms of the body, such as the contraction of smooth muscle, the secretion of glands, the regulation of the heart rate, and the maintenance of blood-pressure. It has two portions, the **sympathetic** and **parasympathetic systems,** both of which are under the general control of a part of the brain called the **hypothalamus** (p. 108).

The sympathetic system

The **sympathetic** nervous system is the more widespread, and innervates visceral muscle throughout the body. Its cells of origin lie in the spinal cord between the first thoracic and the second lumbar segments, and the whole of the **sympathetic outflow** emerges from these segments. Since the sympathetic is a motor system, its fibres come out in the ventral nerve roots and run into the spinal nerves; as they leave the grey matter they acquire a myelin sheath in the usual way, and as they pass into the nerve root a sequence of Schwann cells is added. As soon as the spinal nerve is well clear of the intervertebral foramen, the sympathetic fibres leave it to run to a small ganglion situated on the neck of the corresponding rib. Since the fibres are myelinated, the bundle which they form looks white, and is called the **white ramus communicans**. In the ganglion many of the fibres relay, and the fresh neurones immediately pass back into the spinal nerve. These neurones are non-myelinated, the bundle looks grey, and is referred to as the **grey ramus communicans** (Fig. 34). After reaching the spinal nerve again, the sympathetic fibres run in it to the visceral muscle in the distribution of its branches.

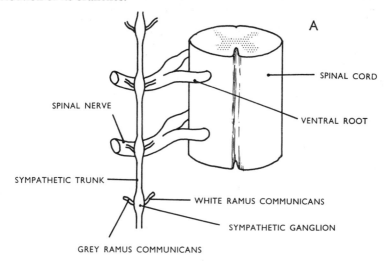

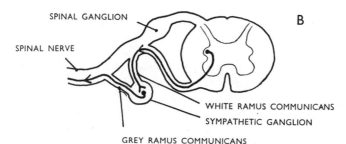

Fig. 34. The sympathetic trunk and rami communicantes. A: surface view from in front. B: horizontal section showing the pathway of sympathetic fibres and the relay in the sympathetic ganglion.

Instead of running straight back into their own mixed spinal nerve, the sympathetic fibres may travel up or down a longitudinal chain formed by bundles of fibres running from one sympathetic ganglion to another. This **sympathetic trunk** extends well beyond the limits of the sympathetic outflow; for example, there are three ganglia in the neck, and a variable number in the abdomen and pelvis, terminating in the mid-line **ganglion impar** in the region of the coccyx (Fig. 35). From all these ganglia grey rami communicantes pass to the cervical, lumbar and sacral nerves, but no white rami pass into the ganglia, since all the fibres they receive come to them via the sympathetic trunk.

Chapter 8

The sympathetic trunk has branches of its own; fibres may run up and down it for some distance, and after relaying may leave it to pass into one of these branches instead of entering a spinal nerve at all. In this way the thoracic and abdominal viscera are supplied. The **splanchnic nerves** arise from the ganglia in the lower part of the thorax and run down into the abdomen, where they join a series of autonomic plexuses from which the supply of the gut is largely derived. The chief of these is the **coeliac plexus,** and lies on each side of the upper part of the back wall of the abdomen.

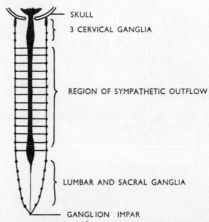

SKULL

3 CERVICAL GANGLIA

REGION OF SYMPATHETIC OUTFLOW

LUMBAR AND SACRAL GANGLIA

GANGLION IMPAR

Fig. 35. Diagram of sympathetic trunks, showing the outflow from the spinal cord.

Closely related to the lateral part of the coeliac plexus on each side is a **suprarenal gland**. This twofold endocrine gland consists of a cortex and a medulla. The **medulla,** or core of the gland, is functionally part of the sympathetic system, and produces noradrenaline (p. 76) as well as the closely related **adrenaline**. When the medulla of the suprarenal is stimulated by impulses reaching it through the coeliac plexus, noradrenaline and adrenaline enter the blood stream and circulate round the body, alerting smooth muscles everywhere, and 'preparing the body for fight or flight'. The blood vessels in the skin and the wall of gut constrict, and those in the muscles, heart and brain dilate, so that the blood is diverted to the organs which need it most in an emergency. The heart rate increases, the blood-pressure rises, the pupil of the eye dilates, the muscles propelling the food along the alimentary canal relax and the sphincter muscles contract, so that no energy is wasted in non-essential activity. The muscles of the respiratory passages relax, so that more air can be taken into the lungs. The resemblance of the general picture to the effects of emotion is well marked, and in fact many emotional disturbances produce their bodily effects through the liberation of adrenaline. Fear, rage, horror, etc., in some way all affect the hypothalamus, which in turn responds by stimulating the sympathetic system. If emotions outlast the need for them, as when fear turns to an 'anxiety state', an excess of adrenaline may circulate in the blood over long periods. It is thought that through this mechanism long continued mental and emotional conflicts may manifest themselves as apparently unconnected bodily disturbances.

The parasympathetic system

The **parasympathetic** system is antagonistic to the sympathetic. In contrast to the sympathetic system, the parasympathetic has a **cranial outflow** in four of the cranial nerves and a **sacral outflow** in the second, third, and fourth sacral segments of the spinal cord. Fibres run in the **oculomotor** nerve to the vicinity of the eyeball, where they leave it to relay in the small **ciliary ganglion**. The secondary neurones enter the eyeball to supply the muscles which constrict the pupil and focus the lens (pp. 118, 119).

Parasympathetic fibres also emerge in the **facial** nerve and run to a series of ganglia in relation to the branches of the trigeminal nerve. The second order neurones from these ganglia supply the salivary glands (p. 171) and the lacrimal gland (p. 125); they are assisted by a few fibres which emerge in the **glossopharyngeal** nerve. But the most important part of the cranial outflow runs in the **vagus** nerve to the viscera of the thorax and abdomen and reaches as far down as the lower part of the large intestine. The pelvic viscera are innervated by the sacral parasympathetic outflow, the fibres of which mingle with the sympathetic plexuses in the pelvis, and are then distributed with them by running in the walls of the blood vessels.

The parasympathetic has not such a wide distribution as the sympathetic, for there are no parasympathetic fibres running to the limbs. It differs, too, in that its fibres do not relay soon after leaving the central nervous system, but defer their synapse until they are close to or actually in the organ to be innervated. Again, the parasympathetic endings, like the cerebrospinal endings, act by liberating acetylcholine and not noradrenaline. The parasympathetic has two outflows, not one, and it has no system of generalized stimulation such as the sympathetic possesses in the suprarenal medulla.

All the viscera have a dual nerve supply, from both sympathetic and parasympathetic, and the effects of stimulating each are diametrically opposed. The parasympathetic causes the salivary and lacrimal glands to secrete; the sympathetic, by constricting their blood-vessels, dries them up (the dry mouth is a familiar symptom of fear). The parasympathetic constricts the pupil, slows the heart, contracts the wall of the gut and the bladder and relaxes their sphincters, thereby moving on the contents.

Both parts of the autonomic system are motor. However, the sensory fibres coming from the viscera take the line of least resistance and travel along the same pathway as the autonomic fibres, usually choosing the sympathetic fibres as their guide. They thus enter the central nervous system at the same level as the point at which the sympathetic supply of the organ in question leaves it.

The sensory supply of the viscera is not so rich as the sensory supply of the skin, and in consequence many stimuli can be applied to them without causing pain. Once the abdominal cavity has been opened under a local anaesthetic, a viscus like the stomach can be cut or burned or clamped without any discomfort to the patient. The pangs induced by unripe fruit or other similar indiscretions are caused by spasm of the wall of the gut, which constitutes a large enough stimulus to arouse the sparsely distributed sensory endings.

The operation of **sympathectomy** consists in cutting the sympathetic supply to a limb. The blood vessels, thus divorced from the nerve supply on which their tone depends, dilate passively under the influence of the blood pressure, and in such conditions as Raynaud's disease, where there is a spasm of the blood-vessels in the fingers, the operation may be extremely beneficial by improving the circulation.

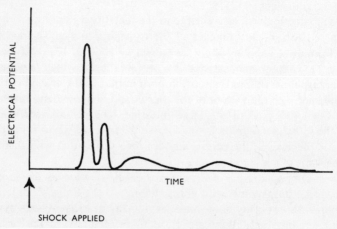

Fig. 36. Electroneurogram.

Structure of peripheral nerve

A peripheral nerve contains nerve fibres which vary in diameter from 15–20 μm down to less than 1 μm. The largest fibres come from the sensory endings in somatic muscle, and most of their diameter is made up by their coating of myelin; among the thinnest fibres are those belonging to the autonomic system. When an electric shock is fed into a nerve trunk, the fibres conduct the stimulus along the nerve at speeds which vary with their thickness. Recording from the nerve trunk at a distance from the point of stimulation gives a picture called an **electroneurogram** (Fig. 36), which shows a number of peaks. The first peak is due to impulses which have travelled along the thickest fibres, and successive peaks are due to groups of smaller and more slowly conducting fibres. In this way we can get an idea of the general grouping of fibres of different size which the nerve contains, and can compare the **fibre spectrum** of one nerve with that of another.

Each nerve fibre is surrounded by delicate connective tissue known as **endoneurium,** and is wrapped into a bundle or **funiculus** with some of its fellows by rather more dense **perineurium.** Between the funiculi run blood vessels supported in still more connective tissue, and the whole structure is surrounded by a tough fibrous **epineurium** (Fig. 37). In some nerves there are many funiculi, and in others most of the fibres are bundled together into one large funiculus. The arrangement of funiculi is not constant in the corresponding nerve in different patients, and one cannot safely predict the position of the fibres going to a given muscle or coming from a given area of skin. The amount of connective tissue in different nerve trunks varies widely, and in some nerves, such as the sciatic, may account for as much as four fifths of the cross-sectional area. This fibrous tissue gives the nerve its tensile strength, which is considerable. At operation it is possible to lift the whole lower limb from the table by a finger hooked under the sciatic nerve.

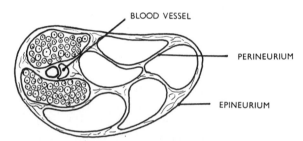

Fig. 37. Diagrammatic cross-section of a small peripheral nerve. Two funiculi are shown in detail. The axis cylinders show as dots surrounded by vacant spaces, for the fatty myelin dissolves out in the process of preparing the section.

Effects of nerve injuries

Peripheral nerves

Peripheral nerves pursue well-defined pathways to their destinations, running through the surrounding structures in a **course** which differs little from one individual to another. The structures in the immediate vicinity of the course of a nerve constitute its **relations,** and these may be a source of danger to it. For instance, a nerve in close contact with a bone may be injured by the fragments should the bone be broken, and one passing close to a joint may be damaged if the joint is dislocated. A nerve which comes close to the surface may be injured by a blow or cut.

Motor paralysis: If there is a complete interruption of all the fibres in the nerve there will be a **motor paralysis** of all the voluntary muscles supplied by its branches. The muscles are incapable of contracting, for the line of communication from the central nervous system has been cut. They are flaccid, having lost their nerve supply (p. 57), and because their reflex arcs have been interrupted the deep reflexes in the distribution of the nerve are abolished (p. 77). This picture of flaccid paralysis with loss of deep reflexes characterizes the **lower motor neurone lesion**, which can be produced by injury to the motor side of the reflex arc anywhere along its length, from the motor cells in the spinal cord (brain, in the case of cranial nerves) to the terminal muscular branches of the peripheral nerves.

The muscular paralysis interferes with voluntary movements. If all the muscles which combine to perform a certain movement are supplied by the cut nerve, then there is a complete paralysis of that movement. If the movement is performed by one or more muscles supplied by the nerve in conjunction with others having a different nerve supply, it is weakened but not altogether lost.

The flaccid muscles can no longer oppose the pull of their healthy antagonists, and the joints on which they act may be pulled into an abnormal position. This **postural deformity** may be severe (Fig. 163, p. 333). In addition, a **configurational deformity** develops as the denervated muscles waste (p. 21), and the contours of the affected part come to differ from those of its healthy counterpart.

Finally, there is a loss of tone and power in the smooth muscle supplied by the sympathetic fibres in the nerve trunk, so that the blood vessels in the cutaneous distribution of the nerve dilate under the blood pressure. For a time after the injury the area remains flushed and red, but later the absence of normal voluntary movements in the part leads to an inefficient clearance of blood from the local veins. In consequence the flow of blood through the territory of the nerve becomes sluggish, the slowly-moving blood loses its oxygen, and the part becomes blue and cold. The loss of sympathetic fibres also means that no sweating can occur in the territory of skin which the nerve supplies.

Sensory paralysis: When a peripheral nerve which supplies skin is cut, the autonomous zone (p. 83) becomes completely anaesthetic to all forms of stimulation, and in the overlap zones which surround it sensibility is partly lost; the annulus of partial sensibility in turn shades off into regions of normal sensibility. In addition to the disturbance of skin sensation, the loss of deep sensory innervation causes a loss of proprioceptive information (p. 71). The brain is inadequately informed about the position of the joints and the degree of tension of the muscles in its distribution; the result is that the patient is unaware of the position of the denervated part until he looks at it.

Trophic paralysis: The skin becomes thin and unhealthy, smooth and shiny. Any nails in the territory of the nerve become dry and brittle and may fall out, together with the hairs. The affected muscles waste and shrink, and ultimately their bellies become replaced by fibrous tissue. These nutritional changes are certainly partly due to disuse, and are described as **disuse atrophy**, but impulses coming down the nerve fibres exercise some sort of nutritional effect, perhaps through the chemicals known to be liberated at their terminals, and this is thought to be an additional factor.

Disability: These changes must all be considered in relation to the total effect on the patient. This depends on the particular nerve severed, and also upon his occupation. A small patch of cutaneous anaesthesia on the dorsum of the foot is not a great trouble to the average person, but if the same size of patch occurs on the pads of the thumb and first two fingers of the right hand it is a catastrophe. An inability to move the fingers separately may not greatly disturb the earning capacity of a navvy, but it at once finishes a musician's career. Again, a man can keep an eye on the joints of his little finger if they have lost their proprioceptive innervation, and no great harm will ensure if he does not, but if his ankle joint has lost its sensory supply he may put his foot down wrongly and sprain his ankle, or perhaps break his leg. Brief accounts of the individual results of paralysis of the main nerve trunks in the upper and lower limbs will be given later (e.g. pp. 330, 379).

Nerve roots

Cutting a ventral nerve root produces a lower motor neurone paralysis in which few muscles are completely paralysed and many are partially paralysed. The characteristic myotome paralysis differs in its grouping from the paralysis induced by interruption of a peripheral nerve, and this serves to distinguish between damage to a nerve and damage to a root. Because of the extensive overlap of contiguous dermatomes injury to a single dorsal nerve root seldom gives rise to an easily detectable sensory paralysis. The patient is, however, conscious of an alteration in his sensory capacity.

Degeneration and regeneration

When a peripheral nerve is cut across, the axons and dendrites within it are cut off from their parent cell bodies, and cannot continue an independent existence. For a time the peripheral part of the nerve remains capable of conducting electrical impulses fed into it, but this rapidly fails, and the nerve becomes physiologically inert. The axis cylinders break up into little fragments like the dots and dashes of the Morse code, the chemical constitution of the myelin alters, and it coalesces into droplets of unsaturated fatty acids. The myelin and axis cylinder débris is soon scavenged up by macrophages, so that eventually all that is left is the framework of Schwann tubes, cleaned and empty, waiting for the nerve cells to attempt to send out processes along them again. Degeneration extends in the proximal stump up each nerve fibre as far as the nearest node of Ranvier, but no further; the axis cylinder beyond this point is still part of the parent cell and does not die.

After some time sprouts appear at the end of these intact axis cylinders, and wriggle slowly down the Schwann tubes to the gap in the nerve. This gap is full of blood clot, which may have become converted into scar tissue. Through it the regenerating fibres have to find their way to the waiting tubes of the peripheral stump. Many lose their way, and run off into the surrounding tissues, or are blocked by trying to grow into firm scar tissue. The more closely the two ends of the nerve are apposed, and the less bleeding there has been, the easier it is for the fibres to cross the gap, and the better chance there is of subsequent functional recovery. If a sprouting nerve fibre finds its way into one of the empty peripheral Schwann tubes, it grows down it at a rate of from 1–3 mm a day. The growing end is sensitive, and tapping over it produces a tingling sensation. This is **Tinel's sign**, and indicates how far the growing ends have travelled. When a sprouting motor fibre reaches the periphery—perhaps after growing for two years or more—it may, if all goes well, establish a functional connexion with a motor end-plate. If the muscle has been kept exercised by physiotherapy and not permitted to atrophy it once more starts to respond to messages from the central nervous system, and a degree of functional recovery takes place. Similarly sensory fibres may connect with sensory endorgans, and sensation may gradually return to the denervated skin.

Recovery is always imperfect, and usually grossly so, for several reasons. The growing fibres often find their way into the wrong type of tube, so that a motor fibre is conducted down to a sensory ending, or a sensory fibre to a motor end-plate; if this happens, they give up the struggle and again degenerate. Even if they do find an appropriate tube, it may not conduct them to their accustomed destination. Particularly if there has been any twisting of the peripheral stump relative to the proximal stump, a motor fibre which has formerly helped to control (say) the flexor muscle of the little finger may now find itself innervating the adductor of the thumb. The brain has been accustomed to send messages along this fibre when the patient desired to move his little finger, and it is disconcerting to find that the result of such messages is now an adduction of the thumb. Similarly, sensory fibres may find themselves misdirected, so that a message received at the brain may mean that the back of the little finger has been touched, instead of the medial side of the palm, as such messages have always previously meant. The patient, in short, has to be re-educated in the meaning of the impulse-patterns which his peripheral nerves conduct to his central nervous system, and in the regrouping of the motor impulses which he sends down to his muscle to produce a desired movement. This is very hard for any adult, but it is relatively easy for an infant, who has not already learned fixed sensory and motor patterns. The best recoveries from nerve injuries are thus made when the patient is only a few years old, and the functional result may occasionally be indistinguishable from normal. Nothing like this can ever be expected in an adult.

Regeneration does not occur in the central nervous system; the regenerating fibres appear to need the support and guidance of the Schwann tubes, which do not exist inside the central nervous system. For this reason also cut dorsal nerve root fibres cannot re-establish their central connexions. Any destruction of fibres in the central nervous system or in the dorsal nerve roots thus produces permanent disability.

Nerve blocks

The territory of a nerve may be rendered anaesthetic by cooling or by pressing on the nerve for a period of twenty minutes or more, and so temporarily abolishing its ability to conduct impulses. The familiar experience of a limb 'going to sleep' is the result of compressing one or more of its nerves. If cold or pressure are too long continued the paralysis becomes irreversible. A drunken patient who goes to sleep with his arm over the back of a chair may fail to wake up in time to rescue one of his main nerves from lasting paralysis. But the most important way of producing a nerve block is by injecting a **local anaesthetic** round the nerve trunk. A great range of local anaesthetics is available, the most important of which is **xylocaine.** The drug penetrates to the axis cylinder at the nodes of Ranvier, and there interferes with the ionizing mechanism for a variable period. The addition of adrenaline to the solution constricts the blood vessels and prevents the drug from being removed too rapidly from the scene of action.

Local anaesthetics may also be used to block the spinal nerve roots inside the vertebral canal. If this is done outside the protective coverings of the spinal cord it is known as an **extradural block,** if inside, it is called a **spinal block** (p. 101).

Referred pain

Any stimulus strong enough to excite impulses when applied directly to a nerve trunk may cause pain to be felt, not in the nerve itself, but in the skin it supplies. The sensation is **referred** to the peripheral distribution of the nerve because the brain cannot distinguish between impulses aroused in a nerve fibre by stimulation of its endings and those aroused by direct stimulation of the fibre itself.

The cut ends of the nerves in an amputation stump may be irritated by the pressure of the developing scar tissue, or by infection. As a result, the patient may be fully conscious of his absent limb, and may feel it performing various movements, some of which cause him pain. Such **phantom limbs** are very common in the early stages after amputation, but the patient often hesitates to . speak of them, because he thinks he will be laughed at. Later the limb may appear to shorten, and eventually may disappear entirely. Some people may have very severe pain in their phantom limbs.

Pain is also often referred from one part of the body having a poor sensory innervation to another which has a good one. For example, the pain of heart disease is felt in the chest wall in front of the heart, or in the left arm or left side of the head and neck. After much argument, the mechanism of this is still not settled. Most references of pain occur within the compass of the distribution of a single spinal segment, but some jump over several segments; very seldom is referred pain left throughout the whole of any one segmental distribution.

The spinal cord

Structure and function

The thin cylindrical spinal cord lies in the vertebral canal. In the embryo it extends the full length of this canal, but it does not grow as fast as the vertebral column does, and so becomes relatively shorter as growth proceeds. In the adult it is about 45 cm long, and extends from the foramen magnum of the skull, where it is continuous with the brain, to the level of the first lumbar vertebra, or sometimes slightly lower (Fig. 30). Its lower end is a conical swelling known as the **conus medullaris**, and to the apex of this is attached a thin glistening thread, the **filum terminale**, which anchors the cord to the front of the coccyx. The filum terminale is surrounded by the lower spinal nerve roots, forming the cauda equina.

In the cervical region there is a **cervical enlargement**, which corresponds to the segments innervating the upper limb. In this enlargement the cord is wider from side to side than antero-posteriorly, and here it attains its maximum diameter of about 1·4 cm. Another swelling, the **lumbar enlargement**, lies above the conus medullaris and corresponds to the segments supplying the lower limb; the cross-section here is almost circular.

In the mid-line of the anterior aspect of the spinal cord there is a narrow linear **anteromedian fissure**; posteriorly a linear groove marks the position of the **posteromedian septum** (Fig. 38). The dorsal nerve roots enter the cord along a posterolateral groove on each side, while the ventral nerve roots emerge on the anterolateral aspect of the cord.

The grey matter of the spinal cord is embedded in the middle of the white matter, like the pattern in a stick of rock. This pattern varies at different levels, but always has the form of a capital H in cross-section. The cross-bar of the H is called the **grey commissure**; in its centre is the tiny **central canal**. The uprights of the H run from the region where the dorsal nerve root enters to near the spot where the ventral nerve root emerges. The two ends of each upright are the **anterior** and **posterior horns** of grey matter; in the thoracic and first two lumbar segments of the cord there is a **lateral horn** on each side, lying between the other two and opposite the grey commissure. The tip of the posterior horn is covered by specialized grey matter called the **substantia gelatinosa**.

On each side the portion of the white matter between the posterior horn of grey matter and the posteromedian septum is the **posterior white column**, the portion between the anterior horn and the anteromedian fissure is the **anterior white column**, and the rest of the white matter constitutes the **lateral white column**. In front of the grey commissure the **white commissure** connects the two anterior white columns. The white matter decreases as the cord is traced distally, for at the lower levels not so many sensory fibres have entered the cord and most of the motor fibres have left it. The grey matter is most abundant in the two enlargements of the cord, where it has to accommodate the nerve cells

which control the limbs. The cells in the anterior horn of the grey matter innervate somatic muscle, and those of the lateral horn are the cells of the sympathetic outflow. In **acute anterior poliomyelitis** (infantile paralysis) the anterior horn
cells are damaged or killed, and in their distribution a lower motor neurone
paralysis results. The posterior horn receives the axons of the sensory cells
stationed in the dorsal root ganglia, and the grey matter of this horn contains
intercalated cells. The typical spinal reflex arc may thus be represented as
shown in Fig. 39.

The white matter close to the grey matter is made up of fibres which emerge
from the grey matter, run up or down the cord for a variable distance, and sink
again into the grey matter. They put one segment in communication with

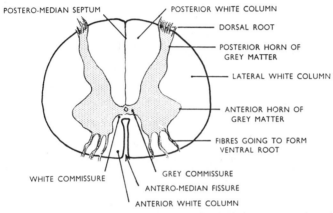

POSTERO-MEDIAN SEPTUM

POSTERIOR WHITE COLUMN

DORSAL ROOT

POSTERIOR HORN OF
GREY MATTER

LATERAL WHITE COLUMN

ANTERIOR HORN OF
GREY MATTER

FIBRES GOING TO FORM
VENTRAL ROOT

WHITE COMMISSURE

GREY COMMISSURE

ANTERO-MEDIAN FISSURE

ANTERIOR WHITE COLUMN

Fig. 38. Cross-section of the spinal cord in the cervical region. Note the central canal in
the grey commissure.

another, and make up the **intersegmental tracts**. The white matter further away
from the grey core is occupied by fibres which may run up or down for much
longer distances, and sometimes traverse the whole length of the spinal cord.
Different **tracts** are defined according to the destination of the fibres which they
contain. For example, in each posterior column there is a **fasciculus gracilis**,
with a **fasciculus cuneatus** lateral to it (Fig. 40). Both consist of fibres ascending
to the brain. Another important ascending tract lies just lateral to the anterior
horn; it is called the **lateral spinothalamic tract**. Close to it on its lateral side is
the **anterior spinocerebellar tract**, and behind this is the **posterior
spinocerebellar tract**; in front of the anterior horn is the **anterior spinothalamic
tract**. Their boundaries are not nearly so clear cut as was once supposed, and it
does not follow that because a given fibre lies in a given region of the cord it will
pursue a given path. Nor are the descending pathways much better defined. The
most important are the **lateral corticospinal** (crossed pyramidal) **tract**, in the
posterior part of the lateral white column, and the **anterior corticospinal** (direct
pyramidal) **tract** close to the anteromedian fissure. Anterior to the lateral

corticospinal tract is a mixed collection of descending fibres called the extrapyramidal tracts (Fig. 40).

Destruction of the posterior white column of one side by disease or injury results in the patient being unable to tell the position of the limbs of that side unless he is watching them. If his toes are moved passively he cannot say whether they are being bent or straightened unless he is watching. The posterior columns are therefore necessary for proprioception on the same side of the body. After a transverse cut through the anterolateral part of the white matter the patient can no longer appreciate pain, warmth or cold on the *opposite side* of the body below the cut. Touch sensibility is only slightly affected, and

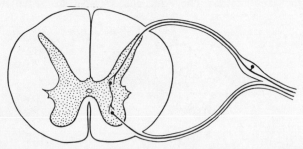

Fig. 39. Diagram of a typical spinal reflex arc. The incoming axon ends round an intercalated cell in the base of the posterior horn; this cell in turn synapses with an anterior horn cell whose axon passes out into the ventral nerve root. (The presence of a lateral horn of grey matter indicates that the section is from the thoracic region or the first two lumbar segments).

proprioception remains practically intact. This operation of **anterolateral cordotomy** is therefore sometimes performed to relieve pain due to an incurable cause. The clinical results indicate that the fibres in this region are necessary for certain forms of exteroceptive perception, and that they must cross over to the opposite side of the cord before running up to the brain. This crossing takes place for the most part in the white commissure, and is not horizontal; the fibres take two or three segments to incline gradually over to the other side.

A spinal injury involving the lateral corticospinal tract produces a motor paralysis of the *same side* of the body below the lesion. The paralysis produced by interrupting the motor pathway from the brain to the anterior horn cells differs markedly from the paralysis produced by destruction of lower motor neurones (p. 92). In **upper motor neurone** paralysis there is, naturally, a loss of voluntary power; the brain is unable to transmit its commands to the muscles concerned. But there is no flaccidity, as when the lower motor neurone is cut; on the contrary, the tone of the muscles is increased, and the paralysis is said to be **spastic** because of the characteristic 'spasm' of the limb muscles. The deep reflexes are not abolished, but exaggerated (p. 77). Just how these results are brought about is not known, but the evidence suggests that the extrapyramidal fibres are as much concerned as the pyramidal ones. Spastic paralysis with increased deep reflexes is typical of injury to the motor pathway at any point between its origin in the brain and the anterior horn cells of the spinal cord. For example, an injury to the brain at birth can produce the condition known as **cerebral palsy,** or congenital spastic paralysis. In such cases the tone of the muscles is increased, and the strongest groups of muscles pull the limbs into postural deformities (in this case it is paralysed muscles which do the pulling, not healthy ones overcoming paralysed antagonists). An exactly similar type of paralysis results from the vascular injury to the upper motor neurones in the brain which is known as a 'stroke'.

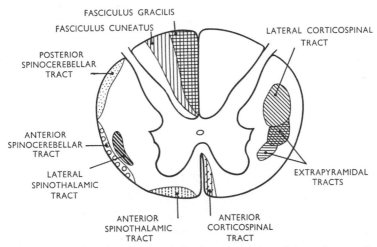

Fig. 40. Cross-section of spinal cord indicating the position of the main ascending pathways (left side of figure) and descending pathways (right side of figure).

Spinal meninges

The spinal cord is closely covered by a thin membranous coat, the **pia mater,** which contains the blood vessels supplying the cord. It dips into the anteromedian fissure, and gives a covering to the anterior and posterior nerve roots as they leave the cord. Outside the pia there is a space full of fluid, so that the cord is held suspended in a water bath which cushions any sudden shocks applied to the vertebral column. This **cerebrospinal fluid** (CSF) is contained in a tube formed by another thin membrane called the **arachnoid mater**; the CSF is said to occupy the **subarachnoid space.** The arachnoid is closely applied to the inner surface of the third and last coating of the cord, the dense and tough **dura mater.** The dura is attached to the margins of the foramen magnum, and continues down as far as the level of the second piece of the sacrum, forming, with the arachnoid, a strong bag which envelops the spinal cord and the upper part of the cauda equina in CSF (Fig. 41). From each side of the spinal cord, midway between the lines of attachment of the ventral and dorsal nerve roots, a series of about twenty ligaments projects into the subarachnoid space to become attached at intervals to the dura-arachnoid coating. These **ligamenta denticulata** occur in the intervals between successive pairs of spinal nerves (Fig. 41), and suspend the spinal cord in its tube of fluid. The covering membranes of the spinal cord are collectively called the **spinal meninges**. Each spinal nerve root

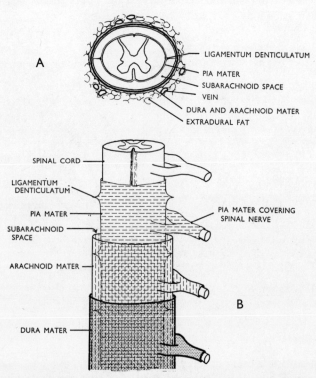

Fig. 41. A: cross-section of vertebral canal showing the spinal cord stabilized within the tube of dura mater and arachnoid mater by the ligamenta denticulata. B: scheme showing how the membranes clothe the cord and are continued over the nerve roots. From below up successive layers have been stripped off until the naked spinal cord is reached.

receives a covering from each of the three meninges; the outermost is naturally derived from the dura, and becomes continuous with the epineurium of the spinal nerve. In this way the dural tube is anchored to the intervertebral foramina, and similarly the filum terminale gets a covering of dura-arachnoid which fixes the lower end of the tube to the coccyx.

Outside the dural tube the remainder of the space within the vertebral canal is packed with semi-fluid fat, which affords an additional soft protective covering for the cord. In this **extradural space** ramify the branches of venous plexuses which drain the cord and the vertebral bodies, and across it pass the nerve roots on their way to the intervertebral foramina. Below the level at which the dural tube terminates, the whole of the vertebral canal is filled up in this way.

Lumbar puncture

If a hollow needle is pushed into the vertebral canal between the third and fourth or between the fourth and fifth lumbar spines it can be made to enter the subarachnoid space, and the CSF will drip out at a steady rate. Samples of the fluid can be taken for analysis, and drugs can be injected through the needle. Antibiotics can be given to control an acute **meningitis,** or a local anaesthetic may be used to produce the condition of **spinal anaesthesia.** By replacing an equivalent amount of CSF with anaesthetic solution the spinal roots are blocked within the arachnoid tube.

It is also possible to block the roots as they pass from the dural tube across the extradural space to leave the vertebral canal, and this procedure is known as **extradural** or **epidural anaesthesia.** The solution will not spread up and down so freely as in spinal anaesthesia, and it may be necessary to make several injections. In the region of the sacral nerves there is no subarachnoid space in which to put an injection, and **sacral epidural blocks** are used in midwifery.

The brain

The central nervous system is essentially a thick-walled tube of grey and white matter. In the spinal cord its construction is relatively simple, and the central cavity is a tiny canal just visible to the naked eye (Fig. 38). The upper part of the tube, however, undergoes a number of contortions as it develops, and becomes much more complicated. The central canal opens out into a series of interconnected caverns called **ventricles** which form a major feature of the anatomy of the brain. The brain is divided into **forebrain, midbrain** and **hindbrain** (these names are given from their relative positions in the embryo). The forebrain has developed two large **cerebral hemispheres** which conceal most of the rest of the brain when it is viewed from above or from the side (Fig. 42). Behind and below the cerebral hemispheres lie the two hemispheres of the **cerebellum,** and buried under the cerebrum and the cerebellum lie the relatively small midbrain and hindbrain. The whole brain weighs about 1·3 kg, and it has long been a source of satisfaction to male medical students that the male brain weighs rather more than the female. Unfortunately this difference is simply due to the larger size of the male body as a whole, and there is no evidence to show that size is in any way related to performance.

The hindbrain

The spinal cord is continuous with the **medulla oblongata** (this name for the posterior part of the hindbrain is often contracted to **medulla**; there is little chance of confusion with the **medulla spinalis**, which is the official name for the spinal cord, because this name is seldom used). The medulla is about 2·5 cm long, and its thickness increases as it is traced proximally. On its anterior surface are two swellings, the **pyramids**, between which the anteromedian fissure of the spinal cord is continued upwards; the fissure is partly obscured by fibres passing from one pyramid to the other, and so forming the **decussation of the pyramids**. Lateral to the pyramids lie a pair of smaller swellings called the

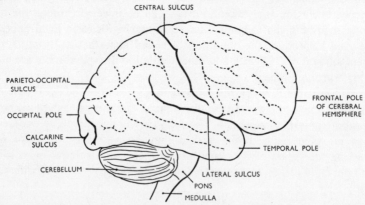

Fig. 42. Lateral surface of the brain. The great cerebral hemisphere overlaps the hindbrain and completely conceals the midbrain.

olives, and lateral to these again the medulla is overlapped by the cerebellar hemispheres; the dorsum of the medulla is completely covered by the middle part of the cerebellum. Between the pyramid and the olive of each side emerge the rootlets of the **hypoglossal** nerve (12th cranial), and those of the **accessory** (11th), **vagus** (10th) and **glossopharyngeal** (9th) nerves come out between the medulla and the cerebellar hemisphere (the nuclei containing the cells of origin of all these nerves lie in the substance of the medulla).

At the superior end of the medulla a large band of fibres sweeps transversely across the mid-line to connect the two cerebellar hemispheres (Fig. 43). The elevation produced by these fibres is the **pons**, which is the second component of the hindbrain. The pons contains the nuclei of the **facial** (7th) and the **abducent** (6th) nerves and the greater part of the nucleus of the **trigeminal** (5th) nerve. The abducent nerve appears on the surface between the pons and the medulla, and the facial more laterally, in the cerebellopontine angle. The large trigeminal nerve leaves the substance of the pons anterolaterally. Close to the facial nerve the fibres of the statoacoustic (8th) nerve enter the pons. Their further travels will be considered later (p. 129).

In the substance of the medulla the central canal of the spinal cord inclines dorsally and widens out into a shadow diamond-shaped cavity which is prolonged superiorly into the pons. This is the **fourth ventricle** (the ventricles are numbered from above downwards), and it has a pyramidal roof, rather like a tent (Fig. 52). The superior part of the roof is covered over by cerebellum, but the posterior and inferior part is very thin, and is perforated by a **median aperture**, or **foramen of Magendie**, and two **lateral apertures**, or **foramina of Luschka**. Deep to the floor of the fourth ventricle are clustered the nuclei of the last seven cranial nerves.

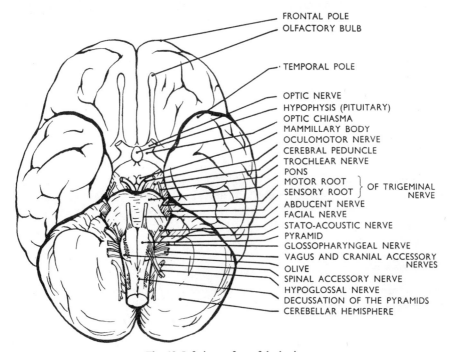

FRONTAL POLE
OLFACTORY BULB

TEMPORAL POLE

OPTIC NERVE
HYPOPHYSIS (PITUITARY)
OPTIC CHIASMA
MAMMILLARY BODY
OCULOMOTOR NERVE
CEREBRAL PEDUNCLE
TROCHLEAR NERVE
PONS
MOTOR ROOT ⎱ OF TRIGEMINAL
SENSORY ROOT ⎰ NERVE
ABDUCENT NERVE
FACIAL NERVE
STATO-ACOUSTIC NERVE
PYRAMID
GLOSSOPHARYNGEAL NERVE
VAGUS AND CRANIAL ACCESSORY NERVES
OLIVE
SPINAL ACCESSORY NERVE
HYPOGLOSSAL NERVE
DECUSSATION OF THE PYRAMIDS
CEREBELLAR HEMISPHERE

Fig. 43. Inferior surface of the brain.

The largest component of the hindbrain is the **cerebellum** (Fig. 43), which consists of two hemispheres joined by a middle portion related to the medulla and pons. The hemispheres have a core of white matter enveloped by a thin **cortex** of grey matter which is scored and furrowed by closely spaced parallel grooves and fissures. Embedded in a white matter is the **dentate nucleus**, a curious crumpled structure shaped like a sack with an open mouth. All the efferent fibres from the cerebellum arise from the inside of this nucleus and pass out through the 'mouth' of the sack. Afferent fibres reach the cerebellum along three large bundles or **peduncles** which connect the cerebellum to the medulla, pons and midbrain. The **inferior** peduncle contains, among other things, the fibres of the **posterior spinocerebellar tract** and fibres from the **vestibular** apparatus which lies in the temporal bone and has an important function in relation to balance and posture (p. 129). The fibres of the **middle** peduncle, which form a great part of the pons, come from the cerebral hemisphere to the cerebellum, mostly relaying on the way in a number of scattered **pontine nuclei**. The **superior** peduncle is largely efferent, and contains fibres coming from the dentate nucleus which run upwards to the midbrain, and so ultimately to the **thalamus** and the cerebral hemisphere. In addition, the superior peduncle contains the **anterior spinocerebellar** tract, whose fibres have climbed to the midbrain only to descend again to the cerebellum. The fibres reaching the cerebellum are thus mainly derived from proprioceptive sources, and indeed one of the main functions of the cerebellum is the co-ordination of posture and the maintenance of balance (see p. 227). If it is damaged, there is giddiness, loss of muscle tone, and a consequent tendency to fall over. Each hemisphere is responsible for its own side of the body. Thus, damage to the right hemisphere causes loss of tone on the right half of the body, and an inability to perform co-ordinated movements on that side. This capacity to group together related muscles as a team in order to carry out a complex movement is the second prime function of the cerebellum. The **cerebrum** (the term given to the two cerebral hemispheres) issues the orders, but the details of carrying them out are left to the cerebellum, which adjusts the degree of contraction of the muscles necessary to maintain posture while the movement is carried out. The instructions framed by the cerebellum are distributed entirely by the efferent fibres coming from the dentate nucleus, and many of these return to the cerebrum before descending to the spinal cord.

The midbrain

The midbrain is the smallest subdivision of the brain. It is about 2 cm in length, and is continuous below with the pons, while above it divides, like the trunk of a tree, into two large bundles called the **cerebral peduncles**, which then sink into the cerebral hemispheres. The peduncles carry the main tracts passing up and down between the cerebrum and the rest of the nervous system. Dorsal to, and between them, is the **tectum** of the midbrain, which is tunnelled through by a small channel connecting the fourth ventricle to the third ventricle in the forebrain. This canal is the **cerebral aqueduct**, or **aqueduct of Sylvius**. The tectum is composed of four rounded swellings called **colliculi** arranged in two pairs. The **inferior** pair are connected with the mechanism of hearing, and the **superior** pair with the mechanism of sight (pp. 129, 121). In the midbrain are the nuclei of two cranial nerves which emerge from its substance, the **trochlear** (4th) dorsally and the **oculomotor** (3rd) anteriorly.

The forebrain

The forebrain consists of two cerebral hemispheres, joined to each other across the midline by a series of bands of fibres called **commissures,** the most striking of which is the **corpus callosum.**

Each hemisphere consists of a core of white matter enveloped in a thin rind of grey matter, the **cerebral cortex.** The cortex is furrowed by fissures called **sulci,** and the areas between the sulci are known as **gyri.** The pattern formed in this way is variable from one individual to another, but its general features are fairly constant. Looked at from the side, each hemisphere has a characteristic shape. The anterior end, or **frontal pole,** is blunt and rounded, and occupies the region behind the forehead. The posterior end, or **occipital pole,** is rather more pointed; it lies deep to the external occipital protuberance (Fig. 114). In addition there is a **temporal pole** deep to the temple (Fig. 42).

There are two important sulci on the lateral aspect of the hemisphere. The **lateral sulcus** runs almost horizontally backwards superior to the temporal pole. Approaching the posterior end of this sulcus, but not quite meeting it, is the **central sulcus,** the upper end of which cuts the superior border of the hemisphere (Fig. 42).

Superiorly, the hemispheres are separated by a deep cleft called the **longitudinal fissure,** occupied by a double fold of dura mater known as the **falx cerebri** (Fig. 51). The medial aspect of the cerebral hemisphere can only be properly seen by dividing it from its fellow (Fig. 44). The cut passes through the corpus callosum and through other less important commissures; it also runs through the middle of the thin slit-like median cavity of the **third ventricle,** which is expanded continuation of the aqueduct of Sylvius (Fig. 52). The medial surface of the hemisphere is flat, and the upper end of the central sulcus can be seen appearing over its superior border. Another important sulcus is the **parieto-occipital** sulcus, which runs obliquely upwards and backwards from a horizontally running fissure which it divides into two parts; this is the **calcarine** sulcus.

The inferior surface of the hemisphere is more irregular. In the centre, between the two hemispheres, lies the midbrain dividing into the cerebral peduncles, and in front of this, between the two temporal poles, can be seen the two **optic** nerves entering into the formation of an X-shaped crossing called the **optic chiasma** (see p. 121), from which spring the two **optic tracts.** In the area between the peduncles and the chiasma are two small **mamillary bodies,** and in front of these the stalk of the pituitary gland (Fig. 43). All these structures lie inferior to the third ventricle. Further forward, and stretching almost to the frontal pole on each side, lie the **olfactory tracts,** terminating in the expanded **olfactory bulbs,** which receive the **olfactory** nerves. Overlapping the olfactory tracts at their termination, close to the point where the optic tracts arise from the chiasma, is a folded-over region of the temporal pole called the **uncus.**

Some of the surface features of the cerebral hemispheres are used to divide them up into arbitrary **lobes** for the purpose of simplifying description. The **frontal** lobe is all that part of the hemisphere anterior to the central sulcus; the **parietal** lobe lies between the central sulcus and the parieto-occipital sulcus; the

occipital lobe is the part of the hemisphere posterior to the parieto-occipital sulcus, and the **temporal** lobe lies inferior to the lateral sulcus (Figs. 42, 44).

The cerebral cortex

The structure of the cortex varies from place to place, and it is possible to make a histological map of the surface of the hemisphere. It is also possible, by observing the results of injury, disease, or electrical stimulation, to obtain an idea of the functions of different regions of the cortex. Sometimes histology and clinical observation agree in marking out an area of distinct structure and definite function; thus, the region round the occipital pole and the calcarine

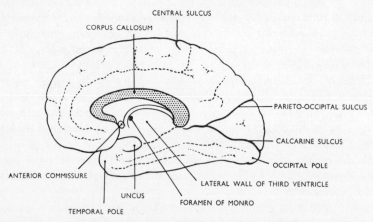

Fig. 44. Medial surface of the right cerebral hemisphere.

sulcus has a characteristic band of fibres in the cortex called the stria of Gennari, and the presence of this band accurately delineates the area of the brain concerned with vision. In other areas, however, differences in structure appear to have little functional significance.

Clinical work has established certain well-defined functional areas (Fig. 45). The **precentral gyrus** is the great **motor** area of the brain, and the body is 'represented' upside down in this gyrus—that is, an injury to the superior part of the gyrus leads to paralysis of the leg, while damage to the lower part paralyses the head and tongue. Behind the central sulcus, in the **postcentral gyrus**, lies the **sensory** area, and here also the representation of the body is upside down. The motor and sensory areas of one hemisphere are in charge of the opposite (contralateral) side of the body, so that damage to the right motor cortex will produce paralysis on the left side of the body. This situation contrasts with the conditions in the cerebellum, where each hemisphere controls the same (ipsilateral) side of the body.

The **visual** region of the cortex is perhaps the best demarcated of the functional areas; both eyes are represented on each side. The **auditory** area lies in the **superior temporal gyrus** just within the lips of the lateral sulcus, and the **olfactory** area occupies the uncus. Surrounding all these **primary areas** like a fringe are other **secondary** or **association areas** which are also concerned with the same function in a more interpretative way. Thus, damage to the auditory association area may lead to disturbances of speech, writing or reading, which go under the generic name of **aphasia**. For example, a patient may be able to hear someone talking but the words appear to him meaningless. Or he may be

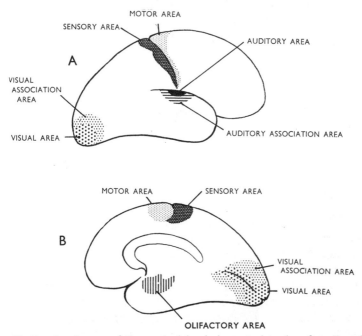

Fig. 45. Functional areas of the cerebral hemisphere. A: lateral surface. B:medial surface.

able to write, but be unable to read. Or he may understand what is said to him but be unable to speak himself, though the motor mechanism of his tongue and larynx may be unaffected. The area controlling speech is known as **Broca's area**, and is largely unilateral. It used to be thought that in left-handed people it lay in the right hemisphere and vice versa, but the evidence is not satisfactory. A similar idea is that one hemisphere 'dominates' the other, and that this accounts for the 'handedness' of the individual. Training and habit certainly account for a great deal of 'handedness' in children, but animals are just as much left or right-handed as children, and the concept of the dominant hemisphere has something to recommend it.

Apart from the primary and secondary areas which they contain, the functions of the parietal and temporal lobes are still largely mysterious. A complicated system of connexions involving a structure known as the **hippocampus** on the under-surface of the temporal lobe, a flat band of fibres inside the hemisphere called the **fornix**, and the **cingulate gyrus** on the medial aspect of the hemisphere, is concerned with emotional reactions, and possibly also with memory. This system, which has connexions with the hypothalamus (see below), is often called the **limbic system**, or **limbic lobe** of the brain.

A glance at Fig. 45 will show that large regions of the hemisphere do not appear to have any specific function, and these are sometimes called the **silent areas**. For example, the region of the frontal pole has many times been destroyed without producing any apparent change in the individual concerned. However, the under surface of the frontal lobe has a special significance, for by cutting the fibres which lead to and from it personality changes may be produced in the patient. The operation of **frontal** or **orbital leucotomy** has sometimes been used in psychiatry for the treatment of violent or grossly disturbed patients, and in successful cases these patients have been able to return to satisfactory everyday life with their families. Leucotomy can also be effective in cases of long-continued pain. After operation the patient 'feels' the pain as much as ever, but simply ceases to pay any attention to it, so that it does not upset him.

The basal ganglia

In the depths of the white core of each hemisphere, close to the region where the cerebral peduncle enters, are masses of grey matter called the **basal ganglia**. The most important of these are the **thalamus**, the **caudate nucleus** and the **lentiform nucleus**. The thalamus is a large ovoid structure close to the peduncle; it is a relay station for afferent fibres on their way to the cortex. The two thalami lie very close to each other, separated only by the third ventricle, and occasionally they may fuse with each other across the midline, so obliterating part of the ventricle. The posterior part of the thalamus overhangs the superior colliculus of the midbrain, and develops two small swellings, the **medial** and **lateral geniculate bodies**. The lateral aspect of the thalamus is covered by the sensory fibres which enter and leave it.

The region just inferior to the thalamus is the **hypothalamus** (Fig. 48), which contains nuclei controlling the general pattern of the autonomic responses of the body. These nuclei have extensive connexions with the limbic system, and functionally the hypothalamus mediates the bodily reaction to emotional stimuli (p. 88). It also controls the activity of the pituitary gland (p. 198), the temperature of the body, and feeding behaviour; it is said to have many other more speculative functions.

Also below the thalamus, and reaching down into the peduncle of the midbrain, lie various other nuclei with a supposed motor function. Some of these, such as the **substantia nigra**, may be important; others, like the **red**

nucleus, are important in animals like the cat, but probably have less significance in the human being. These nuclei are nowadays grouped together with other nuclei in the pons and midbrain under the name of the **reticular formation.** Some of the fibres which arise from the reticular formation run down the spinal cord as part of the extrapyramidal pathway.

The reticular formation is an immensely complicated network which can alter or even suppress incoming messages, and is concerned also with the maintenance of attention and consciousness. Damage to this system may result in prolonged unconsciousness, as is sometimes seen after severe brain injury.

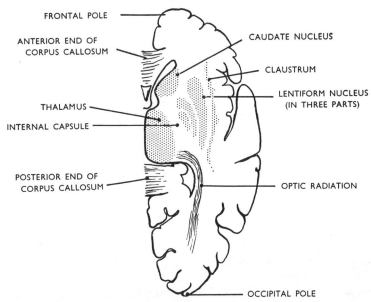

Fig. 46. Horizontal section through right cerebral hemisphere, showing the internal capsule. (The lentiform nucleus is composed of three interconnected parts, and the claustrum is a mass of grey matter of unknown function).

Anterolateral to the thalamus lies the **caudate nucleus,** which is shaped like a comma set on its side, with a long tail curving posteriorly and downwards. The body of this nucleus is fused with the body of the **lentiform nucleus,** and the two form a single functional unit which receives fibres from the motor area and sends out efferents to many parts of the nervous system. Unfortunately little is known about these fibres. It is thought that the condition known as Parkinsonism, in which there are tremors, disturbances of muscle tone and movements, and other manifestations of disturbed motor function, is due to disease of the caudate and lentiform nuclei, but direct evidence is scanty.

The lentiform nucleus lies lateral to the thalamus in the angle between it and the caudate nucleus (Fig. 46). Between the three main ganglia lies a narrow strait, the **internal capsule**, through which the fibres of the main sensory and motor pathways running to and from the cortex must pass. If these fibres are suddenly destroyed by bleeding which ploughs up the soft brain tissue, the disaster is known as a **stroke**, and leads to a sensory and motor paralysis which, like that resulting from destruction of cortex, is contralateral. This is because the sensory fibres cross over in the spinal cord, and most of the motor fibres cross over in the decussation of the pyramids, while the remainder (travelling in the direct pyramidal tract) cross over near their termination in the cord. A paralysis of one half of the body is called a **hemiplegia**.

As the afferent fibres leave the constriction of the internal capsule, they spread out to run to all parts of the cortex. Those concerned with general sensation form a part of a twisted fan-shaped pathway called the **corona radiata**; this also contains the motor fibres converging on the internal capsule as they descend from their broad origin in the precentral cortex.

In the middle of the white matter of the hemisphere lies the **lateral ventricle** (since there is one on each side, the ventricle between the thalami is called the third and not the second ventricle). The lateral ventricle (Fig. 52) has a body, an anterior horn, a large inferior horn, and a small posterior horn opening off it. In the floor of the body are the thalamus and the tail of the caudate nucleus; the inferior horn extends into the temporal lobe and almost reaches the temporal pole. At the junction of the body and the anterior horn each lateral ventricle opens into the third ventricle through an **interventricular foramen (foramen of Monro).**

The electroencephalogram

It is possible to record the electrical activity which occurs on the surface of the brain by placing electrodes on the scalp, and the changes in potential between the electrodes (Fig. 47) produce the **electroencephalogram**. By using many pairs of electrodes it is possible to detect the exact situation of abnormal electrical discharges, and the tracing may thus be used to localize intracranial disease.

When the eyes are shut, the normal tracing shows a periodicity called the **alpha rhythm**. If the subject opens his eyes, or has to perform some mental task, such as memorization or calculation, the rhythm disappears and the tracing becomes irregular. It has been suggested that the electrical activity of the cortex plays the part of a scanning mechanism which examines incoming patterns of impulses and 'recognizes' familiar ones, much as the scanner of a television set converts patterns of electrical impulses into a recognizable picture on the screen. When no impulses other than the familiar background of proprioception are being received, the scanning mechanisms synchronize their activity to produce the alpha rhythm, but as soon as activity begins this regularity is disturbed. This idea is derived from **cybernetics**, a subject in which the working of the brain is compared to the working of computers—which are, indeed, often called electronic brains. No analogy of this kind can possibly be exact, but it is sometimes useful to regard the brain in this light. Another illuminating cybernetic concept is that of **negative feedback**. This means that part of the output of a machine is fed back into it to control its stability. Thus, the governor of a steam engine is driven by the engine itself. If the governor rises too high it automatically cuts out the drive until this allows the governor to fall again. Applied to the brain this idea means that any source of impulses should have some of these impulses fed back into it to control its activity. For example, the cerebral cortex sends messages down to the cerebellum, and some of this activity ought to be returned to the cortex from the cerebellum. This is confirmed experimentally, and the principle seems to be one of general applicability. There is also (in the white matter of the hemisphere) a vast system of **association fibres** which connect different parts of the cortex and help it to work as an integrated whole rather than as an aggregation of disconnected parts.

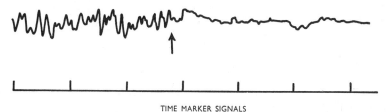

TIME MARKER SIGNALS

Fig. 47. Electroencephalogram. The left-hand part of the tracing was taken while the eyes were shut, and shows alpha rhythm; at the arrow the eyes were opened, and the rhythm is destroyed.

Summary of main fibre pathways

1. *The motor pathways*

The main motor pathway is the corticospinal, or **pyramidal tract**, which originates in the precentral cortex. Its fibres run down through the corona radiata, the posterior limb of the internal capsule, and the anterior part of the cerebral peduncle into the pons, where they are split up by the transversely running fibres of the middle cerebellar peduncle. After entering the ventral part of the medulla, four-fifths of them cross over to the opposite side in the decussation of the pyramids to form the crossed pyramidal tract (lateral corticospinal). These fibres run down in the lateral white column till they reach the appropriate level, when they enter the grey matter and end round the anterior horn cells. The remaining one-fifth of the fibres continue down in the anterior white column as the uncrossed pyramidal (anterior corticospinal) tract until they reach the appropriate level, when they cross over in the white commissure to end round the anterior horn cells of the opposite side (Fig. 48).

The **extrapyramidal** fibres originate in such places as the caudate nucleus and lentiform nucleus, the substantia nigra and the other nuclei of the reticular

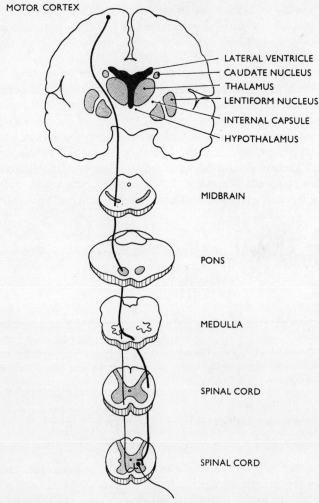

Fig. 48. Diagram of the motor pathway. The crossed pyramidal (lateral corticospinal) tract is indicated by the thick black line; the uncrossed pyramidal (anterior cortico-spinal) tract by the thin line: both eventually cross to end round the anterior horn cells of the opposite side.

formation, and the tectum of the midbrain. They all run down in the lateral white column and are presumed to cross over to the opposite side at some point in their course.

2. *The exteroceptive sensory pathways*

Many of the fibres entering the spinal cord through the dorsal nerve roots relay in the posterior horn of grey matter. The second order neurones cross over

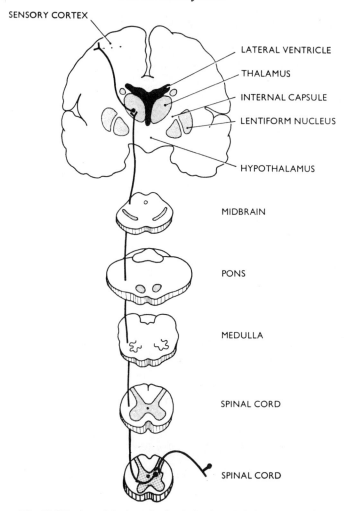

SENSORY CORTEX

LATERAL VENTRICLE

THALAMUS

INTERNAL CAPSULE

LENTIFORM NUCLEUS

HYPOTHALAMUS

MIDBRAIN

PONS

MEDULLA

SPINAL CORD

SPINAL CORD

Fig. 49. Diagram of the lateral spinothalamic system (exteroceptive).

gradually in the white commissure to the other side of the cord. Most of them go to the lateral spinothalamic tract of the opposite side, which travels up near the surface of the medulla, passes through the pons and midbrain, and ends in the ventrolateral part of the thalamus, where a second relay takes place. The third order neurones run up in the internal capsule and the corona radiata to the postcentral gyrus (Fig. 49).

Other exteroceptive fibres do not relay in the posterior horn, but enter the posterior white column of their own side and run up with the proprioceptive fibres, following a very similar pathway.

3. *The proprioceptive sensory pathways*

Many of the incoming proprioceptive fibres enter the posterior column of the same side of the spinal cord, and ascend in this position to the medulla, where they relay in two collections of nerve cells—the **nucleus gracilis** and the **nucleus cuneatus**. The second order neurones run up to the thalamus, crossing over to

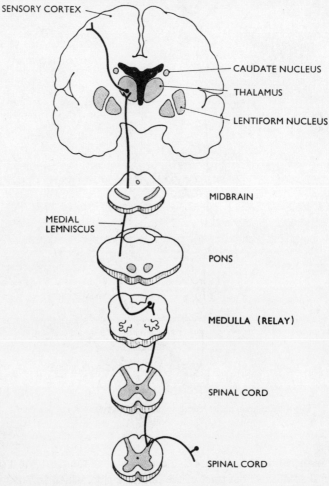

Fig. 50. Diagram of the posterior column/medial lemniscus system (proprioceptive and exteroceptive).

the opposite side in a prominent tract known as the **medial lemniscus** (Fig. 50). The third order neurones pass from the thalamus to the post-central gyrus.

Other proprioceptive fibres relay in the posterior horn and pass to the posterior spinocerebellar tract on the same side. This leads them to the inferior cerebellar peduncle, and they enter the cerebellum without having crossed to the opposite side. Still others relay in the posterior horn and then mostly cross to the opposite anterior spinocerebellar tract, which leads them up to the superior cerebellar peduncle and so to the opposite cerebellar hemisphere.

Cranial meninges

The brain is covered by the same three membranes as cover the spinal cord, but the arrangement is different. The cranial pia mater, like the spinal pia, closely invests the surface of the brain and dips down into all the fissures and sulci, carrying blood vessels to the cortex of the cerebrum and cerebellum, as well as to the midbrain and the hindbrain. The arachnoid, however, unlike the spinal arachnoid, fuses with the pia where it passes over the gyri of the cerebral cortex, though it does not dip into the sulci. In this manner a series of subarachnoid spaces are formed, which intercommunicate to form one continuous space. In some places, for example between the posterior aspect of the medulla and the cerebellum, there are larger subarachnoid **cisterns**. The cranial dura mater is not fused with the arachnoid as is the spinal dura, but separated from it by a very narrow **subdural space**, so that when the skull is opened, the dura can be incised and folded back without dislodging or damaging the arachnoid. The dura is composed of two layers, the outer one being fused with the periosteum clothing the inside of the skull; there is no extradural space. Nevertheless, the dura can be stripped off the skull by bleeding from one of the arteries supplying the bone, and this is referred to as an extradural haemorrhage. The inner layer of dura is in places doubled up to form large folds. One of these hangs down between the two cerebral hemispheres (Fig. 51) as the **falx cerebri**, while another serves as a shelf on which the posterior parts of the hemispheres are supported, so taking their weight off the cerebellum; this is the **tentorium cerebelli**. In some regions the inner layer of dura is separated from the outer by large venous spaces called the **sinuses** of the dura; elsewhere the two layers are firmly attached to each other.

The circulation of the cerebrospinal fluid

Into each of the ventricles of the brain there protrudes a thin vascular fold of membrane containing a **chorioid plexus** of blood vessels. The fluid filtered off from the blood by this plexus (cf. p. 119) enters and fills the ventricular system (Fig. 52). CSF formed in the lateral ventricles escapes through the foramina of Monro into the third ventricle, where more is added by the chorioid plexus of that ventricle; the contents then pass along the aqueduct of Sylvius to the fourth ventricle, where still more CSF is added, and escape through the foramina of Magendie and Luschka into the **cerebellomedullary cistern** of the subarachnoid space. The CSF works along the ramifications of the space until it eventually reaches the wall of one of the venous sinuses of the dura. Here the arachnoid is attached to the dura and the CSF passes through both into the sinus and thus returns to the blood stream again.

Some CSF enters the subarachnoid space of the spinal cord, which is continuous with the cerebellomedullary cistern. It escapes in the sheaths of the spinal nerves (p. 91), and is absorbed into the spinal veins. Some of it trickles down among the nerve fibres, and is called the **endoneurial fluid**. Dyes introduced into a peripheral nerve may be detected in the CSF shortly

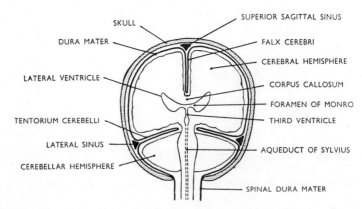

Fig. 51. Diagram of vertical section through skull showing the arrangement of the inner layer of the dura mater. The outer layer, which is fused with the periosteum and in places with the inner layer, is not shown, for the sake of clarity.

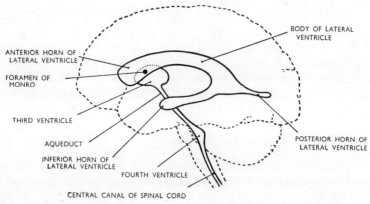

Fig. 52. Diagram of the ventricular system of the brain. Only one lateral ventricle is shown.

afterwards, having diffused upwards against the current. This fluid pathway is one means by which infection can enter the cerebrospinal fluid.

There is thus a continuous circulation of CSF, and if the flow is blocked at any point a catastrophe will result. Blockage of the foramina of Magendie and Luschka can occur following meningitis, the fluid damming up inside the ventricular system. The chorioid plexuses can filter fluid against a fairly heavy pressure, so more and more fluid is still formed, and the pressure built up inside the brain makes it bulge. Since the brain is contained inside a rigid box, this compresses the brain substance, and may be sufficient to interfere with its blood supply; it also forces the hindbrain down into the foramen magnum and so causes an increasing disturbance of its function. The clinical results of this general situation are shown in headache, vomiting, disturbance of vision, coma, and eventually death. The same results may also be produced by interference with the absorption of the fluid.

If CSF is dammed up in an infant, whose skull is not rigid since the bones composing it have not yet fused together, the skull will enlarge very considerably; the condition is known as **hydrocephalus**.

Blood supply of the central nervous system

Two systems of arteries supply the brain substance, **cortical** and **central**. In both systems communications with adjacent arteries are not free enough to provide a 'bypass' for the blood should one of them become blocked (p. 136). Such an occurrence thus destroys brain substance, and if a vital spot is involved blockage of even a tiny vessel may result in paralysis or even death. The central arteries are very thin walled, and may give way under the stress of high blood pressure. This is particularly so in the arteries which supply the region of the basal ganglia and internal capsule, one of which is actually called the **artery of cerebral haemorrhage**. Rupture of this artery gives rise to a **stroke** (p. 110).

The special senses

The eye

Structure and function

The eyeball is roughly spherical and lies in a pyramidal cavity known as the **orbit**. It has a tough fibrous outer coat called the **sclera**, a middle vascular one called the **chorioid**, and an inner sensitive layer containing the visual endorgans and known as the **retina** (Fig. 53).

The **sclera** is everywhere opaque except in the anterior one sixth of the eye, where its structure is modified in such a way as to render it transparent. The transparent portion is the **cornea**, which has a smaller radius of curvature than the rest of the eye, so that it bulges forwards. It has no blood vessels, and is nourished by the tears which flow over it anteriorly, and the **intraocular fluid** which bathes it internally. At the **corneoscleral junction** a small venous channel, the **sinus venosus** of the sclera, runs circularly round the margin of the cornea (Fig. 54).

The **chorioid** coat contains the blood vessels supplying the eyeball, and also a large amount of pigment to prevent light being scattered and reflected inside the eye; the inside of a camera is blackened for the same reason. In the region of the corneoscleral junction the chorioid becomes modified into a disc called the **iris** which is attached all round the sclera just behind the sinus venosus. The iris has a hole in the middle called the **pupil**, the size of which can be adjusted by two visceral muscles. The **sphincter pupillae** (supplied by parasympathetic fibres in the oculomotor nerve) runs circularly round the pupil and diminishes the aperture; the **dilatator pupillae** (supplied by fibres from the cervical sympathetic

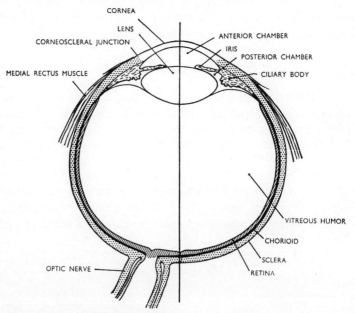

Fig. 53. Horizontal section through the right eye. The straight line indicates the axis of the eye, and passes through the macula lutea.

trunk) radiates like the spokes of a wheel from the margins of the pupil to the corneoscleral junction, and its pull enlarges the hole. The iris regulates the amount of light falling on the back of the eye, and is rendered opaque by the presence of pigment, particularly on its posterior surface. The amount of this pigment gives the eye its colour. Dark brown eyes have much pigment, blue eyes have less, and the iris of an albino has none, so that albinos cannot venture out into bright sunlight without risking damage to their eyes.

The chorioid coat does not extend further forward than the iris, behind which it develops a circular fringe of vascular and pigmented **ciliary processes** which protrude into the interior of the eye. From these processes, and from the adjacent chorioid, a very fine meshwork of fibres runs across and across to form a disc-like transparent **suspensory ligament**, in the middle of which, like an acid drop wrapped in cellophane, lies the elastic and transparent **lens**. The lens is biconvex, the front surface being flatter than the posterior surface. The tight wrapping squashes the lens, and if the pull of the ligament is relaxed the lens will bulge and become more convex. This is done by the radially directed **ciliary muscle,** which pulls the part of the chorioid to which the suspensory ligament is attached forwards towards the iris (Fig. 54). In this way it causes the focal length of the lens to shorten, so that objects near to the eye can be focussed on the retina; this process is called **accommodation.**

About middle age the elasticity of the lens begins to fail, and it no longer bulges so well when the tension on it is relaxed; this is the condition of **presbyopia,** and its victims require convex spectacles to enable them to read.

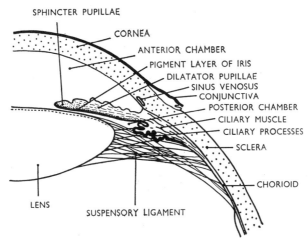

Fig. 54. Section through the ciliary region of the eye.

In old age the lens begins to lose its transparency; at first small opacities develop in it, and later the whole lens may become opaque; this is called a **cataract.** Such a lens can be removed from its sling of suspensory ligament by a small knife introduced through the front of the eye. The removal of the useless lens allows light to enter the eye, but objects cannot be focussed on the retina, and the lens has to be replaced by a convex glass lens placed in front of the eye.

The space between the iris and the cornea is the **anterior chamber** of the eye, and that between the iris and the suspensory ligament is the **posterior chamber;** both are filled with clear **intraocular fluid.** This is formed by filtration from the blood in the ciliary processes, and passes through the pupil into the anterior chamber, where it is absorbed into the sinus venosus. The circulation of the intra-ocular fluid is a continuous process comparable to that of the CSF. If the pupil becomes obstructed after an inflammation of the iris, or if the sinus venosus becomes blocked with inflammatory débris, the fluid becomes dammed up, and the pressure inside the eye rises—a condition known as **glaucoma.** If this is not treated, the sight will certainly be destroyed.

Behind the lens the cavity of the eyeball is filled with a jelly-like material called the **vitreous body**. It is mainly transparent, though it may contain small thread-like opacities.

The **retina** lies internal to the chorioid coat, and extends forwards nearly to the ciliary processes. The **optic nerve** enters the eyeball on the medial side of the axis of the eye, and its fibres immediately spread out in all directions to serve the retina, of which they form the innermost (next to the vitreous) layer. When the point of entry of the nerve is seen from the front of the living eye by means of an opthalmoscope, it stands out as a white circular plate called the **optic disc**, through which the branches of the **central artery of the retina** can be seen entering the eyeball. The retinal receptors are arranged in a layer next to the chorioid, and consist of rod-shaped cells (**rods**) and conical cells (**cones**). From these cells nerve fibres run to ganglion cells in the retina, and after relaying pass into the optic nerve. The light falling on the retina has to pass through several layers of fibres and ganglion cells before it can stimulate the receptors.

At the periphery of the retina receptors are relatively scanty, and there are very many more rods than cones, whereas in the densely populated area in the middle of the retina there is a small patch—on the central axis of the eye—where only cones are found. At this point the other layers of the retina dwindle to nothing, so that there is no obstruction to the light reaching the sensitive cells. This patch has a slight yellowish colouration which gives it the name of the **macula lutea** (yellow spot). It is on this spot that an object is focussed when we look directly at it, and the cones are thus the endorgans of **central vision**. The cones are responsible for visual acuity and colour vision, but they need a relatively strong light to stimulate them. The rods, on the other hand, are colour blind, and cannot form so sharp an image, but they are more sensitive to light. Every girl guide knows that in order to see a faint star she must not look directly at it, but rather to one side of it, so that the image will be formed on the peripheral part of the retina, where the rods are predominant.) Night vision is thus colour-blind and less acute than day vision. The rods require a pigment called **visual purple** which enables them to transduce light into electrical energy, and this pigment is a derivative of vitamin A. Deficiency of vitamin A can thus cause a condition called **night blindness**.

There is evidence that the cones contain three different pigments, deficiencies in which may form the basis of **colour blindness** (8 per cent of men and about 0·5 per cent of women are colour blind in some degree). A common defect is an inability to distinguish red from green, particularly under poor lighting conditions. Since traffic lights and railway signals utilize these colours the significance of red–green blindness is obvious.

Many people are mildly colour blind without realizing it, since they are often able to guess about colour well enough to get along, and it is only when they are confronted by a stringent test such as the **Ishihara figures** (p. 131) that they are lost.

To see anything in detail it must be focussed upon the macula, and the movements of the eyeball are reflexly controlled so that a moving object can be kept focussed in this position. If light falls on the optic disc the image formed will not be perceived because there are no receptors in the optic nerve head. We therefore have a **blind spot** in the **visual field** of each eye. We are not normally conscious of this because the two blind spots do not synchronize, and the gap in one visual field is filled by the other eye. But if we make two crosses about three inches apart on a piece of paper, shut the left eye, and look fixedly at the left-hand cross while moving the paper gradually closer to the eye there will come a stage when the other cross disappears; it is being focussed on the blind spot. If the paper is moved still closer, the cross will reappear because its image is now formed on the retina on the other side of the optic nerve head.

The optic nerve runs backwards through the optic canal into the skull. Shortly after this the two nerves communicate at the optic chiasma in such a way that the fibres coming from the medial parts of both retinae cross over to the opposite optic tract, while the fibres from the lateral halves of both retinae pass into the optic tract of the same side. The optic tract runs posteriorly round the cerebral peduncle, and ends in the **lateral geniculate body** (p. 108). Here a relay takes place, and the new fibres run in the substance of the hemisphere back to the visual cortex (Fig. 55). Other fibres pass from the lateral geniculate body to the superior colliculus of the same side, and so to the underlying oculomotor nucleus. These fibres subserve a number of reflexes. For instance, when light falls on the eye the pupil constricts (the **light reflex**) because the oculomotor nucleus sends impulses to the sphincter pupillae. This occurs in the other eye also, even though no light has been shone into it (**consensual light reflex**).

Optical defects

The focussing mechanism of the eye is frequently at fault. The light rays are bent by the curved surface of the cornea, which is never quite spherical. **Regular astigmatism** is due to the cornea having a greater curvature in one direction than in another, so that objects are not all accurately focussed in the same plane; the condition can be corrected by a cylindrical lens in front of the eye, set at the appropriate angle to the horizontal. **Irregular** astigmatism is more difficult to compensate for, but a contact lens, which is worn inside the lids and can be made to allow for the irregular curvatures of the cornea, is sometimes useful.

Refractive errors may also be due to a disparity between the refractive power of the lens and the length of the eyeball. In **myopia** (short sight) the eyeball is too long for its lens, so that objects lying at a distance are brought to a focus in front of the retina instead of on it (Fig. 56) and are consequently blurred. There is no difficulty about seeing objects close at hand; in fact, less effort is required than in the normal eye, for the ciliary muscle does not have to contract so hard. In **hypermetropia** (long sight) the eye is too short for its lens, and objects are brought to a focus behind the retina. In this condition the ciliary muscle is always working, even when the object is a distant one, and things close to the eye cannot be focussed at all. These two defects are readily corrected by placing in front of the eye spherical lenses of appropriate strength—concave (divergent) for short sight and convex (convergent) for long sight (Fig. 56).

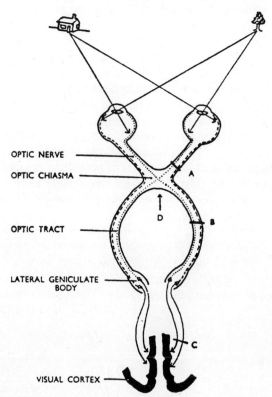

OPTIC NERVE

OPTIC CHIASMA

OPTIC TRACT

LATERAL GENICULATE BODY

VISUAL CORTEX

Fig. 55. Diagram showing the visual pathway. The fibres from the medial parts of both retinae (dotted lines) cross to the opposite side at the chiasma; those from the lateral parts (dashed lines) do not. An injury at 'A' causes blindness in one eye: the other eye continues to see both house and tree. Injury at 'B' or 'C' causes blindness to objects on the left: both eyes continue to see the tree, but the house cannot be seen at all. Injury at 'D' would mean that the right eye could not see the tree and the left eye could not see the house.

Movements of the eyeball

The eyeball is surrounded by fat, which facilitates movement, and special fascial slings and sheaths steady it under the pull of the six extraocular muscles which control it. The four **rectus muscles** (superior, inferior, medial and lateral) take origin from the bone round the optic canal at the back of the orbit, and are inserted into the sclera just behind the corneoscleral junction. The **superior**

oblique muscle arises from the same place, but runs forward along the medial margin of the orbit till it reaches a small fascial pulley, when it makes a hair-pin bend backwards and outwards to be inserted in the upper, lateral and posterior part of the eyeball (Fig. 57). The **inferior oblique** comes from the anteromedial part of the floor of the orbit, and passes backwards and laterally to be inserted close to the superior oblique on the lower, lateral and posterior part of the eyeball.

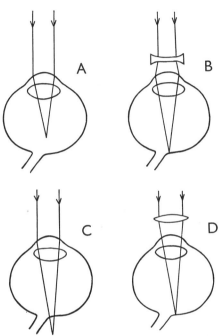

Fig. 56. Myopia and hypermetropia. A: myopic eye focussing parallel rays of light in front of the retina. B: myopia corrected by a concave lens. C: hypermetropic eye focussing parallel rays of light behind the retina. D: hypermetropia corrected by a convex lens.

The lateral rectus pulls the eye directly laterally; the medial rectus pulls it directly medially. The superior rectus pulls the eye upwards and medially; similarly the inferior rectus pulls it downwards and medially. This inward tendency of the muscles which raise and lower the gaze is counteracted by the two oblique muscles. The superior oblique pulls the eye downwards and laterally, and can thus be combined with the inferior rectus to make the eye look straight down; similarly the inferior oblique, which turns the eye upwards and outwards, can combine with the superior rectus to make the eye look directly up.

To keep the images formed by each eye synchronized, the movements of both eyes have to be geared together. The lateral rectus is supplied by the abducent nerve, and the superior oblique by the trochlear nerve; all the others are supplied by the oculomotor nerve. When the eyes follow an object in the horizontal plane, the medial rectus of one side contracts together with the lateral rectus of the other because a tract called the **medial longitudinal bundle** connects the nuclei of the oculomotor, trochlear and abducent nerves. When we look at something near to the eye, the two medial recti contract synchronously, causing the eyes to converge; this is accompanied by constriction of the pupil and contraction of the ciliary muscle (accommodation). All these activities are under control of the oculomotor nerve, and illustrate the close connexion

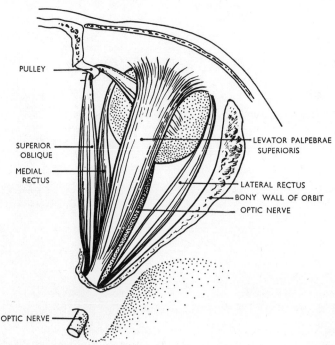

PULLEY

SUPERIOR
OBLIQUE

MEDIAL
RECTUS

LEVATOR PALPEBRAE
SUPERIORIS

LATERAL RECTUS
BONY WALL OF ORBIT
OPTIC NERVE

OPTIC NERVE

Fig. 57. Dissection of the right orbit from above. The levator palpebrae superioris conceals the superior rectus from view.

between the nuclei of either side. The medial longitudinal bundle connects the nuclei of the nerves moving the eyeball to other nuclei. For example, when we look up, the upper lid is pulled out of the line of vision by the **levator palpebrae superioris muscle** (oculomotor), the eyebrow is elevated by wrinkling the forehead (facial nerve), and the head is tilted back by the muscles at the back of the neck (spinal nerves); all these must be co-ordinated.

When the co-operation between the two eyes breaks down, a **squint** may result. If this is due to weakness of one or more muscles, these muscles may be re-educated and strengthened by excerises, or perhaps the balance of power of the active muscles may be adjusted by sliding the insertion of a muscle forwards or back from its natural position, so altering its purchase.

The tears

The eyeball is protected in front by the **eyelids,** each of which is stiffened by a small plate of fibrocartilage. The eyelids are lined by a transparent mucous membrane called the **conjunctiva,** which also covers the front part of the sclera close to the corneoscleral junction (Fig. 58). Into the **conjunctival sac** formed in this way are poured the **tears** secreted by the **lacrimal gland**—an object about 1·5 cm long in the upper lateral part of the orbit. The tears are washed across the front of the eye by the contraction of the muscle which closes the eyelids, and are absorbed through two small canals which open on to the inturned medial ends of the eyelids in little holes called the **puncta lacrimalia.** The

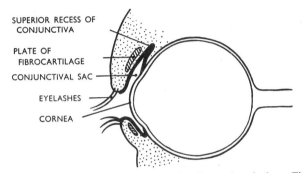

SUPERIOR RECESS OF
CONJUNCTIVA

PLATE OF
FIBROCARTILAGE

CONJUNCTIVAL SAC

EYELASHES

CORNEA

Fig. 58. Diagram of vertical section through eye, showing conjunctival sac. The lids are drawn away from the eyeball; in life they are in contact with it.

lacrimal canals conduct the fluid into the **lacrimal sac,** in the medial wall of the orbit, and eventually they are discharged into the nose (p. 159). The tears nourish the cornea and keep its surface cells from drying up and dying; they also contain substances which prevent the eye from becoming infected in spite of being exposed to the organisms in the air.

The ear

The ear has three parts: the outer, the middle, and the inner ear. The **outer ear** consists of the affair on the side of the head, which is called the **auricle,** and a small tube called the **external auditory meatus,** which leads from this to the eardrum, or **tympanic membrane.** In man the auricle is purely ornamental, though the muscles which move it in other animals are still represented, and can be trained to waggle the ears if so desired. The meatus is about 3·5 cm long; the lateral part is cartilaginous and the medial part is bony. The passage is slightly concave forwards, and it is lined with skin in which there are glands which secrete **cerumen,** or wax. This is a protective device, and, assisted by the bristly hairs which grow just inside the auricle, helps to keep out wandering insects.

At the end of the meatus is the drum, which is set so that its surface looks downwards, forwards and outwards. It is composed of modified skin, and is tightly stretched on the bony **tympanic ring** like embroidery on a circular stretcher. The upper part of the drum is, however, much looser than the rest, and is called the **pars flaccida**. Embedded in the drum is the 'handle' of a little bone called the **malleus**, the first of three **auditory ossicles** (Fig. 59) which convey the vibrations of the drum across the middle ear to the inner ear where the receptors lie. The second is the **incus**, which resembles a miniature molar tooth, and the third is the **stapes**, which looks exactly like a miniature stirrup. The chain of bones stretches across the narrow cavity of the middle ear, being connected by synovial joints, so that when the drum vibrates, the footplate of the stapes is set in corresponding motion.

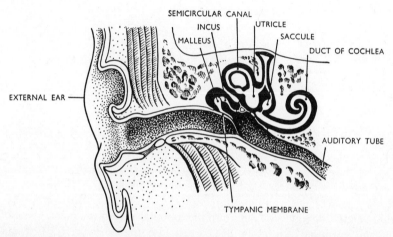

Fig. 59. Diagrammatic vertical section through the ear, showing the middle ear with the chain of auditory ossicles. The relationships of the cochlea and the vestibular apparatus are indicated schematically.

The **middle ear** is an air-filled bony cavity about 1·5 cm high, 1·5 cm from front to back, and 0·5 cm from lateral to medial wall. Anteriorly it communicates with the nasopharynx by means of the **auditory tube**, which is opened during the movements of swallowing, but otherwise remains shut. This tube equalizes the pressures on both sides of the drum. If we go up in an unpressurized aircraft the pressure outside the drum rapidly drops, but the pressure inside remains high. The drum bulges outwards, becomes stretched, and may be very painful. Swallowing allows the air in the middle ear to rush down the tube into the pharynx, the tension is relieved, and the pain subsides. When the aircraft comes down, swallowing causes air to pass from the pharynx to the middle ear, and the drum continues to work freely. If we have a cold in the head, the tube may fail to open, and a flight in an unpressurized aircraft is then a very unpleasant experience.

Posteriorly the middle ear opens into a cavity called the **tympanic antrum**, and this in turn communicates with a series of **mastoid air-cells** in the mastoid process of the temporal bone (p. 259). The whole arrangement would be admirable were it not for the fact that between the pharyngeal openings of the two auditory tubes lies the **pharyngeal tonsil** or **adenoid**—a mass of lymphoid tissue (p. 149) which is frequently infected. Infection can spread up the tubes to the middle ear, where fluid may accumulate, as in **catarrhal otitis media,** or pus may form, as in **suppurative** otitis media. The pressure on the drum may be such as to cause it to burst outwards, so producing a **perforation**, and the mastoid cells may become infected, leading to an **acute mastoiditis.**

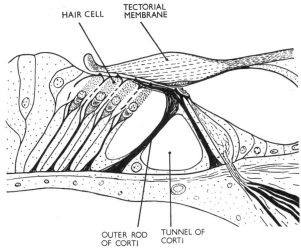

HAIR CELL   TECTORIAL MEMBRANE

OUTER ROD OF CORTI   TUNNEL OF CORTI

Fig. 60. A section through the organ of Corti (an enlargement of the arrowed portion of Fig. 61).

The **inner ear** lies in the **petrous** part of the temporal bone, which, as its name suggests, is extremely dense and hard. The **cochlea** exactly resembles a snail shell, being a curled-up tube with two and a half turns in it. The tube is wound round a central pillar called the **modiolus**, and it is divided into three passages, one on top of the other. The upper passage is called the **scala vestibuli,** and contains fluid which communicates with the fluid in the lower passage (the **scala tympani**) through a small hole in the top of the modiolus. The first turn of the cochlea lies in the medial wall of the middle ear, and at this point the stapes fits into the **fenestra vestibuli**, an oval hole in the scala vestibuli. The stapes works out and in like the plunger of a syringe, compressing the fluid in the scala vestibuli. The pressure waves in the fluid are communicated to the fluid in the scala tympani, at the lower end of which there is a round hole (**fenestra cochleae**) closed by the **secondary tympanic membrane.** When the stapes moves in and compresses the fluid in the inner ear, the secondary tympanic membrane bulges out; when the stapes moves out, the membrane is sucked in.

In between the scala vestibuli and the scala tympani is the third passage, the **duct of the cochlea**, and here lie the auditory receptors. The fluid in the duct does not communicate with the fluid in the other two passages, but is set in motion by the vibrations in them. The duct is bounded by the **basilar membrane** inferiorly; this stretches from the bony **spiral lamina** of the modiolus, which may be compared to the thread of a screw, to the lateral wall of the cochlea. The fibres in it are of different lengths and tensions, and they are selectively set in

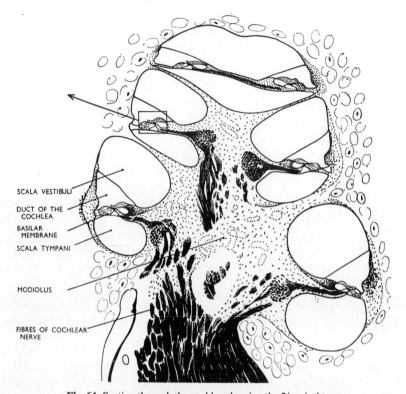

SCALA VESTIBULI

DUCT OF THE COCHLEA

BASILAR MEMBRANE

SCALA TYMPANI

MODIOLUS

FIBRES OF COCHLEAR NERVE

Fig. 61. Section through the cochlea showing the $2\frac{1}{2}$ spiral turns.

motion according to the pitch of the vibrations occurring in the fluid—just as a given string may be made to **resonate** by singing the appropriate note into a piano. Resting on top of the 'strings' of the basilar membrane is the complicated spiral **organ of Corti**, consisting of 'hair-cells' supported on a cartilaginous stretcher called the **tunnel of Corti**. The 'hairs' of the hair cells are embedded in the filmy **tectorial membrane**, and their cell bodies are connected to fibres of the **cochlear nerve**, which run through the bony spiral lamina (Figs. 60, 61).

The sequence of events resulting in the experience of hearing a note is as follows. A disturbance in the air makes the drum vibrate; this sets in motion the chain of auditory ossicles; the movement of the stapes causes pressure waves in the fluid surrounding the duct of the cochlea; this leads to a particular part of the basilar membrane being set in motion by resonance; this makes the corresponding hair cells move up and down in relation to the tectorial membrane, and so stimulates the hair cells by pulling on their hairs; finally this causes electrical disturbances to pass up the branch of the cochlear nerve concerned. Low notes are distinguished from high notes because they cause different parts of the basilar membrane to vibrate, and so activate different branches of the cochlear nerve. The impulses set up in the nerve pass with it into the brain, where they join a tract called the **lateral lemniscus**. This leads them up to the **inferior colliculus** and the **medial geniculate body**. From here the fibres are projected on to the auditory cortex, there being some evidence that the high notes are represented further forward in the superior temporal gyrus than the low notes.

The vestibular apparatus

Also in the petrous temporal bone, and sharing the fluid of the cochlea by means of a small connecting tube, lie two small cavities (Fig. 59), each with a little oval patch of hair cells in its wall; one patch is vertical and the other horizontal. On the hairs are stuck small particles of chalk, which are acted on by gravity, and stimulate the cells in different patterns according to the position of the head. The cavities are the **utricle** and the **saccule**, and they are very important 'static' organs—that is, they tell the brain which way up the head is. Such information can be vital to a pilot who is lost in a thundercloud and cannot see the horizon; if his utricle and saccule are not working he may well emerge from the cloud upside down.

The 'dynamic' organs of posture are the **semi-circular canals**, which are connected to the utricle and saccule. Each canal has a bulge at one end, called the **ampulla**, in which there is a clump of hair cells with chalk particles, exactly as in the utricle and saccule. The three canals in each temporal bone are arranged so that they are all at right angles to each other, and thus respond to movements in any direction. When the head is moved, the walls of the semicircular canals of course move with it. But the fluid which they contain has a certain inertia, and lags behind the movement, so dragging on the hairs and stimulating impulses in the fibres of the **vestibular nerve** (which supplies both static and dynamic organs). In this way the brain learns which way the head is moving, and this information, which is relayed to the cerebellum, is co-ordinated with information from the eye, the static organs, and the proprioceptors throughout the body in order to assist in the maintenance of posture.

Disease of the vestibular apparatus causes very distressing symptoms; the patient may be unable to move his head without becoming intensely nauseated and giddy. A lesser manifestation of the same sort of thing is observed in seasickness, in which the symptoms have been attributed to conflict between the visual and proprioceptive information received by the brain.

Taste and smell

On the surface of the tongue and the soft palate are small endorgans which under the microscope resemble minute oranges, from the way in which the cells are arranged. These are the **taste buds**, which can detect different substances in solution. Different parts of the tongue are sensitive to different kinds of taste; the back responds particularly to bitter tastes and the front and sides to sweet tastes, but there is no evidence to connect this with any structural differentiation of the endorgans. The taste buds are extremely sensitive to certain chemicals; for example saccharin can be detected when diluted with several million parts of water. A substance called phenylthiocarbamide is used in anthropology because certain people cannot taste it, though to the majority it has an intensely bitter taste. This condition of 'taste-blindness' is useful in mapping the relationships of different peoples. The taste buds are supplied by branches of the facial, glossopharyngeal and vagus nerves.

Most of what is commonly called 'taste' is really smell, and food is 'tasteless' when the nose is blocked by a bad cold. The olfactory mucous membrane lies right up at the top of the nose, partly on the nasal septum, and partly on the lateral wall of the nose. It is faintly yellow and differs structurally from the neighbouring respiratory epithelium (p. 159). It is supplied by the olfactory nerves, which perforate the base of the skull to end in the olfactory bulbs (p. 105). And there our definite information ends. We do not know how we appreciate different smells any more than we know how we taste different tastes. Smell is even more sensitive than taste, and substances like mercaptan or musk may be detected in fantastically minute quantity. 'Smell-blindness' also exists, for some people cannot smell hydrocyanic acid at all, while most can detect it instantly; such a condition can be dangerous. Smoking greatly reduces the sensitivity of both taste and smell mechanisms.

Outline of the clinical examination
of the nervous system

The nervous system is complicated, and a neurological examination must be lengthy and detailed if nothing is to be missed. It is only possible here to outline the procedure.

Intellectual functions

While the history of the patient's illness is being elicited some idea may be obtained of his memory, insight, orientation and comprehension, his capacity to recognize common objects, his powers of producing speech, and his ability to understand spoken and written material.

Exteroceptive functions

General sensation

Touch is tested with a nylon thread mounted on a handle and stroked across the skin, and thermal sensibility by tubes containing hot or cold water. Pain is investigated by pinprick (superficial pain), by pressing the skin hard with a blunt object, or by bending a joint beyond its normal range (deep pain). The thresholds of the various sensations are compared with normal thresholds in other patients or in other areas of the body. The best idea of the extent of an area of disturbed sensibility is obtained by passing the hand over the skin and asking the patient to indicate where the sensation becomes abnormal. Vibration sense is tested by placing a vibrating tuning-fork on a subcutaneous bone, such as the tibia; this normally gives rise to a buzzing sensation. The integrative powers of the cortex may be tested by giving the patient a common object—such as a pencil—to identify with his eyes shut.

Special senses

Visual acuity is tested by the familiar test types of the optician or by other typefaces designed to be read at ordinary reading distance. Colour vision is tested by the ability to distinguish coloured lights, or by a series of **Ishihara figures**, in which a coloured letter or figure is embedded in surrounding camouflage of a different colour. The visual fields are mapped by an instrument called a **perimeter** in which a small disc is slowly moved along a curved metal arm marked in degrees of angle. The patient looks straight in front, and the point at which he first sees the disc as it moves centrally is taken as the limit of the visual field in that direction.

Hearing is tested by a tuning fork or a whisper at a given distance from the ear. An electronic **audiometer** is used to measure the range of frequencies which the patient can hear. Smell and taste are very crudely tested by giving the patient various substances to smell or by dropping solutions on the tongue with a glass rod. His reactions can be roughly compared with those of normal people.

Proprioceptive functions

The patient stands with his feet together and parallel and shuts his eyes (Romberg's test). If proprioception is deficient he will sway and fall. He will also have difficulty in touching two fingers together with his eyes shut, or in touching his finger to his nose. He will be unable, with his eyes shut, to put one limb into a position duplicating the position of the corresponding limb. Nor can he tell whether a joint is being flexed or extended. To assess the functioning of the vestibular apparatus the patient shuts his eyes and is then turned round quickly two or three times by the doctor; his capacity to stand up straight with his eyes shut is then estimated.

Motor functions

The general muscle tone can be assessed by handling the limbs; this will readily detect undue flaccidity or spasticity. Gross paralysis may be assessed by comparing the grips of both hands or the capacity to raise the knee off the bed against resistance. Tremor is detected by making the patient put out his tongue as far as it will go, or by asking him to stretch out his hands at shoulder level in front of him, with the fingers spread. Muscular co-ordination is tested by making the patient perform simultaneous duplicate movements with both hands—for example, rotation of the wrists clockwise. Failure may indicate a cerebellar deficiency. The patient's gait may afford valuable evidence of paralysis.

Paralysis of individual muscles is tested by observing and feeling the muscle when a movement in which it is a prime mover is performed. The technique of testing has to be rigidly followed for each individual muscle, otherwise errors creep in through 'trick movements' (p. 60). Paralysis of the extraocular muscles is usually detected by the fact that the patient sees two images (diplopia); the relationship in space of one image to the other indicates which muscle is affected.

Reflexes

The ocular reflexes include the light reflex (constriction of the pupil when a light is shone into it), the consensual light reflex (constriction of the contralateral pupil as well), the convergence reflex (turning in of the eyes when a near object is suddenly looked at), accommodation (constriction of the pupil when a near object is suddenly focussed), and the corneal reflex (blink in response to a touch on the cornea).

Representative of the tenson reflexes which may be tested are the knee jerk (tap on patellar tendon), the ankle jerk (tap on Achilles tendon at the back of the ankle) and the triceps jerk (tap on triceps tendon at the back of the elbow). The most important of the superficial reflexes are the plantar reflex and the abdominal reflexes (p. 78).

9 · The circulatory system

The blood vessels

The blood is circulated throughout the body in a closed system of blood vessels by the pumping action of the heart. The vessels which conduct the blood away from the heart to the periphery are called **arteries**, and those which bring it back again are **veins**. It is necessary to stress this definition, since two sets of arteries, the pulmonary arteries (p. 142) and the umbilical arteries (p. 197) contain what is usually called venous blood.

The vascular system is lined throughout by a thin pavement epithelium called the **tunica intima**, except that where it forms the lining of the heart it is called the **endocardium**. It is exceedingly smooth, and this helps to prevent the blood clotting in the vessels.

Outside the intima there are two other coats which vary in size and in composition in different parts of the circulation. The **tunica media** contains varying quantities of elastic fibres and visceral muscle, and the outer **tunica externa (adventitia)** is made of fibrous tissue (Fig. 62).

The largest artery is the **aorta**, which leaves the left ventricle of the heart (p. 143) and is about 3 cm in diameter. Like all arteries, the aorta has **collateral** and **terminal** branches. The collateral branches come off at intervals along the length of the main trunk, and the terminal ones arise from the end of the trunk. In some cases, to facilitate topographical description, an artery may change its name at an arbitrarily defined point, much as a street may change its name at an intersection. In such instances the named vessel simply becomes another named vessel, and has no terminal branches. The branches of an artery in turn give off branches of their own, and in this way the blood is conducted to almost every part of the body (a few structures, such as the cornea, contain no blood vessels, and are nourished indirectly).

The aorta takes the full impact of the blood forced out by the contraction of the heart. For this reason it has, like all the large arteries, a very strong adventitial coat to prevent it from bursting under the strain. The media contains large amounts of elastic tissue disposed in circumferentially arranged plates. This ensures that the aorta, stretched by the impulse of the blood entering from the heart, recoils again during the rest period of the heart muscle. This prevents

the pressure in the aorta and the other large **elastic arteries** from falling
dramatically in between heartbeats, and so maintains the diastolic pressure (p.
145).

As the succesively branching arteries get smaller the proportion of elastic
tissue in the tunica media becomes progressively reduced, and the relative
amount of visceral muscle increases; this is the stage of the **muscular artery**,
such as the radial artery (p. 338). The pressure inside these vessels can be varied
by alterations in the tension of their muscle coat. Occasionally a muscular
artery can, following injury, go into such severe spasm as to prevent any blood
flowing through it, but the job of closing or opening pathways in the normal
circulation is left to the next stage of arterial branching.

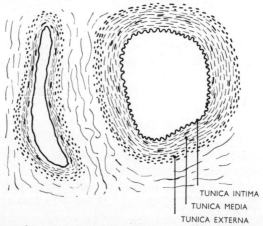

TUNICA INTIMA
TUNICA MEDIA
TUNICA EXTERNA

Fig. 62. Diagram of a cross-section of a large artery (on the right) and a vein. Note the
greater thickness of the arterial walls and the semi-collapsed condition of the vein. The
corrugated appearance of the lining of the artery is due to the elastic tissue, which
contracts when the pressure of the blood inside the vessel is withdrawn.

This reached after successive divisions have reduced the calibre of the
vessels—their **lumen**—to about 150 μm or less, which is small in relation to the
thickness of the media. At this distance from the heart, the pressure inside these
arterioles has fallen considerably, and their muscular wall is therefore strong
enough to close off the flow in the lumen by contracting in response to impulses
reaching it through the sympathetic nervous system (p. 88). Arterioles
therefore afford a means of regulating the flow through the tissues according to
their needs.

After the arteriole stage the vessels grow progressively smaller, and the
muscular coating disappears entirely. The adventitia becomes very delicate and
fine, and is eventually represented only by a few strands of connective tissue.
The final stage, that of the **capillary**, consists of a tiny tube some 8 to 10 μm in
diameter, with only one effective coat, the intima. Certain of the white blood
cells are able to squeeze through the thin capillary wall, particularly when it is

inflamed, and can then wander about as scavengers in the connective tissue spaces (p. 155). The lumen of the smallest capillaries is so small (p. 153) that the erythrocytes have to proceed in single file. In this way the flow becomes slow enough for the necessary exchange of oxygen, food materials, carbon dioxide, and waste products to take place between the blood and the tissues.

The capillaries are arranged in networks, the entrance to which is controlled by **precapillary sphincters** of visceral muscle. If several sphincters of this kind are closed, the blood flows through the network in relatively few **preferred paths** formed by larger capillaries. This situation occurs when the tissue is not active and does not need much blood; when it starts work, the sphincters relax, and blood flows through the whole network. With every subdivision of the arterial tree the total cross-sectional area of the branches becomes greater, and the **capillary bed** (the sum of all the capillary networks) has a capacity enormously greater than that of the aorta.

The **venules** which drain the other side of the capillary network are simply rather larger vessels which have again acquired a recognizable tunica media and tunica adventitia. They join each other to form **veins**, which differ from arteries in having much thinner and less muscular walls (the pressure inside is very much lower). In the larger veins, however, there is a definite muscle coat, and the largest ones of all, the **inferior** and **superior venae cavae** (pp. 272, 286), have some elastic tissue as well. The need for elastic tissue in veins is naturally negligible in comparison with the need in the large arteries, for the thrust of the heart on the blood has largely been exhausted by the time the blood reaches the veins. However, the great veins are stretched by the movements of inspiration (p. 270), and the elastic tissue may be there to allow them to recoil satisfactorily during expiration.

The veins contain valves (Fig. 63) arranged so as to allow the blood to pass onwards to the heart, but not back to the periphery. Normally the flow of blood towards the heart is assisted by the contraction of the muscles among which the veins lie. Both muscles and veins are usually enclosed in a compartment formed either by deep fascia or by deep fascia and bone (p. 22), and the contraction and consequent bulging of the muscle causes a rise of pressure within this compartment, squashing the vein and milking the blood along it because the valves will only allow the blood to pass in one direction. Superficial veins, such as those of the saphenous system (p. 384), lie outside the deep fascia, and muscular action cannot help. They are therefore prone to dilate and become **varicose** under the hydrostatic strain. If this happens, the valves tend to leak, and the condition thus makes itself worse.

The return of blood into the heart is assisted by the respiratory movements. When the diaphragm descends and the ribs rise (p. 270) the pressure inside the closed box of the chest falls as its dimensions are increased, and this sucks blood up against gravity into the heart. The process is also assisted by the fact that as the diaphragm contracts the pressure in the abdomen rises, and this forces blood from the abdominal part of the inferior vena cava into the thorax.

From the general features of the blood vessels and their adaptation to functional demands we must now turn to consider certain aspects of the circulation in more detail. The smaller muscular arteries usually communicate with their neighbours to form an **anastomosis**, particularly in the region of joints (Fig. 64). This acts as a safety device, for if an artery should be occluded blood can still enter the vessel distal to the block by passing along one of the collateral channels, and the tissue supplied by the blocked artery can survive. However, the collateral channels take some time to dilate sufficiently to convey a similar volume of blood. This time lag may be as much as several days, during which the region supplied has to get along with a reduced (but steadily improving) oxygen supply. In the limbs collateral channels are numerous, the oxygen demands of the tissues, providing they are kept at rest, are relatively

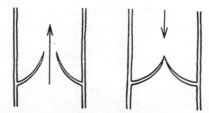

Fig. 63. Diagram of a venous valve, open to allow blood to pass towards the heart, and shut to prevent blood surging back in the opposite direction.

low, and an effective **collateral circulation** is rapidly established (Fig. 65). But certain organs, such as the brain and heart, are so dependent on oxygen that deprivation of blood supply for even a few minutes may lead to the death of tissue in the distribution of the artery. It is unfortunately precisely in these organs that collateral channels tend to be relatively inadequate, and the arteries supplying the heart and brain are often referred to as **functional end-arteries**, meaning that while they have a collateral circulation, it is not good enough to prevent the tissue dying. In other parts of the body there are true end-arteries which have no collateral circulation; the best example is the central artery of the retina (p. 120).

In certain places special distributing arrangements are formed by communication between muscular arteries. For example, the blood supply to the brain is derived from a circular anastomosis at its base called the **circle of Willis** (Fig. 118). The arteries which feed this are of very different calibre, and it has been suggested that the circle is not a safety measure, but a device for equalizing the pressure in the branches which spring from the circle. Again, the radial and the ulnar arteries (p. 338) form two **palmar arches** from which springs the blood supply to the fingers, and there is a comparable arrangement in the foot.

Direct communications between arterioles and venules are called **arteriovenous anastomoses**. They are numerous in the skin and in the wall of the alimentary canal, and blood reaching them passes along the line of least resistance without going through the capillary bed. The anastomosing channel

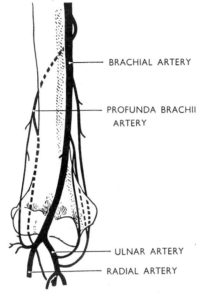

BRACHIAL ARTERY

PROFUNDA BRACHII ARTERY

ULNAR ARTERY

RADIAL ARTERY

Fig. 64. The anastomosis round the elbow joint.

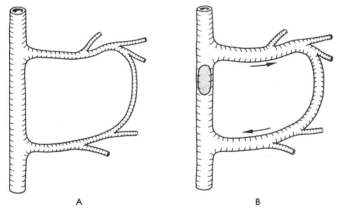

A   B

Fig. 65. Collateral circulation. A: normal arrangement. B: the main artery is blocked, but blood finds its way round the bypass formed by the anastomosing branches.

has a thick cuff of richly innervated visceral muscle, contraction of which shuts down the anastomosis and forces the blood to pass through the capillaries (Fig. 66). The purpose of arteriovenous anastomoses is not altogether clear; they are found particularly in tissues which have irregular periods of activity and when these are resting it would be uneconomical to force through them as much blood as they require when they are active. Nevertheless, the controls provided by the ordinary arterioles and precapillary sphincters would seem perfectly adequate for this purpose.

Capillaries are not the only kind of blood vessel through which exchanges of foodstuffs, gases, and waste materials can take place. In certain organs, such as the liver, they are replaced by **sinusoids**. These, unlike capillaries, have an irregular lumen and in their walls are fixed phagocytes, some of which seize and destroy worn out erythrocytes.

In the mucous membrane of the nose, and also in the substance of the penis the circulation opens into large spaces which often have small sphincters on the venules which drain them. In such **cavernous tissue** blood can be dammed up by shutting the outlet sphincters, and in this way the tissue swells. This is the mechanism of erection of the penis, and also the reason why the nose becomes readily blocked when it is infected by the common cold.

Special circulations

Several organs have specialized circulatory arrangements to serve specialized functional requirements. In the **spleen** some of the vessels open freely into the space among the cells of the splenic pulp, there being no capillary walls. Blood percolates among the cells, and is drained back into the circulation via venules which also open freely into the pulp. In this way the blood is brought into direct contact with the cells of the splenic pulp, and this may facilitate the removal of worn out erythrocytes from the blood stream by the local phagocytes. It also necessitates the presence in the pulp of cells which produce heparin (p. 21), since the blood would otherwise clot as soon as it found itself outside the vessels.

In the **kidney**, filtering devices are interposed in the arterial side of the circulation. These **glomeruli** (p. 188) are small twisted balls of capillary nets, each fed by an **afferent arteriole**. In the glomerulus fluid and waste products are filtered off into an excretory tube, but the blood does not lose its oxygen. The vessel which drains the glomerulus is called an **efferent arteriole**, for it also is provided with a thick muscular wall. This in turn leads to an ordinary capillary network drained by venules back into the venous circulation (Fig. 66). When the afferent arteriole is open and the efferent one closed, the pressure in the glomerulus rises and forces fluid out through the capillaries into the capsule of the glomerulus; when the afferent arteriole is contracted and the efferent one relaxed, pressure in the glomerulus is low, and little filtration takes place.

The **portal** system of veins, which has no valves, drains the blood from the absorptive portion of the digestive tract into the **portal vein**, which brings the products of digestion from the gut to the liver, where they can be dealt with. At its destination the portal vein splits up like an artery into branches, which eventually fill a system of sinusoids. These also receive blood from the

hepatic artery, so that what supplies the liver cells is a mixture of venous and arterial blood. The other ends of the sinusoids lead into venules, which eventually coalesce to form **hepatic veins** draining into the inferior vena cava just before it enters the heart. The portal circulation thus differs from the general circulation in having a network of minute vessels interposed on the venous side. A similar arrangement is found in the blood supply of the pituitary gland (p. 198).

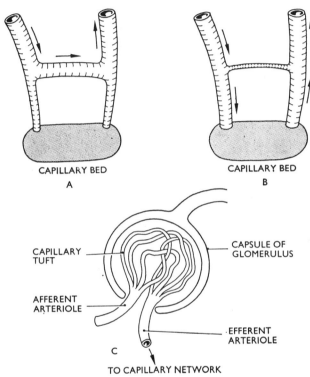

Fig. 66. A: arteriovenous anastomosis, open. Blood flows from the arteriole to the venules without passing through the capillary net. B: arteriovenous anastomosis, closed. Blood is forced through the capillaries. C: the arrangement in a glomerulus.

The blood supply of the heart muscle and of the lungs also exhibits peculiarities, but these will be discussed later (pp. 143, 168).

Innervation of blood vessels

Blood vessels throughout the body receive their innervation from the autonomic system, which supplies the muscle fibres in the tunica media. The effects of sympathetic and parasympathetic stimulation on the redistribution of blood have already been mentioned (p. 88), but blood vessels also have a sensory supply to all three coats of their walls. Irritation of this sensory supply, for example, by injecting certain drugs into arteries, may cause a prolonged reflex spasm of the arterial wall, with potentially disastrous results.

Results of injury to blood vessels

If a blood vessel is torn across by an injury, several things happen. The blood escapes from the vessel, causing a drop in pressure inside it, and proceeds to clot. This clot, once it becomes firm, is an important factor preventing further escape of blood. The tunica media contracts reflexly and helps to prevent any more blood from reaching the site of the injury; in this it is helped by the elastic fibres within it, but will be severely hindered if the adventitia is closely adherent to surrounding dense connective tissue. This is the case in the vessels of the scalp, and explains why wounds of the scalp usually bleed very freely and for a long time. If the vessel is not fixed in this way to its surroundings, the elastic tissue in its walls will cause the torn ends to retract away from each other, a point of considerable importance to the surgeon who is trying to find and tie off the bleeding points.

If the injury is incomplete, as when one side of an artery is 'nicked' by the surgeon's knife, the elastic tissue in the vessel wall operates against the closure of the wound by holding the edges open, a condition known as 'button-holing'.

The heart

The heart is a hollow organ the size (and roughly also the shape) of the clenched fist and weighing about 300 g. It lies towards the front of the chest, most of it being on the left side, and its position and shape alter to some extent as it beats, with breathing, and according to the position the body takes up (it is considerably lower when the body is upright than when it is lying down). The **base** of the heart lies uppermost, and the **apex** is opposite the fifth or sixth left intercostal space (p. 275). The heart is lined by smooth pavement epithelium continuous with the tunica intima of the blood vessels; here it is called the **endocardium**. The **myocardium** corresponds to the tunica media of the blood vessels, and is formed of cardiac muscle (p. 51); it is covered by a thin fibrous **epicardium**. Outside this again is the **serous pericardium**, which, like a synovial sheath (p. 43) is arranged in two layers, a **visceral** layer closely attached to the epicardium, and a **parietal** layer firmly bound to a strong fibrous enclosing bag, the **fibrous pericardium**. As the heart beats, the visceral layer slips smoothly over the parietal layer, so obviating friction between the heart and the fibrous pericardium, which is tethered to the diaphragm and the sternum, and blends with the adventitia of the aorta and the pulmonary artery.

The heart has four chambers arranged in two pairs. Each pair consists of a thin-walled **atrium**, which receives the blood, and a thicker-walled **ventricle**, which pumps it away from the heart again. The **right atrium** forms the right border of the heart (Fig. 67) and receives the two venae cavae which return the venous blood from the tissues (Fig. 69); it also receives venous blood from the heart muscle itself through the opening of the **coronary sinus**, which is the main vein of the heart wall. Most of the wall of the atrium is smooth, but in the **right auricle**, a small appendage to the main cavity, the endocardium is ridged by underlying bars of muscle. The right atrium is separated from the left atrium by the thin **interatrial septum**; normally this is intact in the adult, but in the embryo it is perforated by the **foramen ovale**, through which most of the blood passes directly from the right atrium into the left atrium (p. 143).

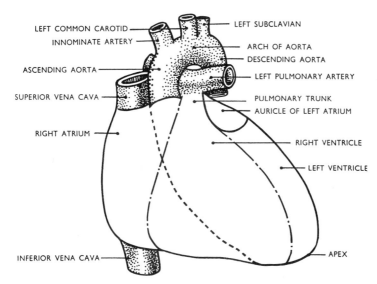

Fig. 67. The heart from in front. The position of the cavity of the right ventricle is indicated by the interrupted lines; the right margin of the left ventricle, which lies behind this cavity, is shown by the dotted line.

The right atrium opens into the **right ventricle**, which forms most of the front of the heart, by the **right atrioventricular (tricuspid) valve**. As its name suggests, this valve has three **cusps** which allow the blood to pass freely through into the ventricle when the atrium contracts, but which are forced together when the ventricle contracts, so preventing the blood from surging back into the atrium. Each cusp is attached to a small **papillary muscle** derived from the myocardium of the ventricle, so that when the ventricle contracts the cusps become taut and cannot be turned inside out (Fig. 68). The wall of the right ventricle is much thicker and stronger than that of the right atrium; the atrium has only to push the blood through the valve into the ventricle, but the ventricle has to drive the blood round the lungs in the pulmonary circulation.

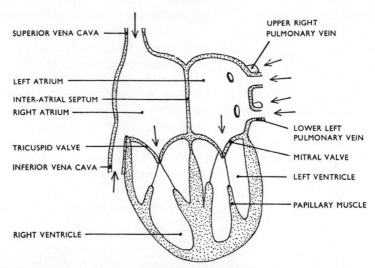

Fig. 68. Diagram of oblique section through the heart. The outlets from the ventricles and the third cusp of the tricuspid valve are not shown. Notice that the cusps are attached to the papillary muscles by their fibrous bands.

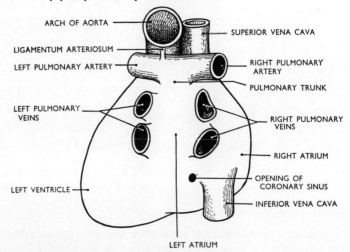

Fig. 69. The heart from behind. The ligamentum arteriosum is the fibrous remnant of the ductus arteriosus.

The blood leaves the ventricle by the **pulmonary artery** (p. 273), the opening being guarded by the small **pulmonary valve**, which has three cusps arranged so as to prevent reflux of the blood back into the ventricle (Fig. 70).

These cusps, like those of the atrioventricular valve, are composed of dense connective tissue, clothed on either side by pavement epithelium; they have no blood supply. Each pulmonary cusp is thickened at the point where all three meet to form a **nodule**; the rest of the cusp is known as the **lunule**.

In the lungs the branches of the pulmonary artery divide progressively until a capillary bed is formed. In these capillaries the blood exchanges its carbon dioxide for oxygen and so returns refreshed to the heart along the four **pulmonary veins** which open into the **left atrium** at the back of the heart (Fig. 69).

The left atrium has a small **auricle** similar to that of the right atrium, but its walls are otherwise smooth and thin. It opens into the left **ventricle** by the **left atrioventricular (mitral) valve**. This valve has only two cusps, and its appearance fancifully suggests a bishop's mitre. The mitral cusps, like those of the tricuspid valve, are held taut by papillary muscles during contraction of the ventricle. The left ventricle forms most of the left margin of the heart as well as its apex. Its wall is much thicker than that of the right ventricle, for it pumps blood all over the body and not just around the

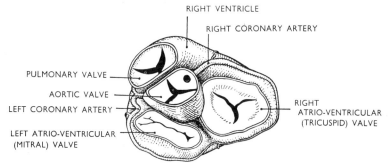

Fig. 70. The valves of the heart viewed from above. The atria have been removed to display the upper surfaces of the atrioventricular valves.

pulmonary circulation (Fig. 71). The opening of the aorta from the left ventricle is guarded by the **aortic valve**, which is exactly similar in construction to the pulmonary valve.

Just above the cusps of the aortic valve there arise from the aorta two **coronary arteries** (Fig. 70), which run over the surface of the heart, supplying the myocardium and the conducting system (p. 144). These arteries fill with blood when the heart beats, but this blood cannot enter the muscular wall of the heart while the muscle fibres are contracted; the coronary arteries therefore have to dilate to accommodate the blood forced into them, and are only emptied of this blood when the heart muscle relaxes after a beat and can allow the blood to pass through the capillaries between the muscle fibres. The coronaries *dilate* in response to sympathetic stimulation, in accordance with the principle that the sympathetic system prepared the body for flight or fight (p. 88). They are also functional end-arteries (p. 136). A **coronary occlusion** may be rapidly fatal, but not necessarily so; the heart may continue to function fairly well in spite of the fact that it has in its wall a patch of fibrous tissue, representing the dead area of muscle.

In the embryo, the blood reaching the heart is oxygenated blood from the mother, and there is no point in sending it round the lungs, which are not functioning at this stage. The circulation is therefore routed from the right atrium through the foramen ovale to the left atrium, and so into the left ventricle and round the body. Such blood as does pass into the right ventricle escapes from the pulmonary artery into the aorta through a by-pass channel called the **ductus arteriosus**; after birth this channel is converted into fibrous tissue and is then known as the **ligamentum arteriosum** (Fig. 69). Either the foramen ovale or the ductus arteriosus may fail to close normally, leading to circulatory defects which may have a serious effect on the child. Other congenital defects of the heart are not uncommon, and many of them permit blood to avoid passing through the lungs. The patient then has a dusky appearance because a proportion of the blood in the arterial tree is not oxygenated.

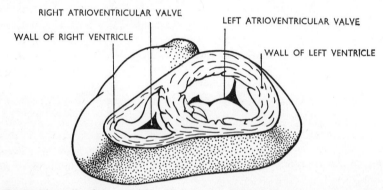

RIGHT ATRIOVENTRICULAR VALVE

LEFT ATRIOVENTRICULAR VALVE

WALL OF RIGHT VENTRICLE

WALL OF LEFT VENTRICLE

Fig. 71. The walls of the ventricles. The heart has been sliced open and viewed from below.

The cardiac cycle

The normal heart rate is usually taken as 72 beats per minute, but in many people it is slower than this. The rate is usually ascertained by feeling the **pulse** in an easily accessible artery; the most convenient is the **radial** artery at the wrist (p. 338). The pulse is a wave of contraction passing along the vessel wall.

Each heart beat is initiated in the right atrium by a small mass of nervous tissue and specialized muscles fibres called the **sinuatrial node**. This lies in the wall of the atrium at the point where the superior vena cava enters it, and it is controlled by **cardiac plexuses** derived from the vagus nerves and the thoracic and cervical sympathetic chains.

The contraction of the atria spreads through both chambers almost simultaneously, and eventually reaches the region of the atrioventricular valves. Here the arrival of the muscular contraction stimulates the **atrioventricular node**, which lies close to the opening of the coronary sinus, and sends an impulse down into the ventricles along the **atrioventricular bundle**. This specialized bundle of tissue containing nerve cells and large pale muscle cells called **Purkinje fibres** conducts the impulse to the apex of the heart, which therefore begins to contract first. In this way the blood in the ventricles is forced upwards, towards the aortic and pulmonary valves, instead of being forced down towards the apex, as would be the case if the contraction wave spread directly from atria to ventricles.

The period of contraction of a given chamber of the heart is called its **systole**, and the period of its relaxation is called **diastole**; the whole process from one ventricular systole to another is called the **cardiac cycle**. Atrial systole lasts about 0·1 seconds; ventricular systole some 0·3 seconds. The electrical changes produced by the active cardiac muscle during the cycle can be recorded as an **electrocardiogram.** The tissues round the heart act as conductors, and the recording electrodes can be applied to the hands and feet. The three 'standard' leads are: right and left hand; right hand and left foot; left hand and left foot. Usually another lead is taken from the chest. The four traces obtained in this way show several distinct waves of electrical activity, which are labelled 'P', 'QRS', and 'T'. Interference with the normal conduction of the cardiac contraction is shown up by distortion or suppression of part of the normal pattern of waves. The type of distortion is often characteristic of a particular disease process.

The cardiac cycle also produces the **heart sounds**. These can be heard by a stethoscope placed over the chest wall, and in the normal heart two sounds may be distinguished. The first is a quiet one likened to a person saying 'luub', and the second is a sharp slapping sound produced by the closure of the valves and usually represented as 'dupp'. Alterations of the normal sounds may afford information as to the state of the heart.

The blood pressure and the control of the circulation

The blood, propelled by the heart muscle, exerts a pressure against the walls of the vessels which contain it, and this is known as the blood pressure. It decreases continuously in the direction of the flow, and actually becomes negative in the great veins as they open into the right atrium. The blood pressure is measured by encircling the arm with a rubber bag contained in a nonstretchable cover. This armlet is inflated with air until the radial pulse is obliterated. The pressure in the armlet is now released gradually until a rhythmic thumping noise is heard through a stethoscope placed over the brachial artery at the bend of the elbow; the pressure in the armlet at this point is taken as the **systolic pressure**—that is to say, the maximum pressure in the artery, caused by the systole of the heart. As the armlet pressure falls still further, the sound disappears; this stage is taken as the **diastolic pressure**—i.e. the pressure in the artery between the heart beats. In arteries near to the heart there is a considerable difference between the systolic and diastolic pressures, but further away this difference—the **pulse pressure**—becomes less because the elasticity of the great arteries cushions the force of the cardiac systole (p. 134). In the arm the systolic pressure is usually about 120 mm of mercury in a healthy young adult at rest, and the diastolic pressure about 80 mm of mercury. With increasing age both pressures tend to rise.

The blood pressure depends on two factors, the rate and force of the heart beat, and the resistance offered to this by the state of contraction of the peripheral arterial tree. The peripheral resistance depends mainly on the arterioles. Normally only a proportion of these are relaxed at any one time, and if they are all suddenly opened there is not enough blood to fill the enormous capillary bed brought into play. The pressure in the arterial tree thus falls catastrophically, and the brain is deprived of its normal blood supply; this is one of the causes of fainting, or **syncope**. Conversely, if more arterioles than usual are contracted, the same volume of blood is now held in a smaller container, and the blood pressure must rise. The tone of the arterioles is controlled by the sympathetic system (p. 88), and is varied in such a way as to keep the blood pressure fairly constant. Emotion may raise the pressure, and in a nervous patient the mere act of taking the blood pressure may be enough to send it up. Intense emotional stress may thus rupture one of the fragile arteries in the brain, resulting in a **stroke** (p. 110).

The sinuatrial node is controlled by the autonomic fibres of the cardiac plexuses; stimulation of the vagus slows the heart, while stimulation of the sympathetic accelerates it. Normally the two effects balance each other. The outgoing impulses are affected by incoming impulses from the **arch of the aorta** (Fig. 67) and the **carotid sinus**. The latter is a dilatation on the common carotid artery at the point where it divides into internal and external carotids (p. 265). If the pressure within the arteries at these points rises, stretch receptors in their walls send up impulses which eventually reach the vagus nucleus. As a result the parasympathetic supply to the heart is stimulated, and the heart slows, allowing the pressure to drop (**Marey's reflex**). If the drop is too great, the inflow of messages from the sensitive areas is reduced, the vagus is not so thoroughly stimulated, and the opposing sympathetic system is allowed to take command; as a result the heart rate increases, and the pressure rises again. The whole mechanism is thus self-regulatory, and an excellent example of the principle of negative feedback (p. 111).

If the peripheral resistance of the circulation should fall, as in syncope, the fall in pressure in the carotid sinus and the arch of the aorta allows the sympathetic to gain the upper hand and increases the heart rate. The onset of unconsciousness causes the patient to fall down; this brings the head level with the heart, which now does not have to pump the blood up against gravity. As a result the circulation in the brain is improved, and consciousness returns.

Another important factor in the control of the circulation is **Bainbridge's reflex**. The stretch receptors for this lie in the right atrium, and respond to distension by sending impulses up to the vagus nucleus. But these impulses have the opposite effect to those arriving from the carotid sinus and arch of the aorta, for they *inhibit* the vagus (see p. 78) and so the heart rate increases because of the unopposed influence of the sympathetic supply. The result is that an increase in the volume of blood delivered to the right atrium results in an increase of the heart rate.

Emotion increases the heart rate; everyone is familiar with the thumping, rapid, sympathetic-stimulated heart which accompanies us on the way to an important interview. Some people can control the heart rate—up to a point—at will, but this is uncommon. Other reflexes operate on the action of the heart; for example, it will beat faster in response to pain, lack of oxygen, or carbon dioxide excess. Some of these reflexes are mediated through the carotid body (p. 169), and others, such as the cardiac confusion which results from stepping into a cold bath, are mediated through cutaneous nerves.

Tissue fluid and the lymphatic system

As the blood flows along a capillary, fluid passes in and out of the vessel through the intact capillary wall. The pressure in the capillary bed, which is equivalent to about 30 mm of mercury, tends to drive fluid out of the vessel into the tissue spaces. But the osmotic pressure of the proteins and salts in the blood tends to keep fluid in the capillary, though this is opposed by the osmotic pressure of the materials in the **tissue fluid** (p. 19). Under normal circumstances, the net result of all these opposing forces is a positive pressure of about 10 mm of mercury tending to drive fluid into the tissue spaces.

The tissue fluid is thus continually being added to, and the blood capillaries are unable to reabsorb it all, particularly if it contains protein which has managed to escape through the capillary walls. A supplementary system of vessels, with walls more permeable to large molecules, is provided by the **lymphatic capillaries**, a system of closed tubes which have no connexion with the blood capillaries, but which, like them, have walls formed only by pavement epithelium. Into them drains the excess tissue fluid, which, once it has gained entry, is called **lymph**. The lymphatic capillaries join together to form **lymphatics**, which again unite to form larger vessels very similar in structure to veins, though the largest one of all, the **thoracic duct**, is not more than about 6 mm in diameter. The larger lymphatics have numerous valves, and this gives them a beaded appearance; their lymph is eventually poured back into the venous system. The thoracic duct opens into the left brachiocephalic vein at the point where it is formed (p. 266). It drains the lymph from the whole body except the right arm, the right side of the head and neck, and the right side of the chest; the lymph from these regions enters the venous system at the corresponding point on the right side of the body through a series of separate vessels.

The lymphatics which drain the alimentary canal have an additional importance, for they are the chief means by which fat absorbed from the intestine is carried into the general circulation. After a fatty meal the lymph which they contain is milky with fat droplets; it is called **chyle**, and the vessels, because of their milky appearance, are called **lacteals**.

The flow of blood in the vascular system depends primarily on the heart beat; lymph has no such direct propulsive force, and the flow of lymph from a given region may not be continuous when the body is at rest. Lymph is 'milked' along the lymphatics by the contraction of surrounding muscles, and 'sucked' into the chest by the movements of respiration, in exactly the same way as blood in the veins (p. 135). Where lymphatics and blood vessels lie within a tight sleeve of fascia, the pulsations of adjacent arteries may assist in the 'milking' process. The flow of lymph from a limb can be increased (by gravity) if the limb is elevated, and decreased if the limb is prevented from moving by a splint or a plaster. Immobilization of an infected limb thus helps to minimize the dissemination of toxic material throughout the body.

At intervals along the course of the lymphatic vessels (except the main terminal trunks) are interposed a series of filters, for lymph capillaries can actively engulf any particles which the tissue fluid may contain. In the absence of filters the particles might reach the general circulation and be distributed all over the body. The filters are small rounded bodies, the largest being not more than about 1·5 cm in length, and are known as **lymph nodes**. If some coloured ink is injected into the skin, as in tattooing, the particles of ink tend to lodge in the next filter on the way to the thoracic duct, and methods of this kind are often used to determine the pathways followed by the lymph drainage of a given area. For the lymphatics have a more sinister connotation than tattooing; they are involved in the spread of acute bacterial infections, and are also the pathways along which certain types of cancer tend to spread from one part of the body to another. When a patient has an untreated infected finger, the lymph nodes in the armpit, which attempt to hold up the further progress of the bacteria brought to them, become swollen and painful; a secondary abscess may form in one of them.

Each lymph node is enclosed in a capsule which sends down partitions into the interior of the node; between these partitions is packed the **lymphoid tissue**, which consists of small round cells. These cells give rise to lymphocytes (p. 155), which are carried into the veins along the larger lymphatics. The lymph stream in the node flows over and through the lymphoid masses along marginal channels (Fig. 72). Each lymph node receives several lymphatics on its convex surface (the nodes are usually kidney shaped) and despatches one efferent vessel from its concave **hilum**. In the channels running through the node lie phagocytic cells ready to snatch any foreign material out of the lymph stream and retain it in the node.

The lymph nodes are not the only representatives of lymphoid tissue in the body. Much of the **spleen** (p. 154) consists of very similar tissue, and it is presumed that lymphocytes are formed there also. In addition there is a system of **epitheliolymphoid tissue** in relation to mucous membranes. The most familiar examples of this system are the **tonsils** and the **adenoid** (p. 161), but there are large patches of similar material in the walls of the ileum (p. 181), and smaller ones throughout the whole length of the gut. The functions of epithelio-lymphoid tissue are unknown; lymphocytes appear to be formed there and are shed into the alimentary tract. The tonsils and the adenoid frequently become infected, and for many years they were removed as a kind of ritual, but nowadays surgeons are reluctant to remove something which is suspected of having a protective function, unless they are manifestly diseased. Epitheliolymphoid tissue certainly differs from the ordinary lymphoid tissues of the lymph nodes and the spleen, for it is unaffected by certain diseases which attack them.

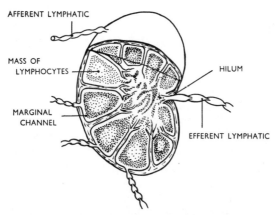

Fig. 72. Section of a lymph node.

Immunity

Lymphoid tissue has a significance beyond the filtration and destruction of material brought to it in the lymph, for it contains many 'B' lymphocytes and plasma cells (p. 21) which produce **antibodies**. Antibodies are proteins which are manufactured to react specifically with the proteins of foreign material introduced into the body; these invading proteins are called **antigens**. The antibodies are not formed immediately; the cells involved have to 'learn' how to make them in response to each different invading protein. But they remain in the body for a prolonged period and can be called up for use immediately a second invasion by the same organism takes place. Antibodies are extremely specific; for example, those formed in response to an injection of sheep's erythrocytes will have no effect on erythrocytes from another species. They act in several ways; sometimes by directly 'neutralizing' the antigen, sometimes by encouraging more rapid phagocytosis of an invading organism, and sometimes by destroying its plasma membrane.

The anatomical basis of the theory of immunity heavily involves the lymphoid tissues. An important part is played by the **thymus**, a lymphoid structure which occupies the region above and in front of the heart in the infant and child, but which regresses at puberty and in the adult is more or less converted into a strip of fibrous tissue. It is suggested that lymphocytes which are formed in, or pass through the thymus are shed into the blood stream, migrate to other lymphoid tissues, and settle down there to help to build up the local lymphocyte population. These special 'T' lymphocytes are thought to convey some sort of 'instructions' regarding the recognition of protein which does not 'belong' to the body.

The body is said to be immune when it can recognize and destroy a foreign protein which gains entry to it. This is sometimes of great benefit, as when invading bacteria or viruses are defeated and the body remains protected against subsequent invasions by the same organisms in the future. But it is also an enormous difficulty when we wish to transplant an organ or tissue from one patient to another; the immunity mechanism is so sensitive that it immediately detects and tries to destroy such material, even though it may come from a close blood relative. One method of getting round this is to paralyse the lymphoid tissues temporarily, either by giving them a dose of radiation, or by means of antilymphocyte serum. This may allow the body to accept the transplant, but of course renders it vulnerable to all kinds of infections which might be acquired during the period of temporary lack of protection. Hence the need for the strictest precautions against infection of all kinds.

Outline of the clinical examination of the circulatory system

The patient's **colour** is important; excessive pallor may indicate anaemia, and a bluish tinge (**cyanosis**) means inadequate oxygenation of the blood, which may arise from several causes, such as a failing circulation. This may also produce breathlessness (p. 169) and swelling of dependent parts such as the ankles. This **oedema** results because if the heart weakens, the blood passes more slowly along the capillaries, more time is allowed for fluid to leak out into the tissues, and the vascular and lymphatic capillaries may be unable to cope with the excess. Feeling the **pulse** will reveal its rate, regularity and volume as well as giving some indication as to whether the blood pressure is raised or the wall of the artery diseased. If no pulse is palpable the blood pressure may be very low, the artery may be blocked, or the walls of the vessel may be so thickened and rigid as to be unable to respond to the pulse wave. The blood pressure can be measured as described on p. 145.

The state of the circulation in the extremities may be assessed by pressing the thumb firmly on the skin and noting the time taken for the area to fill up with blood again. The skin temperature of a limb, taken by special apparatus, may indicate the efficiency of the circulation in it. The condition of the retinal arteries can be directly observed with an ophthalmoscope (p. 120).

The apex beat of the heart is often visible, and can usually be felt. The size of the heart may be estimated by the method of percussion (p. 239) or by taking a radiograph; the heart is considered enlarged if on the X-ray plate it occupies more than half the diameter of the chest. By the same means the shape of the heart can be made out, so that any abnormal dilatation of one or more chambers can be detected. The cardiac cycle can be investigated by listening with a stethoscope (auscultation) or by making a record of the electrical changes during it (p. 145).

Finally, the best test of the efficiency of the heart is its response to exercise. The patient is usually asked to step on and off a stool perhaps twenty times; at the end of this his pulse rate and respiration rate are again observed. In an 'exercise tolerance' test such as this the pulse ought to return to normal again after about a minute.

10 · The blood

The body contains about 6-8 litres of blood, which transports oxygen and food materials to the tissues and takes away from them carbon dioxide and other waste materials to be excreted. It also helps to distribute the body heat evenly throughout the tissues and to convey the internal secretions of the endocrine glands.

The plasma

The liquid part of the blood is the **plasma**, a clear straw-coloured fluid containing three kinds of protein—**serum albumin**, **serum globulin**, and **fibrinogen**—and many inorganic salts in solution. The osmotic pressure of the proteins and salts helps to prevent the water of the plasma from leaking out through the walls of the capillaries. The plasma fibrinogen is concerned in the mechanism of blood clotting, and the globulin in the process by which the body acquires immunity.

In **surgical shock** there is a diminution in the volume of the circulating blood: this may be due to loss of blood, to a pooling of the blood in the capillary bed, or to a leakage of plasma through abnormally permeable vessel walls. If the patient is to recover the loss of circulating blood must be made good by giving a transfusion of whole blood or of plasma, which is more suitable if there has been no great haemorrhage. Plasma can be obtained by adding citrate to fresh blood taken from a donor. The solid elements are allowed to settle, and the overlying plasma is drawn off. If it is then dried, it will keep for long periods, and can be reconstituted by merely adding water. The extensive use of plasma and blood on the battlefield was one of the great surgical advances of World War II.

If no citrate is added to the blood it will **clot** to form a porous mass of **fibrin**, which is derived from the plasma fibrinogen by a complex process involving the platelets (p. 156) and the calcium salts in the plasma. The filaments of fibrin gradually contract, squeezing out from the clot a thin fluid called **serum**, which is identical with plasma except that it contains no fibrinogen. Clotting occurs in from three to ten minutes—faster when the temperature is kept high or when the blood is in contact with a surface such as skin. When a blood vessel is wounded, the cessation of bleeding depends largely on mechanical factors; if the pressure in the vessel is too great any attempt at a clot will be washed away, and haemorrhage will continue. If the pressure is low the clot forms properly and bleeding will rapidly cease. The clot gradually becomes permeated with fibrous tissue, so sealing off the torn ends of the vessel. In **haemophilia** and other 'bleeding diseases' there is a defect in the complicated chain of chemical reactions leading to clotting, and bleeding from even the most trivial injury may continue for a very long time.

After surgical operations, particularly those involving the pelvis and lower limb, the blood flow in the veins of the part may be considerably slowed down, and clotting may actually take place within them. Subsequent movement of the limb may then dislodge the clot, which will travel to the heart and be discharged into the pulmonary circulation (p. 143), where it may stick, blocking that particular segment of the lung from receiving venous blood. This accident, which may be fatal, is known as **pulmonary embolism**, and to avoid its occurrence patients are given exercises to perform after operations, so that the circulation may not be allowed to stagnate.

The solid elements

The solid elements in the blood are of three kinds, erythrocytes, leucocytes, and platelets. The **erythrocytes** (red blood corpuscles, red blood cells) have the form of a biconcave disc (Fig. 73) about 7 μm in diameter. This shape allows them to bend and distort when passing through the capillaries, which are just big enough to allow them free passage (p. 135), and also means that no point within the corpuscle is far from the plasma membrane. The erythrocytes have no nuclei, and this allows the whole of the cell to be packed with the respiratory pigment **haemoglobin** (p. 158), which gives them their red colour, and is the basis of their function as carriers of oxygen. The lack of a nucleus means that they are unable to divide or to effect running repairs, but they survive the buffeting of the circulation very well in spite of this handicap (p. 5).

Erythrocytes are heavier than the plasma, and are present in it in great numbers (5 to 6 million per cubic millimetre of blood in men and 4 to 5·5 million per cubic millimetre in women). Haemoglobin is a protein containing iron, and without a sufficiency of iron in the diet the cells no longer contain their normal complement of haemoglobin; the result is a **hypochromic anaemia**. The skin and mucous membranes become pale because of the lack of red pigment in the blood stream, and the patient becomes breathless on exertion because the blood cannot carry an adequate supply of oxygen to the tissues (p. 158). In other types of anaemia the number of erythrocytes may fall, or their structure and shape may become abnormal. In all cases the capacity of the blood to carry oxygen is interfered with, and the symptoms are similar.

In the fetus and for a short time after birth, erythrocytes are manufactured in the liver, but in the adult the main site of erythrocyte formation (**erythropoiesis**) is in the marrow of the bones, particularly the bodies of the vertebrae and the flat bones. Each erythrocyte traces its ancestry back to the stem cells of the bone marrow through a series of precursors which at each successive division lose some of their nuclear material and gain some haemoglobin. When there is an urgent need for large numbers of erythrocytes in the bloodstream, the marrow may pour out into circulation some of the immediate precursors of the mature red cell, there being no time to complete their development. These **reticulocytes** still retain traces of nuclear material in the form of a reticulum of RNA in the cytoplasm.

The response of the marrow to a sudden loss of erythrocytes takes time to develop, and after a severe haemorrhage a transfusion of whole blood may be called for to tide the patient over until he can form enough blood of his own.

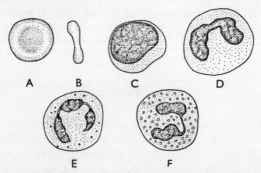

Fig. 73. Blood corpuscles. A: an erythrocyte. B: an erythrocyte in profile. C: a lymphocyte. D: a monocyte. E: a neutrophil granulocyte (polymorph). F: an eosinophil granulocyte.

Under normal conditions the erythrocytes formed in the marrow remain in the circulation for about four months, after which they are seized upon by the cells of the macrophage system (p. 20) and destroyed. The marrow thus has to provide a continuous supply of new cells to replace this steady drain, and about ten thousand million are formed every hour.

The main site of erythrocyte destruction is in the **spleen**, a soft organ with a capsule of smooth muscle which lies in the upper part of the abdomen on the left side, behind the stomach. The large splenic artery pours enormous quantities of blood through the loose 'pulp' of the organ, in which are many macrophage cells which in some way select the superannuated red cells and devour them. In the splenic pulp there are also islands of lymphocyte formation (p. 148) known as **Malphigian corpuscles**.

The **leucocytes** (white cells) are much less numerous than the erythrocytes; there are about 4 to 11 thousand of them in each cubic millimetre of blood. There are two main groups. The **granulocytes** have granular cytoplasm and a nucleus with several lobes (Fig. 73). Some 70 per cent of all leucocytes are **neutrophil granulocytes**, in which the granules stain purple with certain dyes. **Eosinophil** granulocytes have coarse granules which stain red, and **basophil** granulocytes coarse granules which stain blue with the same dyes, but these types together account for less than 4 per cent of the total number of leucocytes. The granulocytes are often called **polymorphs** because of the variety in their nuclear structure.

The **hyaline** leucocytes have clear cytoplasm free from granules. The most important are the small **lymphocytes**, which account for some 20–25 per cent of the total number of leucocytes. They are smaller than the polymorphs, and have a relatively large round nucleus which stains dark purple and cytoplasm which stains pale blue with the usual dyes. A few hyaline cells are larger, with big round nuclei, and are called **monocytes**.

The granulocytes are formed in the bone marrow, while the lymphocytes are formed both in the marrow and in the lymphoid tissues of the lymph nodes, spleen, and thymus. The life span of the granulocytes is not known, and the difficulty of finding it out is enhanced by the fact that most of them are constantly crawling in and out of the circulation; it has been estimated that fewer than 5 per cent of the total number of granulocytes are actually circulating in the blood stream at any given time. However, it is thought that their life is probably less than two weeks, most of which time is spent in the tissues.

It is still more difficult to estimate the life span of the lymphocytes, and it may be that there are two groups, a short-lived one in the bone marrow and thymus, and a longer lived one in the lymph nodes and the lymph; the lymphocytes in the blood are a mixture of both groups. Enormous numbers of lymphocytes enter the lymph every day from the lymph nodes and are poured into the blood stream via the thoracic duct (p. 147); from there they are carried back to the lymph nodes where they settle down for a time before returning again to the bloodstream via the lymph. Other lymphocytes are formed in the epitheliolymphoid tissues and many of these appear to be lost through the surface of these patches into the lumen of the alimentary canal, where they are probably destroyed.

Lymphocytes are concerned in the mechanism of immunity (p. 149), and may also be stem cells producing plasma cells (p. 21) and certain types of phagocytes, though they are not phagocytic themselves.

The neutrophil granulocytes too play an important part in the defence of the body against invasion by micro-organisms. Their motility allows them to chase foreign material such as bacteria, and the fact that they are accomplished phagocytes enables them to ingest and destroy their prey. The number of leucocytes, particularly granulocytes, is increased when infection is present, and the condition is called **leucocytosis**; the total count may rise up to 50,000 or so per cubic millimetre. The pus which forms in an abscess is composed largely of dead granulocytes and dead bacteria.

The term **leucopenia** refers to a reduction in the leucocyte count; it may occur in some forms of poisoning or after exposure to radiation. If the white cells are deficient, bacterial invaders will have a relatively easy task, and this is usually what kills such patients.

The leucocyte count fluctuates during the day, and may be increased following exercise or severe emotional strain, probably because the secretion of adrenaline (p. 88) in some way calls the leucocytes out, ready for action, from the tissues in which they have been resting.

Monocytes are also phagocytic, and are apparently better at dealing with large particles; eosinophils are relatively numerous in the blood stream in allergic asthma, and are also found in normal people in the wall of the gut, in the respiratory system and in the skin. They are thought to be concerned with breaking down and removing protein. Virtually nothing is known about the basophil leucocytes.

The remaining solid elements in the blood are the **platelets**, which are small irregular scraps of protoplasm present in varying numbers; they are derived from **megakaryocytes**. These large cells have processes which penetrate the walls of the sinusoids in the bone marrow, and come to protrude into the bloodstream; they then break off and become platelets. The platelets play a key part in blood coagulation. When a blood vessel is injured (p. 140) its lining epithelium becomes altered in some way, and platelets are attracted to the injured area and adhere to it. A change in their structure results in their becoming fused together into an amorphous mass, and this mass then liberates a substance which has a considerable influence on subsequent events in the sequence of clotting. A deficiency in the platelets thus leads to defects in the clotting of the blood, and is one form of haemorrhagic disease.

Blood transfusion

It is possible to transfuse blood from a given person into certain individuals without causing any apparent damage, but if the same blood is transfused into someone else a severe and sometimes fatal reaction may occur. In this case the blood of the donor is said to be "incompatible" with that of the recipient, and the erythrocytes of the donor become clumped together in the circulation of the recipient. This incompatibility is simply an example of the immune reaction (p. 149). The antigens are carried on the surface of the erythrocytes of the donor, and the antibodies are carried in the globulins of the recipient's serum (p. 152). They are either inherited genetically, or can be acquired following a previous transfusion.

The interactions of donor and recipient blood are complicated, and many different antigens have been recognized and systematized into schemes of **blood groups**; these enable the results of a given transfusion between individuals to be predicted. In the primary system the antigens are designated A and B, and there are four groups—A, B, AB (both present) or O (neither present). Group O blood (46 per cent of the population) can be transfused into anybody with reasonable safety, and group AB blood (3 per cent of the population) is not safe to give to anybody except another group AB individual. Those with group AB blood are sometimes called 'universal recipients', since they can receive blood belonging to any of the four groups, but cannot give any without risk. Group O is sometimes called the "universal donor" group for similar reasons. Group A (42 per cent) and Group B (9 per cent) can give and receive blood to and from their own groups, and can receive from group O and give to group AB.

However, many other antigens have been discovered since the ABO system became understood, and to be really safe the donor's blood must be matched directly against the recipient's blood before starting the transfusion. By observing the result it is possible to tell without reservation whether the proposed transfusion is compatible.

The most important of these other antigens in clinical medicine belongs to the Rh system, in which some have Rh antigens and others do not. An Rh negative mother who conceives an Rh positive fetus may become immunized to its erythrocytes because they occasionally escape through the placental barrier (p. 197) into her circulation. The antibodies formed in her serum can then cross back again into the fetal circulation where they proceed to destroy the fetal erythrocytes, causing a serious condition called haemolytic disease of the new born.

11 · The respiratory system

Every cell in the body requires oxygen. Cells in which very active chemical processes are taking place may be killed by a lack of oxygen lasting only a few minutes, but others which work more slowly can survive for up to an hour or longer. Each cell uses the oxygen brought to it by the blood stream for its own particular chemical purposes. The waste materials formed by the cell differ, but always include carbon dioxide, which enters the blood. It is thus clearly essential that there should be some means of renewing the oxygen in the blood, and of getting rid of the carbon dioxide. These primary needs are met by allowing the blood to come into intimate relation with the air in the lungs.

Transport of gases in the blood

Two-thirds of the carbon dioxide is transported in the plasma (p. 152), about 5 per cent of it being in solution and the remainder combined to form sodium bicarbonate; the remaining one-third is carried in the erythrocytes. The oxygen, on the other hand, is mainly carried in the erythrocytes (p. 153) combined with **haemoglobin** to form the loose compound **oxyhaemoglobin**. As the blood moves slowly along the capillaries, oxygen is given off from this compound and carbon dioxide is taken up by the plasma. Both processes involve diffusion through the capillary wall. The pressure of carbon dioxide outside the capillary is higher than it is inside, and so carbon dioxide is forced *in*; on the other hand, the pressure of the oxygen inside the capillary is higher than the pressure in the tissues, and so oxygen diffuses *out*. The blood leaving the tissues is thus relatively poor in oxygen and rich in carbon dioxide.

This venous blood returns to the heart and is pumped through the lungs, where again it passes slowly through capillaries. These lie in the walls of innumerable tiny air spaces called **alveoli** (p. 165), which contain air relatively rich in oxygen and poor in carbon dioxide. Here the pressure of carbon dioxide in the capillaries is greater than the pressure in the alveoli, so that carbon dioxide diffuses *out* into the air; conversely, oxygen diffuses *in* from the air to the blood. The blood which leaves the lungs to be distributed by the arteries is therefore rich in oxygen and poor in carbon dioxide. Arterial blood is scarlet, for oxyhaemoglobin is bright red; venous blood is dark red because haemoglobin is a much darker colour.

A third gas, nitrogen, is present in small quantities in solution in the plasma. It is completely inert, and its importance lies only in the fact that if the atmospheric pressure in which a man works is increased, more nitrogen is forced into solution. If the pressure is now suddenly reduced, as when a deep sea diver is quickly pulled to the surface, the nitrogen may bubble out of his plasma as the bubbles emerge from soda water when the stopper of the bottle is removed. This may cause serious or even fatal damage, in particular to the nervous system.

The respiratory passages

The **nose** protects the delicate alveoli in the lungs by filtering, warming, and moistening the inspired air. At the entrance to each nostril, in the part of the nose called the **vestibule**, several small stiff hairs act as a primary defence against grit and dust. This region is lined by stratified squamous epithelium, but, except for the small patches of olfactory epithelium (p. 130) tucked away high in its roof, the whole of the rest of the nasal cavity is lined by pseudostratified ciliated columnar epithelium in which are many mucus glands. The sticky mucus traps small particles of dust, and the cilia then waft the mucus to the back of the nose, where it is swallowed. The rich lymphatic supply to the mucous membrane helps to deal with particulate matter (p. 148).

The air is warmed by being forced through narrow channels bounded by extensive warm surfaces. Deep to the respiratory mucous membrane are numerous arteriovenous anastomoses (p. 137) and cavernous spaces (p. 138) which can adjust the flow of blood.

The nose is divided into two by the **nasal septum** (p. 258), a thin vertical plate composed of bone and elastic cartilage, and into each cavity scroll-like bones called **conchae** protrude from the lateral side. These considerably increase the surface area of the nasal mucosa. They lie very close to the septum, and when we have a cold in the nose, the inflamed vascular tissue swells and blocks the airway. The passages between the conchae are the **meatuses** of the nose (Fig. 74), and into them open the drainage orifices of the accessory air sinuses (p. 258), as well as the **nasolacrimal duct**, which conveys the tears from the eye down to the inferior meatus of the nose. The secretions of the sinuses and the inflow of tears combine with the presence of large amounts of mucus to make the nose a very moist place; the air passing through it is therefore humidified.

The back of the nose opens into the **pharynx** (Fig. 75). The **nasopharynx** is that part of the pharynx which lies above the level of the palate, and is lined by columnar ciliated epithelium; the **oropharynx** lies below this level, and is lined by stratified squamous epithelium because it is a passage common to both the

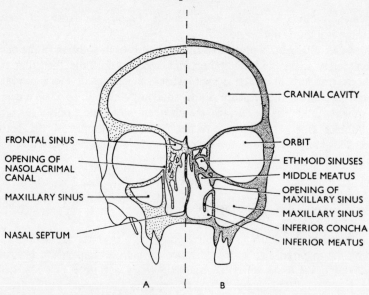

Fig. 74. Sections through the skull showing the nasal cavity. In A, the section passes through the nasolacrimal canal which conveys the tears into the inferior meatus of the nose. In B, which is more posterior, the section passes through the opening of the maxillary sinus into the middle meatus.

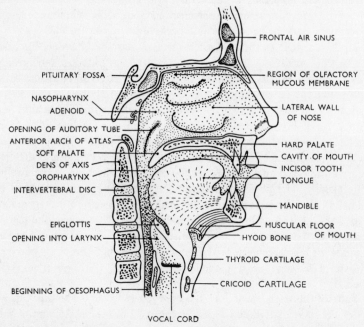

Fig. 75. A median vertical section through the head and neck, showing the cavities of the nose, pharynx, and larynx.

respiratory and the alimentary systems, and thus has to stand a good deal of hard usage. In speech and in swallowing the nasopharynx can be shut off from the oropharynx by the **soft palate**, a muscular flap which can be pulled upwards and tightened. (During swallowing respiration is reflexly inhibited). In the posterior wall of the nasopharynx lies a mass of epitheliolymphoid tissue called the **pharyngeal tonsil (adenoid)**, and at the sides of the opening between mouth and oropharynx are two similar epitheliolymphoid masses called the **palatine tonsils** (Fig. 83, p. 172). These can be regarded as additional protective devices (p. 149) interposed on the pathways of the air stream and the food intake respectively.

Below the oropharynx the respiratory and alimentary systems part company, and columnar ciliated epithelium begins again at the opening of the **larynx** in the anterior wall of the pharynx (Fig. 75). The larynx is yet another protective mechanism, rendered necessary because the crossing of the air and food passages could allow food to pass into the lungs, with disastrous consequences. The whole larynx is raised up under cover of the cartilaginous flap called the **epiglottis** (Fig. 75) every time the movement of swallowing is made. The opening into the larynx, the **aditus**, is thus drawn out of harm's way, and the epiglottis folds down over it like a lid. At the same time the aditus is narrowed by means of a sphincter muscle.

Further down two horizontal folds protrude into the lumen of the larynx. These **vestibular folds** exhibit patches of stratified squamous epithelium, since they undergo friction when they meet in the midline. If a food particle lodges anywhere in the respiratory tract, which is well supplied with sensory nerves, these folds are reflexly brought together, and while they are closed the pressure behind them is built up by a movement of expiration. The folds are then suddenly relaxed, and the resulting violent current of air which we call a **cough** blows away the source of irritation.

Just inferior to the vestibular folds, another pair of **vocal folds** protrude like thin horizontal shutters into the larynx (Fig. 76). These contain the fibro-elastic **vocal cords**, alongside each of which runs a **vocalis** muscle. This muscle is the fine adjustment determining the tension of the cords; the coarse adjustment is provided by other laryngeal muscles, which can also alter the distance between the vocal folds as well as the shape of the orifice between them. The voice is produced by the vibrations of the vocal cords in the air blast, and the quantity of air reaching the lungs is determined by their degree of separation. Because of the hard usage they receive, the vocal folds are covered with stratified squamous epithelium.

The vocal folds and their complex musculature are accommodated and protected by the laryngeal cartilages. The **thyroid cartilage** forms the prominence on the front of the neck commonly called the Adam's apple. It is larger in the male because the whole larynx is larger; the greater length of the vocal cords explains the greater depth of the male voice. The thyroid cartilage is V-shaped in cross-section, the closed end of the V being anterior. It is attached above to the **hyoid bone** (p. 260) by muscles and ligaments, and below it articulates with the **cricoid** cartilage, which lies opposite the sixth cervical vertebra (Fig. 77) and is shaped like a signet ring. The posterior ends of the vocal cords are attached to two **arytenoid** cartilages which rest on top of the 'signet' of the cricoid, and the anterior ends are attached to the inside of the thyroid cartilage at the apex of the V. By moving the cricoid backwards or the thyroid forwards the cords can be stretched, and the opening between them can be altered by moving the arytenoids closer together or further apart.

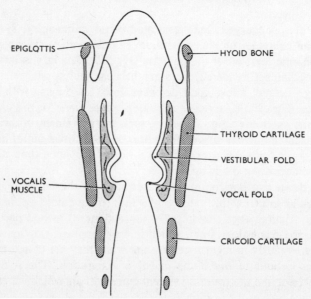

EPIGLOTTIS

HYOID BONE

THYROID CARTILAGE

VESTIBULAR FOLD

VOCALIS MUSCLE

VOCAL FOLD

CRICOID CARTILAGE

Fig. 76. A vertical section through the larynx, viewed from behind, showing the vestibular and vocal folds.

The arytenoids can also pivot round on a vertical axis in response to the pull of the muscles attached to them, and this alters the shape of the aperture between the vocal cords (Fig. 78).

The cricoid cartilage is attached inferiorly to the upper end of the **trachea**, a flexible tube composed of incomplete hoops of hyaline cartilage joined to each other by membranous connective tissue with many elastic fibres, the whole being lined by ciliated columnar epithelium (Figs. 79 and 80). This plan of construction is necessary because the trachea is always being stretched, whether by upward movements of the head, or by elevation of the whole larynx every time anything is swallowed (p. 172). If the trachea were rigid this movement would pull on the lungs and interfere with their functioning. To prevent the trachea being pushed bodily forwards when food passes down the oesophagus (p. 172) which lies directly behind it, the posterior wall of the trachea contains no cartilage, but is formed merely of muscular tissue which 'gives' as the food passes.

In the chest the trachea divides into two primary bronchi, one going to the right lung and one to the left. The right bronchus is the more direct continuation of the parent trunk, and so foreign bodies which find their way past the larynx usually pass into the right lung.

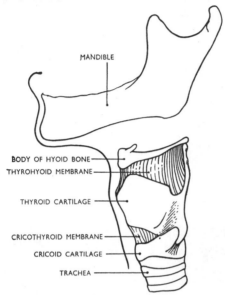

Fig. 77. The larynx from the side.

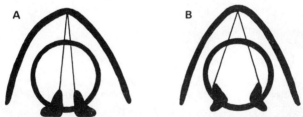

Fig. 78. Scheme showing the movements of the arytenoid cartilages and vocal cords. The circle represents the cricoid cartilage, lying inside the wings of the thyroid cartilage. The two triangular masses are the arytenoid cartilages, to which the vocal cords are attached. A: quiet breathing; B: forced deep breathing; the arytenoids have been pulled apart and rotated on their own axis, so widening the gap between the vocal cords.

Each primary bronchus divides into secondary bronchi, which in turn give off tertiary branches, and so on. The amount of cartilage in the walls of the air passages gradually lessens until it is lost as the tubes reach a diameter of about 0·5–1 mm. The muscular walls of these **bronchioles** can constrict sufficiently to obliterate the airway, and in **asthma** they go into spasm and obstruct the passage of air to and from the alveoli. Sympathetic stimulation or an injection of adrenaline (p. 88) will cause them to dilate, so affording relief to the patient by increasing the ventilation of the lungs and thus the amount of oxygen available.

The bronchioles give off smaller branches, about 0·2 mm in diameter, called **respiratory bronchioles**, which have no muscular coating, and communicate with the alveoli (Fig. 80), of which there are something like 750 million in each

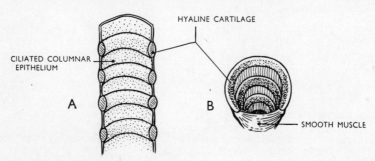

Fig. 79. The construction of the trachea. A: the front wall of the trachea from inside. B: looking down into the trachea.

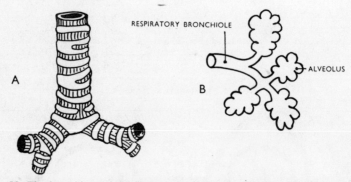

Fig. 80. The bronchi and bronchioles. A: the trachea dividing into the two main bronchi. B: diagram of a repiratory bronchiole and its associated alveoli.

lung. The whole of the bronchial tree down to the respiratory bronchioles is lined with columnar ciliated epithelium; the cilia beat towards the larynx, and dust and dirt are in this way removed in the secretions of the respiratory tract, which together form the **sputum**. In coal miners the strain put on the mechanism is excessive, and dust does penetrate down to the terminal reaches of the respiratory tree, where it may set up an irritation leading the the establishment of disease. The particles are removed as far as possible by lymphatics and are taken to the nearest group of lymph nodes (p. 148). In a newborn baby the lungs are pink, but in a city dweller the cumulative inhalation of filth leads to the lungs becoming pigmented from the particles of dirt which they contain, and in old age they may be almost black.

The respiratory bronchioles, the ducts which lead from them to the alveoli, and the alveoli themselves are called the respiratory portion of the lung, for in these anatomical structures respiratory exchanges go on. The total surface area of the respiratory portion of the lung is over 50 square metres. The alveoli are about 100–200 μm in diameter, and are lined solely by a thin pavement epithelium. They are surrounded by a dense basket-work of capillaries, and the arrangement is such that the blood is separated from the air only by the basement membranes of the two layers of pavement epithelium. This blood/air barrier normally has a total thickness of about 1 μm. The final defence against particulate matter is found in the alveoli, where macrophages called 'dust cells' invade the lumen, clean it out, and return to the lymphatics which richly supply the lung substance.

The lungs and pleural cavities

The lungs have a light spongy framework of very elastic connective tissue. They lie in the chest cavity on either side of the mediastinum (p. 273), and each is covered by a layer of pavement epithelium constituting the **visceral pleura**. This forms one wall of a closed **pleural sac** into which the lung is pushed from the medial side; the outer wall, or **parietal pleura**, adheres to the inner surface of the chest wall (Fig. 81). The two layers are continuous with each other at the **hilum** of the lung, where the blood vessels and bronchi enter the lung substance.

Normally there is no air in the pleural sac, which, like the pericardial sac (p. 140) or any synovial sheath, contains nothing but a few drops of fluid. Atmospheric pressure transmitted down the respiratory passages forces air into the lung and inflates it like a balloon, so thrusting the visceral pleura outwards against the parietal pleura. At the same time the atmospheric pressure outside the chest wall is forcing the parietal pleura inwards against the visceral pleura. Because the pleural sac is moist, a third force tending to make the two layers stick together is the surface tension between them, and the fluid serves the additional purpose of allowing sliding movements between the visceral and parietal pleura during breathing. Acting against these forces are the elasticity of the lung framework and surface tension in the lining of the alveoli, both of which tend to pull the two layers of pleura apart. If the thoracic cavity is made larger by moving the chest wall (p. 270), atmospheric pressure will push more air into the lung as the visceral pleura follows the parietal pleura. Movements which reduce the volume of the chest also reduce the volume of the lung and air is pushed out, assisted by the elasticity of the lung.

If air enters the pleural sac, as may happen after an injury to the chest wall, there is now atmospheric pressure on both sides of the visceral pleura, and the lung collapses because of its own elasticity just as a balloon collapses when the pressure outside it is adjusted to balance the pressure inside. The collapsed lung is about the size of a clenched fist, and fills only about a quarter of the chest cavity. The two pleural surfaces are now widely separated, and the lung ceases to follow the movements of the chest wall. This condition is called a **pneumothorax**, and with one lung out of action the strain of aerating the blood is thrown entirely on the lung of the opposite side; this can be equal to the task so long as no violent demands for oxygen are made by the tissues.

If a lung is placed in water it will float, because of the amount of air it contains, and this test may be used to determine whether a person has met his death by drowning, for if so the lungs will sink because of the water inhaled into them. The lungs from a baby who has never breathed will also sink, for before the first breath the lungs are more or less solid. 'Consolidation' of part of the adult lung may be produced by inflammation in **pneumonia**; there is no air entry to the affected area, which therefore is dull instead of resonant to percussion (p. 239).

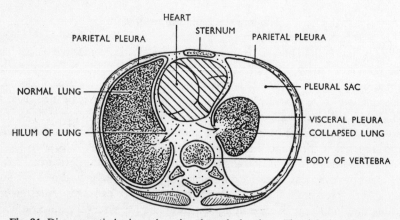

Fig. 81. Diagrammatic horizontal section through the chest. The right lung is shown collapsed following the introduction of air into the pleural sac. The arrows indicate the path of the pulmonary vessels to the hila of the lungs. Most of the back muscles have been omitted for the sake of simplicity.

If a bronchus is inflamed, as in **chronic bronchitis**, the thick sticky secretion poured out into its lumen immediately sets off a cough reflex (p. 161). The chest is forcibly compressed by the violent expiratory efforts, and the pressure is transmitted to the air in the respiratory part of the lung. If this air is prevented from escaping by a plug of secretion, the delicate walls of the respiratory bronchioles and the alveoli may rupture, so forming large intercommunicating cavities which may become further distended as more air becomes trapped within them. This is the condition of **emphysema**, in which the lung is full of useless air. Should the walls of the bronchioles give way at weak points under the strain of coughing, infected sacs full of pus and secretion may be formed, and this condition is known as **bronchiectasis**.

Each lung is divided into two main **lobes** by an oblique fissure, which runs from above downwards and forwards: in addition, the upper lobe of the right lung is partially subdivided by a horizontally running fissure. The visceral pleura dips down to line these fissures. Of greater practical importance are the **bronchopulmonary segments** (Fig. 82). Each segment is the portion of lung tissue supplied by one tertiary bronchus; if that is blocked, the air supply of the segment is completely shut off, and it will collapse. The segments are separated

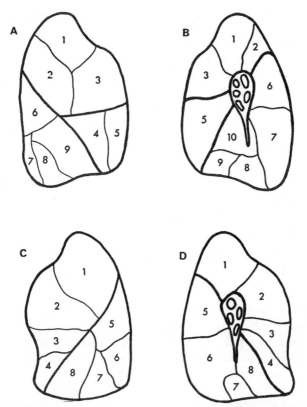

Fig. 82. Bronchopulmonary segments. A and B lateral and medial views of the right lung; C and D lateral and medial views of the left lung. The fissures of the lung are indicated by the thick black lines. Note there is no middle lobe in the left lung.

Right lung

Upper lobe	*Lower lobe*
1. Apical	6. Apical (superior)
2. Posterior	7. Posterior basal
3. Anterior	8. Lateral basal
	9. Anterior basal
Middle lobe	10. Medial basal
4. Lateral	
5. Medial	

Left lung

Upper lobe	*Lower lobe*
1. Apicoposterior	5. Apical (superior)
2. Anterior	6. Posterior basal
3. Superior lingular	7. Lateral basal
4. Inferior lingular	8. Anterior basal

from each other by thin septa which allow the surgeon to remove a diseased segment. The segmental pattern is reasonably constant, and there are usually ten segments in each lung.

Some of the tertiary bronchi turn upwards from the hila to the segments of the upper lobes, some turn backwards to the posterior segments, and others continue downwards to the lower lobes. The direction which is taken to supply a given segment is of importance when it is necessary to drain the fluid out of it by **postural drainage**.

At the hilum, the pulmonary arteries and veins lie close round the main bronchus, but in the lung substance the arteries accompany the branches of the bronchial tree into the bronchopulmonary segments, while the veins return in the septa between the segments. In addition to the pulmonary vessels, which bring to the lung its raw material and take away its finished product, there is a system of **bronchial** arteries and veins. The bronchial arteries spring from the aorta or the intercostal arteries (p. 273), and therefore contain oxygenated blood, which supplies the active tissues in the lung. Curiously, there are numerous anastomoses between the bronchial and the pulmonary circulations, so that the deoxygenated blood brought to the lung by the pulmonary arteries receives a certain amount of arterial blood from the bronchial arteries before it reaches the alveoli.

The control of respiration

At rest under normal conditions some 12–16 inspirations are taken every minute. Each of these draws in a volume of about 0·5 litres which is known as the **tidal air**, and the same amount is expelled during expiration. By taking the deepest possible breath an additional 1·5 litres of **complemental air** can be inspired, and after a normal expiration another 1·5 litres of **supplemental air** can be expelled. The maximal amount of air which can be blown out after taking a very deep breath is thus of the order of 3·5–4 litres, depending on the size of the individual, and is known as the **vital capacity**. This still leaves another litre or so of **residual air** inside the lungs.

There is thus considerable scope for varying the amount of air taken in at each breath, and breathing deepens in response to a demand for oxygen from the tissues. At the same time the respiratory rate rises—perhaps to 30 or 40 a minute. All this greatly increases the oxygen supply to the lungs, and the concurrently increased blood flow through them enables more oxygen to be picked up and more carbon dioxide to be given off. The movements of respiration are under the control of the **respiratory centre**—a group of neurones lying close to the floor of the fourth ventricle of the brain (p. 103). If this region is injured, respiration stops and death occurs from asphyxia. The respiratory centre is partly under the control of the will, but for the most part it functions automatically, being influenced by the pattern of incoming nerve impulses and by the chemical composition of the blood.

When the centre sends out impulses which cause the lungs to inflate, stretch receptors in the lung substance are stimulated, and these send up to the centre, along the vagus nerves (p. 272), impulses which inhibit the outflow of inspiratory impulses. The process of inspiration is thus normally self-limiting. Other impulses arriving along different nervous channels may also influence the centre. Thus if the skin is stimulated by cold, as in those bathers who break the ice to take a New Year's Day swim, the respiration usually becomes gasping or the breath may be held for some time. Similarly, irritation of the respiratory passages leads to the characteristic form of respiration known as coughing (p. 161).

The respiratory centre is sensitive to changes in the reaction of the blood. When the amount of carbon dioxide in solution rises, a certain amount of carbonic acid is formed, and the hydrogen-ion concentration of the blood rises slightly—that is, the blood becomes a little less alkaline. This stimulates the respiratory centre, causing an increase in the ventilation of the lungs so that the excess carbon dioxide is washed away into the alveolar air. During exercise, when the amount of carbon dioxide is increased because of the increased activity of the cells, respiration is stimulated to get rid of it. Similarly, by breathing an atmosphere containing more carbon dioxide than normal, the rate and depth of respiration may be made to go up very considerably. This fact is made use of by the anaesthetist when he wishes the lungs to be kept well ventilated; the patient is given an anaesthetic mixture with 5 per cent carbon dioxide added to stimulate his breathing.

Oxygen lack also stimulates the respiratory centre to increase the rate and depth of breathing, although this is a less sensitive mechanism. People who live at high altitudes develop compensatory alterations, such as an increase in the number of erythrocytes, which diminish the need for increased ventilation of the lungs.

Close to the point where the common carotid artery bifurcates to form the internal and external carotid arteries (p. 265) there is a collection of specialized tissue, lavishly supplied with sensory nerves, which is called the **carotid body**. Oxygen lack and carbon dioxide excess stimulate the endings in the carotid body and the impulses from this organ supplement the direct effect on the respiratory centre itself.

If the lungs are over-ventilated by deliberately breathing very deeply for a time, more carbon dioxide is washed out into the air than is being produced, the hydrogen-ion concentration of the blood falls, and as a result breathing stops. It takes some minutes for the concentration of carbon dioxide to rise high enough to stimulate the centre again, and breathing may be restarted because of the oxygen deficiency before the carbon dioxide level has risen to normal. Deep breathing is practised by pearl and sponge divers who work without helmets, and by this means they are able to hold their breath for as much as six minutes.

Outline of the clinical examination of
the respiratory system

Cyanosis (p. 150) indicates inadequate oxygenation of the blood, perhaps due to deficient ventilation of the lungs. The respiratory rate may be timed over a minute, and any distress in breathing or alteration in rhythm may be noted; for example, prolongation of expiration is common in asthma. The expansion of the chest in response to a full breath may be measured by a tape, and the patient's vital capacity can be recorded by an instrument called a spirometer. The presence or absence of cough may be noted, and also the amount of sputum produced; the sputum may be examined for bacteria or other material such as blood. The patient may be asked to perform an **exercise tolerance test** such as stepping on and off a stool twenty times. If his respiratory and cardiovascular systems are sound this should not alter his respiratory rate very much.

The nervous control of the respiratory system may be tested by asking the patient to hold his breath for a specified period, both before and after deep breathing. Or he may be asked to keep a column of mercury at a given level for a given time by blowing into a tube.

The nose and throat, the larynx, and even the bronchi, are accessible to direct visual examination by appropriate instruments. The degree to which the lungs are filled with air may be estimated by **percussion** (p. 239).

The sounds made by the air entering and leaving the lungs can be heard through a stethoscope, and this may serve to detect alterations which are characteristic of diseases such as pneumonia. The vibrations caused by the patient saying 'ninety-nine' or 'one-one-one' are transmitted through the chest wall and can be felt with the fingers or heard with the stethoscope; here again alterations may suggest disease.

A radiograph of the chest may show up disease of the lung or an accumulation of fluid or air in the pleural sac. The behaviour of the diaphragm during respiration can be watched, as can the way in which the lungs fill and empty.

Finally, samples of blood may be taken from the veins, the arteries and the heart itself, to provide a direct measurement of the amount of oxygen and carbon dioxide in various parts of the circulatory system; from this the efficiency of the oxygenation in the lungs may be estimated. Samples of the inspired and expired air can be analysed to complete the picture.

12 · The digestive system

The digestive system is primarily a long tube, the **alimentary canal**, in which the complicated molecules of the food are split into simpler constituents; these pass through the wall of the canal and are absorbed into the blood stream. The food is first reduced to a pulp, which then travels slowly along the canal, propelled by peristalsis (p. 50), and is acted on by the secretions of numerous glandular cells which lie in the wall of the canal. Other glands which are too bulky to be accommodated on the spot—the **salivary glands**, the **liver**, and the **pancreas**—lie at some distance from the canal, and their secretions are poured into it by means of ducts (p. 16).

The mouth, pharynx, and oesophagus

The process of digestion begins in the mouth. Here the food is bitten into pieces by the incisor teeth, chewed and ground up by the molar teeth, and mixed with saliva into a smooth ball or **bolus** which can be easily swallowed; the tongue plays an important part in the mixing process. The saliva contains much mucus, which helps to make the bolus smooth and slippery, and also the enzyme **ptyalin**, which can convert starch into sugar. This process is limited, because as soon as the food reaches the stomach, the presence of acid puts a stop to the activity of the ptyalin. However, a certain amount of salivary digestion can go on in the stomach until the acid penetrates the centre of the bolus.

There are three salivary glands on each side of the mouth. The largest is the **parotid** gland, which lies in the angle between the ear and the jaw and has a duct which enters the mouth opposite the second upper molar tooth; it is the gland which is usually swollen in mumps, and through its substance run the branches of the facial nerve on their way to the muscles of the face (p. 264). The **submandibular** gland lies under cover of the lower jaw under the floor of the mouth; its duct leads into the mouth below the tongue, where the dentist puts his sucker, and beside the openings of the small ducts of the **sublingual** gland, which lies on the floor of the mouth under the tongue.

The mouth opens posteriorly into the **oropharynx**, which has a muscular wall composed of three overlapping **constrictor** muscles. The wall is deficient

171

Chapter 12

anteriorly, where it is formed by the back of the tongue, the cartilaginous flap of the **epiglottis** which diverts fluid away from the opening into the larynx, and the opening into the larynx itself (Fig. 75). Below this opening, where the air and food passages diverge from each other again, the wall of the pharynx is formed anteriorly by the posterior wall of the larynx, and eventually the muscle fibres of the constrictors meet each other anteriorly at the level of the cricoid cartilage (p. 162), below which the pharynx becomes the **oesophagus**. This narrow muscular tube conducts the food through the neck and the thorax down to the diaphragm, which it penetrates on the left side of the body to enter the stomach. The mouth, oropharynx, and oesophagus are lined by stratified squamous epithelium which is able to stand up to hot coffee and ice-cream, to hard apples and curry powder. The pharynx and oesophagus secrete mucus, but no digestive enzymes.

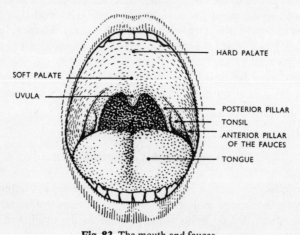

Fig. 83. The mouth and fauces.

During **swallowing** the bolus is pressed upwards and backwards, by the muscles of the tongue and the floor of the mouth (p. 262), against the soft palate and the posterior wall of the oropharynx. This sets up reflexes which result in the breath being held and the soft palate being drawn up to shut off the nasopharynx from the oropharynx. The oropharynx is also shut off from the mouth by the constriction of an incomplete ring of muscle running from the soft palate above to the tongue below. The muscles in this ring form the **pillars of the fauces** (Fig. 83). The constrictor muscles now contract in sequence from above downwards. The food cannot be forced into the nose or the mouth, and so this wave of contraction pushes it down past the opening of the larynx and into the oesophagus (Fig. 75). At the same time the larynx is drawn up under the tongue to prevent the entry of any food particles. About half way down the oesophagus the somatic muscle of the upper end of the alimentary canal gives way to visceral muscle, and voluntary control over the movements of swallowing is lost. At this point the autonomic nervous system takes over, and the wave-like contractions become fully automatic **peristalsis** (p. 50), the wall of the canal relaxing in front of the bolus and contracting behind it, so forcing it onwards.

The greater part of the alimentary canal lies in the abdomen (Fig. 92). After passing through the left dome of the diaphragm (p. 271), the oesophagus almost immediately enters the **stomach**.

The stomach

The stomach is a large distensible bag which stores the swallowed food, and liberates it as required into the duodenum (p. 176). The stomach has a main

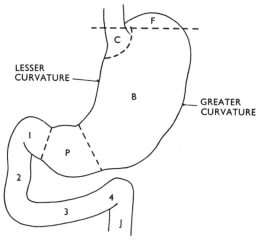

Fig. 84. Subdivisions of stomach and duodenum. B: body, C: cardiac portion, F: fundus, and P: pyloric part of stomach. 1, 2, 3, 4: parts of the duodenum. J: jejunum.

portion, or **body**, and a **pyloric portion**, which differs from the body in its microscopic structure and its functional activity, and leads into a narrow **pyloric canal** (Fig. 84). This in turn terminates in the **pyloric sphincter**, a circular ring of visceral muscle which guards the entrance to the duodenum. The region round the point of entrance of the oseophagus also differs structurally from the rest of the body, and is called the **cardiac portion**: the junction of oesophagus and stomach is the **cardia**.

The stomach has anterior and posterior surfaces, which meet each other in the **lesser** and **greater curvatures.** The part of the body superior to the entrance of the oesophagus is the **fundus**, which is in contact with the diaphragm just under the heart. In the upright position swallowed air accumulates here, but when the patient lies down the bubble moves to the region of the pyloric portion which is now uppermost, in front of the vertebral column.

The stomach is completely covered by pavement epithelium called **visceral peritoneum**, and this enables it to move around without friction, since the inside of the abdominal wall is also lined with **parietal peritoneum** (Fig. 91). The lesser curvature is attached to the liver by a double fold of peritoneum called the **lesser omentum** (Fig. 85), and from the greater curvature a similar fold, the **greater omentum**, hangs down in front of the other viscera and doubles up behind itself to become attached to the transverse colon (p. 183).

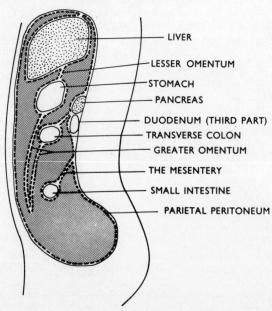

LIVER

LESSER OMENTUM

STOMACH

PANCREAS

DUODENUM (THIRD PART)

TRANSVERSE COLON

GREATER OMENTUM

THE MESENTERY

SMALL INTESTINE

PARIETAL PERITONEUM

Fig. 85. Diagrammatic median section of abdomen showing the scheme of arrangement of the peritoneum; pelvic organs not shown. The peritoneal cavity, like that of the pleura, normally contains nothing but a few drops of fluid, and the wide spaces in the diagram are completely filled by other organs, such as the small intestine, only one loop of which is represented.

The cardia of the stomach represents a more or less fixed point, for the oesophagus is fairly firmly attached to the posterior wall of the thorax. Similarly, the first part of the duodenum, which is about 5 cm long, runs from the pylorus to join the second part (Fig. 84), which is firmly plastered on to the posterior abdominal wall. The junction of the first and second parts of the duodenum is therefore another relatively fixed point, but the pylorus has quite a range of movement because of the mobility of the first part of the duodenum. In between these two points, the stomach is allowed great positional freedom. When it is well filled with food, say after a Christmas dinner, and its owner stands upright, the greater curvature may extend right down into the pelvis (Fig. 86). On the other hand, when he lies down, gravity pulls the stomach up towards the chest, for the most dependent part of the abdominal cavity when the patient is lying on his back is right up under the diaphragm (Fig. 87).

Inside the peritoneal coat of the stomach is a muscular coat with three layers. The outer longitudinal layer and the middle circular layer are continuous with the longitudinal and circular layers of the oesophageal musculature. In addition there is an innermost oblique layer, which is not found anywhere else in the digestive tract. The body of the stomach does not show any well marked peristaltic activity, but during gastric digestion peristaltic waves begin where the body joins the pyloric portion, and sweep downwards over the pyloric canal. The pyloric sphincter relaxes with each peristaltic wave, allowing gastric

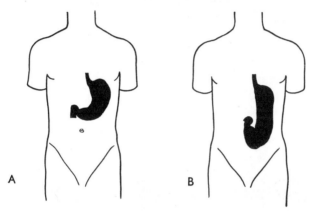

Fig. 86. Diagram showing the position of the stomach as seen after filling it with radio-opaque material (a 'barium meal') and taking a radiograph. A: lying down. B: standing up.

contents to escape into the duodenum. Between waves it acts as a safety mechanism preventing reflux from the duodenum back into the stomach. Although the pylorus does not stop food leaving the stomach, the narrowness of the pyloric canal prevents large pieces of food passing along it until they have been adequately broken up by the churning action of the stomach musculature. Starchy food passes on quickly, and fatty food slowly. Normally food begins to leave the stomach a few minutes after it has been swallowed, and even after a fatty meal the stomach is usually empty within five hours. Emotion can affect the rate of emptying; fear may delay the emptying of the stomach for several hours (p. 88).

The mucous membrane lining the empty stomach is thrown into longitudinal folds called **rugae**, which flatten out when the stomach is full. Inspection of the mucosa with a lens discloses innumerable small tunnels opening on to it; these are the **gastric pits**, into which open the tubular gastric glands secreting the **gastric juice**. The surface epithelium is columnar, and is covered with a protective film of a special kind of mucus which it produces. Some of the cells in the gastric glands produce the enzyme **pepsin**, which can only function in an acid medium, while others produce **hydrochloric acid**. The structure of the acid-producing cells is extremely complex, for they produce a secretion which would instantly kill any other cell in the body. The surface cells are shed at a high rate, and are replaced by cells which are born in the depths of the gastric pits and travel up to the surface; mitotic figures are therefore very common in the pits. The glands of the fundus and body produce all the acid; the simpler glands of the pyloric and cardiac portions produce mainly mucus.

Between them, pepsin and hydrochloric acid convert the protein taken in the food into smaller molecules called **polypeptides** and **peptones**; they have virtually no action on carbohydrates and fats. This seems a surprisingly small result for the immensely complicated glandular structures which the stomach contains, and a good deal more remains to be learned about gastric functions.

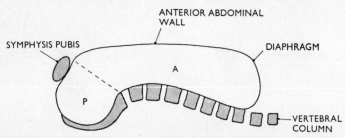

Fig. 87. Scheme to show that abdominal viscera tend to gravitate to the upper part of the abdominal cavity if the patient is lying on his back. A: abdominal cavity, P: pelvic cavity.

Nevertheless, the stomach, unlike the small intestine, can be completely dispensed with, and digestion may proceed surprisingly well in its absence. Nor is absorption much interfered with, for the stomach can only absorb water and alcohol. The chief difficulty, if the stomach is removed, relates to the bulk and consistency of the material which is now fed directly from the oesophagus into the small intestine. The main physiological function of the stomach is therefore to act as a hopper into which are fed, at irregular times, large amounts of food of various consistencies. This hopper converts the food into suitably sized particles and feeds these at regular intervals into the small intestine; the small amount of digestive activity which occurs appears to be a bonus.

The small intestine, liver, and pancreas

The **duodenum**, the first part of the small intestine, is a canal about 25 cm long which forms a C-shaped bend on the back wall of the abdomen (Fig. 88). The first part is fairly mobile, being partly enclosed in peritoneum, but the second part, which travels vertically downwards, and the third part, which runs more or less horizontally and slightly upwards to the left side of the vertebral column, are firmly adherent to the posterior abdominal wall. The short fourth part, about 2·5 cm long, is again a transition zone, for it joins the fixed duodenum to the immensely mobile jejunum (p. 180).

The duodenum, like the rest of the small intestine, has two complete muscular coats, an outer longitudinal and an inner circular, but there is no true peristalsis. Instead there are irregular movements which appear to mix up the

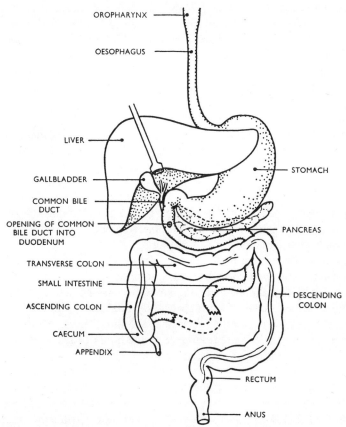

OROPHARYNX

OESOPHAGUS

LIVER

GALLBLADDER

COMMON BILE DUCT

OPENING OF COMMON BILE DUCT INTO DUODENUM

TRANSVERSE COLON

SMALL INTESTINE

ASCENDING COLON

CAECUM

APPENDIX

STOMACH

PANCREAS

DESCENDING COLON

RECTUM

ANUS

Fig. 88. Scheme showing the main features of the alimentary system. Most of the small intestine has been cut away.

contents thoroughly. The mucous membrane, like that of the jejunum (p. 180), is thrown into circular folds which enormously increase the surface area. On these folds are great numbers of **villi** (p. 16), which give the surface a velvety appearance under a lens. The cells covering the villi are columnar, and their free margins, which look into the lumen of the canal, are covered with microvilli (p. 4). All this underlines the absorptive function of the small intestine which requires a very large surface area.

Between the villi are small burrows called **crypts** lined by cells which produce a mucous and enzymatic secretion called the **intestinal juice**. A high mitotic rate in the crypts produces new columnar cells which migrate up the crypts and along the sides of the villi until they are ultimately shed into the lumen of the bowel. In this way the entire lining membrane of the small intestine is renewed every two to four days.

The first part of the duodenum receives the acid material from the stomach, and bears the brunt of any damage this may cause; duodenal ulcers are most commonly found here. Further distally, the acid is neutralized by the alkaline secretions of the liver and pancreas, both of which are poured, usually by a common duct, into the middle of the second part of the duodenum (Fig. 90). The mucous membrane of the duodenum itself secretes mainly mucus, though the first part has some specialized **Brunner's glands**, which have a watery alkaline secretion.

The **liver** (Fig. 88) is the biggest gland in the body, and produces **bile**. In the fetus it manufactures erythrocytes (p. 154), but in the adult its main function is to carry on all kinds of metabolic processes (p. 209). It accounts for something like 2·5 per cent of the body weight in the adult, but more in the child (p. 221), and it has a very rich blood supply, which gives it a semi-fluid consistency; wounds of the liver are very dangerous because of the amount of bleeding. The larger **right lobe** of the liver lies under the right dome of the diaphragm (p. 271), while its **left lobe** reaches across the midline, separated from the heart by the central tendon and left dome of the diaphragm.

The **portal vein** brings to the liver the raw materials absorbed from the digestive tract. This vein branches like an artery and empties into a system of sinusoids (p. 138). Between the sinusoids and bathed by their contents lie the cords of liver cells, and in between the cells are the tiny **bile canaliculi** which drain away their external secretion, the bile. These in turn drain into intrahepatic ducts, which unite to emerge at the **hilum** of the liver as two main **hepatic ducts**, one for the right lobe, and one for the left.

Inside the liver substance, the branches of the hepatic artery and the portal vein and the tributaries of the hepatic ducts accompany each other in the connective tissue between the liver **lobules**. In the centre of each lobule, which is roughly hexagonal in cross-section (Fig. 89), lies a tributary of a **hepatic vein**; blood filters across the lobule to its centre and enters the vein, ultimately leaving the liver through the main hepatic veins, which empty into the inferior vena cava (p. 287) just before it passes through the diaphragm into the chest. The bile passes in the opposite direction to the blood, outwards from the centre of the lobule.

Outside the liver, the two hepatic ducts join together to form the **common hepatic duct** (Fig. 90), which in turn is joined by the **cystic duct** coming from the gall bladder. The result is the **common bile duct**, a fibromuscular tube some 10–12 cm long which opens into the second part of the duodenum. At its lower end it is surrounded by a sphincter of visceral muscle.

The secretion of bile is continuous, but between meals bile discharged into the common bile duct cannot escape into the duodenum because the sphincter is closed. The pressure inside the biliary tree therefore rises, and bile passes along the cystic duct into the **gall bladder,** a small pear-shaped organ holding 30–60 ml of fluid. The gall bladder is lined by columnar absorptive epithelium, and it concentrates the bile and stores it until it is needed.

When food enters the duodenum, the muscular wall of the gall bladder contracts and the sphincter muscle at the lower end of the bile duct relaxes, so discharging the bile. This mechanism is under the control of a hormone liberated from the duodenal mucosa.

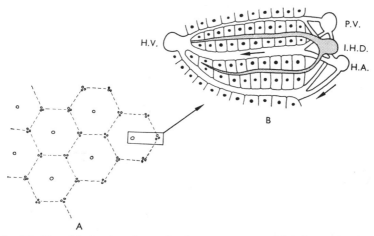

Fig. 89. Liver lobule. A: scheme showing arrangement of lobules, each with a central hepatic vein and having at its angles branches of the hepatic artery and the portal vein, and a tributary of the hepatic duct. B: schematic enlargement of the area indicated, showing the blood passing from the portal vein (PV) and hepatic artery (HA) between cords of liver cells to the hepatic vein (HV). Bile (indicated by stippling) flows in the reverse direction to the intrahepatic duct (IHD).

There are no digestive enzymes in the bile, but the **bile salts** which it contains reduce the surface tension of the duodenal contents, so enabling the watery secretions of the pancreas and the intestinal wall to mix freely with greasy food material and allowing their enzymes better access to the material on which they work. If the common bile duct is blocked by a gall stone, the digestion of fats cannot take place because of the lack of this emulsifying action.

The pancreas is the main digestive gland. It is about 15 cm long, and lies plastered almost horizontally across the posterior wall of the abdomen. It has a head, a body, and a tail. The head is enclosed in the concavity of the duodenum, and the tail makes contact with the spleen (p. 154). The bulk of the pancreas is an exocrine gland, but about 1 per cent of the total mass consists of islets of endocrine tissue which produce the hormone insulin (p. 200). The exocrine secretion drains by a main and an accessory duct into the second part of the duodenum. Usually the main duct shares a single opening with the common bile duct, and the accessory duct opens about 1 cm higher up.

The pancreatic juice contains three enzymes, all of which work best in an alkaline medium; both the juice itself and the intestinal secretions are alkaline. The acid material fed into the duodenum is thus soon neutralized and converted into a suitable environment for the work of intestinal digestion, which occurs mostly in the upper part of the small intestine. The **trypsin** of the pancreatic juice converts protein or polypeptides into aminoacids, which have a simple enough structure to pass through the intestinal wall.

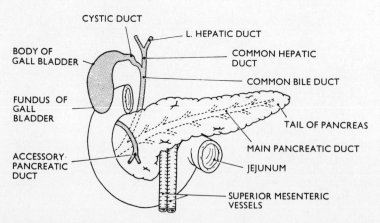

Fig. 90. The extrahepatic biliary system and the pancreas. The accessory pancreatic duct is indicated by stippling although it too is concealed by the substance of the pancreas.

Carbohydrates are broken down into **disaccharides** (sucrose is an example) by the **amylase** of the pancreatic juice. The disaccharides are then in turn broken down into simple sugars (chiefly **glucose**, though some **fructose** and **galactose** are also formed) inside the microvilli of the columnar cells of the small intestine. These simple sugars are then absorbed into the capillaries in the villi.

Finally, the fats in the food are split by the pancreatic **lipase** to **glycerol** and **fatty acids,** which are conveyed through the intestinal epithelium in an emulsion with the bile salts. Inside the epithelial cells the components immediately recombine to form tiny droplets of fat which then pass out of the cells into the central lymphatic vessels of the villi. These vessels are called **lacteals** (p. 148) because of the milky appearance this congregation of fatty droplets gives them when they are examined after a fatty meal. The fat is ultimately delivered into the blood stream via the thoracic duct (p. 147), and reaches the liver by the hepatic artery.

The second part of the small intestine is called the **jejunum,** and the third part is known as the **ileum**. One merges gradually into the other; it is impossible to say where the jejunum ends and the ileum begins. Nevertheless, a typical part of ileum differs from a typical part of jejunum. The ileum is less vascular, and has fewer and less well marked circular folds and fewer villi. On the other hand, it contains more epitheliolymphoid tissue, which is collected in oval areas called **Peyer's patches**. The ileum can absorb water readily.

Stages of digestion

Site	*Reaction of contents*	*Active agents*	*Source*	*Raw material*	*End products*
Mouth	Alkaline	Ptyalin	Salivary glands	Starch	Compound sugars (Disaccharides)
Stomach	Acid	Pepsin Hydrochloric acid	Stomach wall	Protein	Peptones
Small Intestine	Alkaline	Bile Salts Lipase	Liver Pancreas:	Fat	Emulsion
			Intestinal wall	Emulsified fat	Glycerol & soaps
		Amylase	Pancreas	Carbohydrates	Compound sugars
		Sugar splitting enzymes	Intestinal wall	Compound sugars	Simple sugars Glucose Fructose Galactose
		Trypsin	Pancreas	Proteins & peptones	Aminoacids
		Erepsin	Intestinal wall	Peptones	Aminoacids

Mucus is secreted in all sites: it protects and lubricates the mucous membrane.

The total length of the small intestine is variable and difficult to measure, but it usually more than 4 metres. The accommodation in the abdominal cavity is so restricted that the intestine is necessarily coiled up. The whole of the small intestine, except for the greater part of the duodenum, is completely covered with visceral peritoneum (Fig. 91), which allows the coils to slip smoothly over each other during the peristaltic movements of this part of the gut. The small intestine is suspended from the back wall of the abdomen by a double fold of peritoneum known as **the mesentery** (Figs. 85, 91); this permits it considerable freedom of movement. The attachment of the mesentery to the abdominal wall is about 15 cm long, but its free margin (in which the intestine is contained) is the same length as the intestine itself; the mesentery is therefore much folded. It conveys the blood vessels and nerves to the gut, and at the 'root' of the mesentery the visceral peritoneum becomes continuous with the parietal peritoneum. The arrangement is similar in principle to the relationship of a tendon to a synovial sheath (p. 43), but the details are more complicated.

In the right lower quadrant of the abdomen the ileum leads into the **caecum**, the first part of the **large intestine**. The material resulting from a meal begins to arrive here some three hours after it has been swallowed, and the last of the meal passes here about five hours later. Passage through the small intestine is controlled by the **myenteric plexus** between the two layers of muscle in its wall. This plexus receives parasympathetic innervation which hurries the bowel

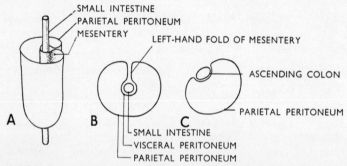

Fig. 91. The relationship of the peritoneum to the alimentary canal. A: the small intestine and its mesentery. The intestine is represented as a straight tube not in direct contact with the visceral peritoneum. In fact the tube is thrown into numerous coils, and the peritoneum is firmly adherent to its walls. B: cross-section of 'A'. C: cross-section showing the peritoneal relationship of the ascending colon.

contents along, and sympathetic innervation which holds them up. In addition to peristalsis the small intestine exhibits intermittent regularly spaced contractions which churn the intestinal contents backwards and forwards, so aiding absorption; these movements are known as **segmentation**.

The opening from the ileum into the caecum is guarded by the **ileocaecal valve**, a horizontal slit-like narrowing of the wall of the gut which prevents reflux from the caecum back into the ileum. It is not a true sphincter, but its lips are passively drawn together by distension of the caecum.

The large intestine

The total length of the large intestine is about 1·5 metres, and it is arbitrarily subdivided for descriptive purposes; the subdivisions do not reflect any great differences in function.

From the caecum, close to the ileocaecal valve, there protrudes a small blind process, the **vermiform appendix**, which varies in length from 1–20 cm; it is about 5 mm in diameter. It usually lies folded up behind the caecum, but may protrude downwards over the brim of the pelvis. It is prone to become infected, and its narrow lumen can readily be obstructed, with the result that an abscess may form and burst into the peritoneal cavity. The appendix has provided a steady source of income for generations of surgeons, but apart from this humanitarian role it has no known function.

The caecum gives place to the **ascending colon** (Fig. 92), which runs up vertically on the right hand side of the abdomen, and under cover of the ribs turns abruptly to the left as the **transverse colon,** which slings in a drooping curve across to a corresponding point on the left side of the abdomen,

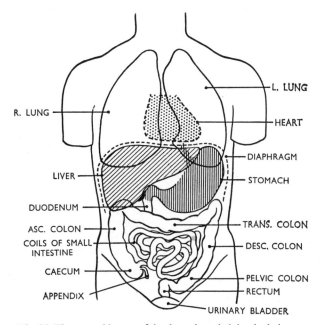

R. LUNG

L. LUNG

HEART

DIAPHRAGM

LIVER

STOMACH

DUODENUM

ASC. COLON

TRANS. COLON

COILS OF SMALL INTESTINE

DESC. COLON

CAECUM

PELVIC COLON

RECTUM

APPENDIX

URINARY BLADDER

Fig. 92. The general layout of the thoracic and abdominal viscera.

suspended by a mesentery of its own from the posterior abdominal wall. At its termination it turns downwards to become the **descending colon,** which, like the ascending colon, has no mesentery, and is more or less firmly fixed to the posterior abdominal wall. At the brim of the pelvis its name changes to become the **pelvic colon,** which, like the transverse colon, has a mesentery of its own. The pelvic colon ends by entering the pelvis to become the **rectum,** perhaps the worst named structure in the body, for it is curved in every direction. It accommodates itself to the concavity of the sacrum (Fig. 98), and has two convexities to the right and one to the left (Fig. 93). The rectum gives place in front of the coccyx to the **anal canal,** which is about 4 cm in length and turns backwards at right angles from the rectum to open at the **anus,** which is guarded by a complex sphincter muscle. The first part of the rectum has peritoneum on both sides and in front; the second part is covered by peritoneum in front, but the third part, together with the anal canal, has no relation to peritoneum at all, lying behind and below the peritoneal cavity.

Most of the large intestine is wider than the small intestine, but the appendix is an exception, and the pelvic colon may also be relatively narrow. Unlike that of the small intestine, the longitudinal muscle coat of the large intestine is deficient, except in the appendix, where it is complete, and in the rectum, where it is nearly so. In the rest of the large intestine the longitudinal fibres form three separate strips called **taeniae**, and these appear to bunch up the gut into bulging **sacculations** (Fig. 88).

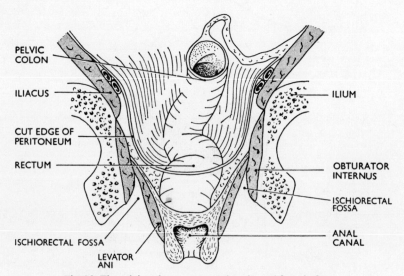

Fig. 93. The pelvic colon, rectum, and anal canal from in front.

Except for the terminal half of the anal canal, the large intestine is lined with columnar epithelium covering small scattered accumulations of lymphoid tissue; these do not become aggregated together except in the appendix.

There are no villi in the large intestine, and the circular folds are relatively few and much less well marked, for the large intestine merely has to absorb water. What passes through the ileocaecal valve is material which is still fluid, though a good deal of water has been absorbed in the ileum. By the time the contents reach the pelvic colon they have attained a solid consistency.

Something like 8·5 litres of fluid are poured out into the average digestive tract every day, made up as follows:

Saliva	1·1 litres
Gastric juice	2·8 litres
Bile	1·1 litres
Pancreatic juice	1·1 litres
Intestinal juice	2·3 litres

If all this volume of water were lost to the body there would have to be an enormous daily intake of fluid to keep the body in water balance (p. 214). However, it is virtually all recovered and fed back into the venous blood and lymph, to be used again. But if for any reason the absorptive part of the gut is thrown out of action there may be rapid and extensive fluid depletion. This may happen in diarrhoea (where the fluid contents of the bowel are hurried past the absorptive surfaces too fast for much recovery of fluid to take place).

The filling of the colon is largely passive, and material passed on through the ileocaecal valve usually reaches the pelvic colon about twelve hours after the meal has been taken, though there are wide variations. The large intestine exhibits a kind of peristalsis, one segment contracting while the succeeding one relaxes, and also shows a churning type of mixing movement. Three or four times a day, however, a strong peristaltic wave pushes the contents of the large gut into the pelvic colon, and, usually about 24 hours after the meal has been taken, the intestinal contents are propelled out of the pelvic colon into the rectum, when there is an immediate desire to defaecate. This 'mass movement' is stimulated by the **gastrocolic reflex** set off by filling the stomach, particularly with the first meal of the day. The rectum usually remains empty except just before defaecation, and the faeces are held up at the **pelvirectal junction**.

The process of defaecation begins with contraction of the colon and relaxation of the involuntary **internal sphincter** of the anus, a thickening of the circular muscle coat of the lower part of the rectum. The **external sphincter** of the anus is a voluntary muscle, and under the control of the will. At the same time the anal canal is lifted upwards over the descending mass of faeces, and evacuation is helped by an increase in intra-abdominal pressure due to contraction of the diaphragm and the abdominal muscles.

The **faeces** consist of the waste material plus dead bacteria whose home is the large intestine, and a quantity of mucus which lubricates the mass. If they have stayed longer than usual in the large intestine they may become dry and hard, whereas if they have been hurried through by violent peristalsis, as happens when the gut is infected, they may be extremely watery. The brown colour of the faeces is due to the products excreted in the bile; if bile is prevented from reaching the intestine the faeces are clay-coloured and fatty. The quantity of the faeces depends on the amount and kind of food eaten, and the regularity of defaecation varies very greatly from person to person. A daily motion does occur in many people, but many others may have perhaps two motions a week and remain in perfect health.

Outline of the clinical examination of
the digestive system

The condition of the tongue, mouth, fauces, teeth, gums and palate is readily inspected.

The respiratory excursion of the abdomen may be restricted if there is inflammation of the viscera, and palpation may reveal protective 'guarding' or rigidity of the muscles of the anterior abdominal wall (p. 279). The only part of the alimentary canal which it is usually possible to palpate in the normal abdomen is the pelvic colon.

Though the liver cannot be felt, the area of dullness it produces can be percussed (p. 239), and this method will also often map out the stomach and the caecum, by virtue of the resonant gas which they may contain. The intestinal movements can be heard through a stethoscope, and a finger inserted into the rectum will discover by touch the state of the lower end of the bowel and of many of the related pelvic viscera.

The interior of the oesophagus can be inspected by means of an oesophagoscope—a long straight metal tube carrying a light—and the interior of the stomach by a flexible tube containing a system of lenses and known as a gastroscope. The lower end of the bowel can be inspected by instruments inserted through the anus.

Samples of the gastric juice may be withdrawn through a tube after the patient has been given a 'test meal', and samples of the faeces may be examined chemically and bacteriologically.

The whole of the alimentary canal may be outlined radiographically by giving the patient a suspension of barium sulphate to drink. The large intestine, however, is usually examined by giving a barium enema. The gall bladder can be visualized by a radiographic technique called cholecystography, which makes use of the fact that certain iodine-containing organic compounds are concentrated in the bile.

13 · The genito-urinary system

The urinary apparatus

The **urine** is a clear yellow fluid which is usually slightly acid. It is the chief channel by which waste products are removed from the body, and variations in its amount and chemical composition play a vital part in the regulation of the quantity and reaction of the fluid content of the tissues. The organs which produce the urine are the **kidneys**, which lie on either side of the back wall of the abdomen (Fig. 94). Each kidney measures about 10 cm long, and is about 5 cm wide and 2·5 cm thick; its weight is about 140 g in a young person, but decreases somewhat with age. The left kidney usually lies a little higher than the right, and its upper pole reaches as high as the eleventh rib in expiration; when the diaphragm descends (p. 271) in inspiration, the kidneys are pushed down to an extent which varies with the individual, but may amount to 4–5 cm. Both kidneys, as befits their importance, are well protected by the thoracic cage and the powerful back muscles.

A section through the kidney (Fig. 95) has a distinctive appearance. The **hilum** of the kidney is found on its medial, concave surface, where the **renal artery** enters the organ and the **renal vein** leaves it.

Behind the vessels there emerges from the hilum the expanded upper end or **pelvis** of the **ureter**, into which the urine is passed from the ducts of the kidney. The hilum leads into a space called the **kidney sinus**, occupied by the branches of the renal vessels and the pelvis of the ureter, which is subdivided into two or three **major calyces**. These in turn give rise to a number of **minor calyces** (usually about 7 to 10). The minor calyces are small cup-shaped hollow protrusions, and into each cup fits a dark coloured **pyramid** of kidney substance, shaped like a cone; the rounded end which fits into the cup is called a **papilla**, and the pyramids collectively form the **medulla** of the kidney. In between the pyramids lie zones of lighter kidney substance; these are the **renal columns**. Surrounding the medulla externally, and continuous with the columns, is the light coloured **cortex** of the organ. The embryonic kidney is made up of about 14 lobes; each may be represented in the adult kidney by a pyramid. Often some lobes fuse together, so that there are usually fewer than 14 pyramids.

The large renal arteries spring directly from the abdominal aorta (Fig. 130.

187

p. 287). Their branches run outwards in the renal columns to form a network of arteries in the zone where cortex joins medulla. From this smaller branches arise, and ultimately the vascular tree bears little bunches of capillaries called **glomeruli** (p. 138). There are about one million of these in each kidney.

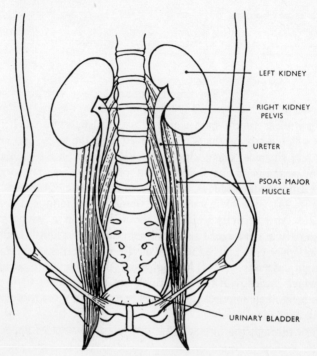

LEFT KIDNEY

RIGHT KIDNEY
PELVIS

URETER

PSOAS MAJOR
MUSCLE

URINARY BLADDER

Fig. 94. The urinary organs.

Each glomerulus protrudes into the highly specialized blind end of a tiny **uriniferous tubule**, which is about 4 cm long. The portion surrounding the glomerulus is the **glomerular capsule**, and receives the fluid filtered off from the capillaries. After leaving the glomerulus, the tubule undergoes several intricate twists and turns. It forms successively a **first convoluted tubule**, which lies in the cortex close to the glomerulus, a **loop of Henle**, which dives into the nearest pyramid and runs centrally towards its papilla before turning sharply back to the cortex again, a **second convoluted tubule**, a very close to the first, and a **junctional tubule** (Fig. 96). Finally, the junctional tubule unites with others to form a **collecting tubule**, and this runs through the pyramid once more to open at its papilla into the pelvis of the ureter. It follows that the pyramids of the kidneys are largely formed by loops of Henle and collecting tubules, while the rest of the apparatus lies in the cortex.

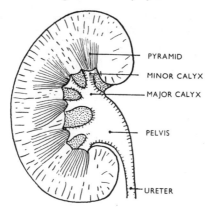

Fig. 95. Diagram showing the pelvis of the ureter. The kidney substance has been cut down the middle, and the blood vessels and fat removed from around the pelvis. The collecting tubules are gathered together in the darker areas called **pyramids**, and they drain into the **calyces** or subdivisions of the pelvis.

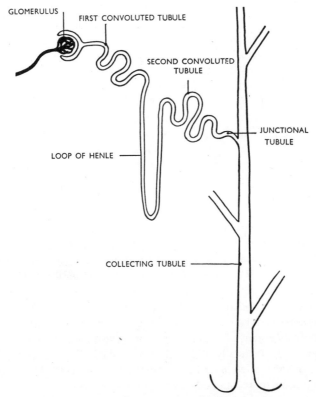

Fig. 96. Diagram of a nephron.

The whole affair, glomerulus and tubule together, is known as a **nephron**, and constitutes the functional unit of the kidney. There is a tremendous reserve of nephrons, and one whole kidney can be removed surgically without evidence of loss of function. If any stress is thrown on the remaining kidney tissue, however, it may prove unable to cope, and this indicates that the numbers of functioning nephrons may vary according to the exigencies of the situation.

The flow of blood through the glomerulus is slow, and at a high pressure. As a result, water and the simpler chemical constituents of the blood plasma are filtered off through the thin capillary wall and the incomplete inner wall of the glomerular capsule. The holes in the capsule are of such a size that they do not in health permit any of the proteins to leave the plasma. If they are diseased, protein may leak through and be lost to the body in the urine; this serious condition is one form of **nephritis**. About 600 ml of glomerular filtrate is formed every five minutes. Yet the total amount of urine passed in a day by a healthy person seldom exceeds 1·5 litres, for there is a great deal of reabsorption of fluid as the filtrate passes down the uriniferous tubule. At the same time valuable materials like glucose, which have passed through the glomerular sieve, are selectively reabsorbed.

Other substances are actively secreted by the tubules, which also play an important part in regulating the hydrogen-ion concentration of the blood (p. 169). This they do by manufacturing varying quantities of ammonia, and by secreting hydrogen-ions into the filtrate. Most of this work takes place in the first convoluted tubule, which has a complex lining membrane. Here the amount of water reabsorbed is a function of the amount of waste material which has to be carried in solution in the urine. In the second convoluted tubule, however, further water absorption is under the control of a hormone secreted by the posterior lobe of the pituitary gland (p. 199), and the amount is variable. By this means the water content of the body can be controlled.

The urine which runs into the collecting tubules is therefore the result of a complicated balance of chemical and physiological forces. The pelvis of the ureter, into which the urine drains, is the expanded upper end of a thin muscular tube about 25 cm long. The muscular coat of the ureter exhibits peristalsis, and propels the urine down it in spurts, one to six times a minute. The two ureters run down the posterior wall of the abdomen into the pelvis, where they narrow as they enter the **urinary bladder** by passing obliquely through its thick muscular walls (Fig. 94), which act as a kind of sphincter muscle.

The empty urinary bladder lies in the pelvis, but when full it rises up above the pubis into the abdominal cavity. It is a muscular organ which can be enormously distended providing the stretching is done gradually. Under normal circumstances a content of about 300 ml of urine elicits a sense of fullness, and the bladder is then usually emptied, but if there is a gradually increasing obstruction to the outlet of urine by an enlarged prostate gland (p. 191), the bladder may come to accommodate as much as 6–7 litres.

Emptying the bladder is an extremely complicated process. In its simplest terms it depends on the stimulation of stretch receptors in the walls of the bladder. These set off an autonomic reflex, whereby the bladder wall contracts and the urethra relaxes; the urine is thus voided in the act of **micturition**. There is no well-marked sphincter muscle guarding the exit from the bladder, and the urine is thought to be held up largely by the contraction of the elastic tissue in the walls of the urethra. The emptying reflex can be brought under conscious control and micturition can be suppressed if the time is unpropitious. Eventually, however, a pressure is reached at which no effort of will can postpone the emptying of the bladder.

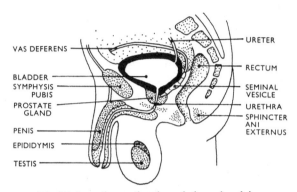

Fig. 97. A median section through the male pelvis.

The **urethra** leads from the bladder to the exterior. In the female it is 4–5 cm long, and runs downwards and slightly forwards to open just in front of the opening of the vagina (Fig. 98). In its first part it is lined, like the bladder and the ureters, by transitional epithelium (p. 14) but lower down the lining comes to resemble stratified squamous epithelium. The male urethra, on the other hand, is about 20 cm long. Its **prostatic** portion runs downwards through the **prostate gland**, a rounded body about 3 cm in its maximum diameter, which has a number of ducts opening into the urethra; it is part of the male sex apparatus (p. 192). Below this is the **membranous** portion of the urethra, which is surrounded by a somatic muscular sphincter. The **penile** portion, about 15 cm long, traverses the **penis** to open at its tip (Fig. 97). The upper portion of the prostatic part is lined by transitional epithelium, the lower end of the penile part by stratified squamous epithelium, and the rest by columnar epithelium.

Because the urinary passages of necessity open to the exterior, there is a risk of their becoming infected, and such an infection may spread, particularly in the female, up to the ureters, a condition known as **pyelitis**. Infected and stagnant urine in the pelvis of the ureter affords suitable conditions for the growth of a **renal calculus**—a 'stone' which is formed by the precipitation of salts from the urine on to debris resulting from infection of the ureteric walls. Such a stone may pass down the ureter to the bladder if it is small enough, but its passage is likely to be very painful. Other stones may form in the bladder itself, and may become quite large, even bigger than a hen's egg, if they are not surgically removed.

The male sex apparatus

The sex glands or **gonads** of the male are the **testes** (testicles), which are contained in a bag of loose skin known as the **scrotum**. Each testicle is an ovoid body about 3 cm long containing a complex system of **seminiferous tubules**. Each tubule is about 75 cm long, but is coiled on itself to take up very little space; there are anything up to a thousand seminiferous tubules in each testicle. Cells which line the upper reaches of the tubules give rise to the **spermatozoa**, or male sex cells, and during the process the number of chromosomes is halved by meiosis. Every spermatozoon thus contains 23 chromosomes instead of 46 (p. 10). Formation of spermatozoa proceeds best at a temperature somewhat below that of the interior of the body, and for this reason the testicles descend down the posterior abdominal wall before birth, and tunnel their way through the anterior abdominal wall on each side just above the groin (p. 282) to emerge into the relative coolness of the scrotum. The tubules are solid until puberty, when they become canalized, and begin activity under the influence of pituitary hormones (p. 198). Quite commonly the testicle does not fully descend for some months or even years after birth, and it is probable that in the vast majority of such cases the formation of spermatozoa proceeds normally. But if the testicle has not descended before puberty, there is some risk that the products of that particular testicle may be sterile.

The spermatozoa are pushed along the seminiferous tubules by the pressure of numbers behind them, and find their way into a network of tubes which ultimately leads them into the **epididymis**, which is closely applied to the back of the testicle. The epididymis is no longer than the testicle, but it contains a tube some 6 metres long. Eventually the tortuous windings of the **duct of the epididymis** straighten out, and at an acute-angled bend it changes its name to become the **vas deferens**, which retraces the path originally taken by the testicle during its descent and enters the front wall of the abdomen obliquely (p. 282). Each vas is about 45 cm long, and runs to the back of the urinary bladder, where it is joined by the duct of a small coiled tube called the **seminal vesicle** to form a **common ejaculatory duct**; the two ejaculatory ducts run through the prostate gland (p. 191) to open into the prostatic part of the urethra.

The vasa, the seminal vesicles and the prostate all have muscular coats, and during **orgasm** the spermatozoa are expelled along the urethra, accompanied by the secretions of the seminal vesicles, the prostate, and the mucus glands of the urethra. The fluid so produced is known as **semen**, and each ejaculation of semen may contain about 200–300 million spermatozoa. Although the spermatozoa have thin whip-like tails, these are not used until the moment of ejaculation, when they begin to lash from side to side during their passage along the vas deferens; when the spermatozoa are shed they are actively motile, and capable of swimming up the female genital tract to meet and fertilize the ovum on its way down (p. 196).

In addition to forming spermatozoa, the testicles manufacture an endocrine secretion produced by the **interstitial cells** which lie between the seminiferous tubules. It is responsible, among other things, for the secondary sexual characteristics which appear at puberty, such as the growth of the hair on the face and the increase in size of the larynx (p. 221).

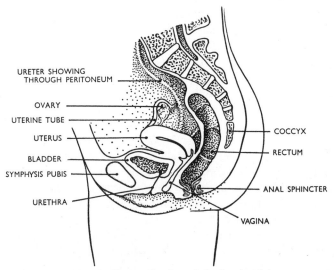

URETER SHOWING
THROUGH PERITONEUM

OVARY

UTERINE TUBE

UTERUS

BLADDER

SYMPHYSIS PUBIS

URETHRA

COCCYX

RECTUM

ANAL SPHINCTER

VAGINA

Fig. 98. A median section through the female pelvis.

The female sex apparatus

The female gonads are the **ovaries**, which lie inside the pelvis (Fig. 98). Each ovary is similar in size to a testicle, but instead of tubules it contains immature egg cells or **oocytes**. Between these there are scattered **interstitial cells** which produce hormones which play an important part in the control of the menstrual cycle (p. 195), and are responsible for the appearance of the secondary sexual characteristics such as the alteration in shape of the pelvis (p. 222) and the deposition of subcutaneous fat.

Although the ovary contains large numbers of oocytes, most of them never develop further. At puberty the ovary is aroused to activity by hormones produced by its own interstitial cells and by the anterior lobe of the pituitary gland. The latter is in control, and it has a cyclic rhythm (p. 218), the result being that, from puberty until the cycle stops, one oocyte (sometimes two) is matured and liberated from the ovaries each lunar month. The exact rhythm varies with the individual; some women have a 28 day cycle, but other cycles may range from 25 to 31 days. The average cycle is longest in girls 15 to 19 years of age, and shortens gradually until the age of 45 or so, after which it becomes longer again until finally it ceases at a variable time round about 50 years of age. There is no necessary alternation between ovaries; the same ovary may produce the ova for several successive cycles. During reproductive life only about 400 oocytes reach maturity, and this number must be contrasted with the immense numbers of spermatozoa capable of reaching maturity in a similar period.

During maturation the selected oocyte becomes surrounded by a group of smaller cells forming a **follicle**, within which the oocyte grows steadily bigger. A fluid-filled cavity appears in the centre of the follicle, and the pressure in this gradually rises, so that the whole follicle expands and finally nears the surface of the ovary. Just before **ovulation** the oocyte begins to undergo the first phase of a meiotic division (p. 10) similar to that undergone by the spermatozoa, so that the mature ovum contains only 23 chromosomes.

The final stages of this division take place just after the follicle bursts, so shooting the ovum through the surface of the ovary and into the peritoneal cavity. The other three of the four cells resulting from the meiotic division have scarcely any cytoplasm, and their nuclear material forms **polar bodies**, which degenerate and are lost. The cavity in the ovary is filled with blood clot, and in the next few days the cells in its walls multiply to form the **corpus luteum of menstruation**, which functions as a temporary endocrine gland. If the ovum does not become fertilized, this yellow body is converted into scar tissue (p. 24) called a white body, or **corpus albicans**. In this way the surface of the ovaries, originally smooth, becomes pitted and scarred as age advances.

The mature ovum enters the mouth of the muscular **uterine tube**, which is about 10 cm long and is lined by ciliated epithelium (Fig. 99). It is propelled along the tube by peristalsis and by ciliary currents, but it takes about three days to make the journey, for the tube has an elaborate labyrinthine system of mucosal baffle plates which slow the ovum down and prolong its opportunity of being fertilized.

The uterine tubes lead to the cavity of the **uterus** or womb, an organ about 8 cm in length possessing very thick walls of visceral muscle lined by a special glandular epithelium. The upper two-thirds of the uterus contains a flattened triangular cavity, and is known as the **body**; the lower one third is narrower than the rest and is called the neck, or **cervix**. The cervix is traversed by the small **cervical canal** through which the cavity of the body opens into the upper end of the **vagina**. This muscular tube, about 9 cm long and lined with stratified squamous epithelium, runs downwards and forwards to open on the exterior at the **vulva**, situated between the anus and the urethra. The urethra is embedded in the anterior wall of the vagina and surrounded by its muscle coat. It is important that there is a direct communication between the peritoneal cavity and the exterior via the uterine tubes, the uterine cavity, the cervical canal and the vagina, for along this channel infection may pass in the reverse direction, so giving rise to ill-health and possible sterility.

The epithelium of the uterus undergoes a series of progressive changes known as the **menstrual cycle** between one ovulation and the next (Fig. 100). After the ovum is shed, the secretions of the pituitary gland, the ovary, and the corpus luteum of menstruation cause the uterine epithelium to become steadily thicker and more vascular. The glands in the epithelium enlarge, accumulate glycogen and fatty material, and begin secreting into the uterine cavity. Eventually, about a fortnight after ovulation, changes in the balance of the endocrine control of the process lead to the beginning of **menstruation**, in which the superficial layers of the epithelium become detached from the basal layer.

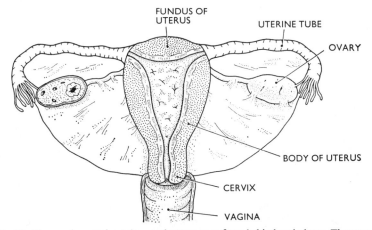

Fig. 99. The ovaries, uterine tubes, and uterus seen from behind and above. The uterus and the left ovary have been sectioned.

The consequent tearing of blood vessels causes a certain amount of haemorrhage which continues for four or five days. After this time the basal parts of the glandular epithelium begin to regenerate, and the raw surfaces become covered with epithelium again; the epithelium then grows until another ovulation takes place, and the whole process begins all over again. This regenerative part of the menstrual cycle is under the control of the ovary and the pituitary, but the corpus luteum has now degenerated and is being replaced by scar tissue.

It is usual to call the day on which menstruation begins the first day of the menstrual cycle. Ovulation occurs roughly fourteen days before the onset of the next menstruation; this fact is the basis of the so-called 'safe period' method of birth control, but unfortunately the time of ovulation is by no means regular, and the time for which the ovum remains capable of being fertilized following ovulation is also impossible to predict.

The first menstrual period occurs on the average about the age of 13 years, though the time varies very considerably in different girls; the event is referred to as the **menarche**. In the early cycles ovulation does not necessarily take place, but on the other hand ovulation may occur without menstruation necessarily following. Menstruation continues throughout reproductive life until about the age of 50 years—again the time is very variable—when it ceases gradually, the event being known as the **menopause**.

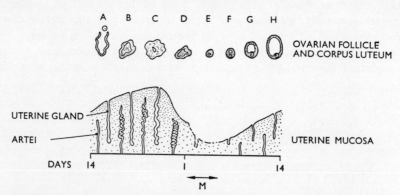

Fig. 100. The ovarian and uterine cycles. A: Ovulation. The thickness of the uterine mucosa is increasing. B: Corpus luteum forming. The uterine vessels are becoming tortuous, and the glands are enlarging. C: Corpus luteum well established. Glands and vessels at their most complicated stage of development. D: Corpus luteum degenerating. Thickness of uterine mucosa decreases and the arteries become spiral. E: New follicle begins to grow. Most of uterine mucosa shed during menstruation (M). F: Follicle growing. Repair of uterine mucosa begins. G: Cavity in follicle now enlarging. Uterine mucosa thickening. H: Mature follicle, meiosis begins.

Notice that the menstrual cycle is timed from the first day of menstruation (day 1) and that ovulation occurs on or about 14 days prior to the beginning of the next cycle.

Pregnancy and childbirth

If a spermatozoon meets the ovum during its passage along the uterine tube it may penetrate the outer covering of the ovum and initiate a series of changes which culminate in the formation of a new individual. The number of chromosomes in the nucleus of the cell is now restored to 46 (23 from each partner), and a series of mitotic divisions (p. 9) is initiated. The fertilized ovum then adheres to the uterine wall and becomes 'implanted' there to begin its growth into a baby. If fertilization does not occur, the ovum gradually disintegrates.

Successful implantation is followed by events designed to establish and nourish the developing fetus. The corpus luteum of menstruation does not now degenerate, but enlarges to form an endocrine gland called the **corpus luteum of pregnancy**, which is important in steering events in the early part of pregnancy. The next menstrual period is suppressed. The walls of the uterus thicken and become more vascular; its epithelium undergoes modification, and the fetus at first takes its nourishment directly from the epithelial cells. Later, the **placenta** is developed, in which the mother's blood and that of the fetus circulate together, separated only by a very thin membrane. Foodstuffs and oxygen can cross this placental barrier into the fetal circulation and waste products and carbon dioxide can be transferred into the maternal circulation to be got rid of by the mother's excretory apparatus. The **umbilical cord,** of variable length, connects the circulation inside the body of the fetus to the fetal side of the placental circulation by means of two umbilical arteries (p. 133) and one umbilical vein.

The breasts are stimulated to grow, and towards the end of pregnancy they show signs of secretory activity (p. 205). During pregnancy stores of fat and other materials are laid down in the mother's tissues, thereby causing a considerable increase in weight (p. 221) over and above that due directly to the fetus and its placenta. Other hormones cause fluid to accumulate in the joints of the pelvis, so that they are able to 'give' a little during childbirth. By the end of the third month of pregnancy the uterus and its contents begin to rise out of the pelvis into the abdomen; the centre of gravity of the body is disturbed by the heavy weight in the lower part of the abdomen, and postural changes and strains on the lower part of the back result (p. 229).

It is not yet exactly certain what triggers off childbirth, but after some 280 days of pregnancy, with a wide margin of error on either side, the uterus begins to contract rhythmically, so pushing the baby's head down against the uterine cervix. This in turn has to dilate to allow the head to pass through, and the process takes some time. When dilatation is complete, the baby is thrust down through the pelvis by powerful contractions of the upper segment of the uterus. Most often the head is the first to appear at the vulva, in which case it is called a **vertex presentation,** but occasionally the baby is born legs first, in a **breech presentation**. The passage of a large child may cause injury to the soft tissues forming the **birth canal,** in particular to the pelvic diaphragm (p. 282) and the perineum (p. 283). Finally the placenta and the membranes which surround it are expelled in the same way.

As soon as the placenta has been delivered, a period of reversion to 'normal' called the **puerperium** begins. The uterus shrinks from a weight of about 1 kg at birth to its normal 60–90 g in a period of six weeks or so. The return of menstruation is usually delayed until breast feeding is finished; if the mother is not feeding the baby herself, menstruation usually returns within six months. The mother's blood volume, which increases by about 30 per cent during pregnancy, falls to the normal value. Her stretched abdominal wall, particularly in a young woman, can return to normal, though in older people it may remain somewhat lax and flabby.

Lactation (p. 205) is originated by hormonal stimulation from the pituitary gland, and maintained by reflex control due to the sucking of the infant. It ceases when the infant is weaned.

14 · The endocrine system

The endocrine glands constitute a chemical system of communication and control. This system is more primitive than, and supplements the activities of, the nervous system, which overlaps it at several points. For example, the liberation of noradrenaline and adrenaline from the suprarenal medulla, an endocrine gland, is part of the defence reaction of the sympathetic nervous system (p. 88), and several bodily functions, such as the secretion of milk, are under both nervous and hormonal control.

The chief endocrine gland is the **hypophysis** or **pituitary gland** (Fig. 43), which lies in a special cavity in the base of the skull called the hypophyseal fossa. The gland is composed of two functionally distinct parts. The **anterior lobe** originally grew up into the skull from the roof of the mouth, and has no direct connexion with the brain. But the **posterior lobe** is attached to the floor of the third ventricle (p. 105) by a stalk called the **infundibulum**. It is really an outgrowth from the brain, and still retains a functional connexion with the hypothalamus (p. 108). Many nerve cells lying in hypothalamic nuclei send axons down through the stalk to end in the posterior lobe, and it is believed that the secretion of this lobe is actually produced by these neurones. In contrast, the anterior lobe and the hypothalamus are connected by a portal system (p. 139) of small blood-vessels. The amount of interaction between the hypothalamus and the anterior lobe mediated by this system is not yet fully established. The whole gland is about the size of a pea, but it elaborates several different secretions. Some of these control other endocrine glands such as the thyroid, the ovary, or the insulin-producing part of the pancreas; for this reason the pituitary has been fancifully called 'the leader of the endocrine orchestra'. Hormones from the anterior lobe are important factors in metabolism, growth, sexual development, pregnancy and lactation. An excess of growth hormone (**somatotrophin**) in a child will produce giantism. By the age of 18–21, however, most of the epiphyses are closed (p. 34) and a similar situation occurring after this age can only produce excessive growth in certain restricted situations, such as the face, hands and feet. The result is known as **acromegaly**. Conversely, an impairment of somatotrophin secretion in childhood may produce a special variety of dwarfism. If the damage is sustained in adult life there may be a regression of secondary sexual characteristics, loss of weight and premature senility. The posterior lobe is concerned, among other things, with the distribution of water in the body (p. 190), and damage to it may lead to the condition known as diabetes insipidus, in which enormous quantities of urine are passed.

In the front of the neck, lying across the thyroid cartilage and closely applied to its sides, is the **thyroid gland**, which has two large lateral lobes connected across the midline by an isthmus, the whole weighing about 25 g. It is related posteriorly to the larynx, with which it moves. It extends vertically from the middle of the thyroid cartilage to the level of the sixth ring of the trachea. The thyroid gland may be swollen if there is an inadequate supply of iodine in the diet; its secretion contains iodine, and the swelling may be regarded as the result of an attempt to make bricks without straw. This **simple goitre** used to be common in some regions of England—hence the name 'Derbyshire neck'—and in Switzerland; at one time it was a sign of beauty, and painters used to take as their sitters women with well-marked goitres. The addition of a very small quantity of potassium iodide to table salt is sufficient for the needs of the thyroid.

The most important secretion produced by the thyroid is a close chemical relation of **thyroxin**, which has been synthesized and found to have the physiological properties of the natural hormone, though not quite the same degree of activity. Thyroxin has a considerable effect on growth and on metabolic activities. An excess speeds up the patient's metabolism, so that he produces more heat; the heart rate increases and he tends to sweat. Other symptoms are tremor, anxiety, and protrusion of the eyes, and the condition is known as **toxic goitre**. When there is a deficiency of secretion two conditions may result. In a child, normal mental and physical development do not occur; the child is called a **cretin**. In an adult the individual becomes slow, cold and lethargic, and may develop characteristic mental symptoms; he is said to suffer from **myxoedema** (the name is given from a typical puffiness of the face, which makes it possible to diagnose some cases of thyroid deficiency at sight). In both cases dramatic improvement and cure may result from the administration by mouth of thyroid extract, but this must be continued for the rest of the patient's life; the thyroid never learns to secrete properly.

Immediately posterior to each lateral lobe of the thyroid, and sometimes embedded in its substance, are two small brownish masses about the size of a small pea; these are the **parathyroid** glands. Their secretion, **parathormone**, regulates the exchange of calcium and phosphorus between the bones and the blood stream (p. 30). The parathyroids and the thyroid secrete another hormone called **calcitonin**, which is also concerned with calcium metabolism, though its effects differ from those of parathormone.

On the posterior wall of the abdomen, above the kidneys, lie the two **suprarenal glands**. The right suprarenal lies behind the liver and on the right of the inferior vena cava, and the left one is posterior to the stomach. Each gland consists of two distinct endocrine glands, one inside the other. The **medulla**, in the centre, has already been discussed in relation to the sympathetic nervous system; it is a factory for the mass-production of adrenaline (p. 88). The **cortex**, which covers the medulla like the peel of an orange, produces a number of hormones essential to life. The best known of these is perhaps **cortisone**, which, with its chemical relatives, is in use to treat conditions in which the suprarenal cortex is thought to be functionally deficient. There is an intricate relationship between the cortical hormones and those produced by the gonads; tumours of the suprarenal cortex result in disturbances of sexual function and changes in the secondary sexual characteristics. Destructive lesions of the cortex, on the other hand, give rise to a condition known as Addison's disease, in which there is a disturbance of the inorganic salt content of the tissues and a characteristic bronzing of the skin and mucous membranes.

The interstitial cells of the **gonads** (testicles or ovaries) have already been described as endocrine glands. Their hormones are concerned in the secondary sexual changes at puberty, and also in the time of closure of the epiphyses (p. 34). Ovarian hormones control the timing of the menstrual cycle, but are in turn controlled by the anterior lobe of the pituitary gland.

The last of the endocrine glands we must consider is scattered in small pieces throughout the substance of the pancreas (p. 179). These are the **islets of Langerhans**, which manufacture **insulin**. Insulin is one of the substances responsible for controlling the level of the blood sugar, and if the islets do not function properly the blood sugar rises above its normal limits after a meal and may remain persistently high (Fig. 103). This in turn leads to deleterious effects on various tissues and systems, resulting in loss of weight, thirst, damage to the central nervous system and the retina, weakness and eventual death. The condition is known as **diabetes mellitus**, and is treated by diet or by giving regular doses of insulin. Unlike the thyroid hormone, insulin does not survive the passage through the stomach, and must therefore be given by injection; the injections must be continued indefinitely, as the islet cells do not recover. The basic cause of diabetes is unknown, but the secretion of the islet cells is controlled in some way by the anterior lobe of the pituitary gland, and possibly the initiating factor is some defect of the pituitary.

15 · The skin

The skin is a much underestimated organ. It measures up to two square metres in area, and weighs 4–4·5 kg. It provides a tough protective coat against dangerous external agencies, such as friction, drying, ultraviolet light, and bacterial invasion. It is the most extensive and versatile of the sensory organs, though its sensitivity is not so great as that of the special senses. It excretes water and salts, and can get rid of various other waste products; it plays a vital part in the regulation of body temperature (p. 216) and in adjusting the fluid content of the body (p. 214). It can synthesize vitamin D (p. 36) when exposed to sunshine, and can store materials like glycogen (p. 210) and lipids (p. 4). It is essential to life, and if more than about one-third of it is destroyed—say by extensive burns—the patient is pretty certain to die.

The skin consists of two distinct layers. The **epidermis** is a stratified squamous epithelium in which four main layers can be recognized. The innermost or **germinative** layer contains the stem cells the progeny of which are pushed up towards the surface by their successors, taking two to three weeks to reach the superficial layer. On the way they undergo degenerative changes which begin in the **granular** layer (Fig. 101), in which the cells start to lose their nuclei and begin to accumulate keratin (p. 15). The next layer is called the **clear** layer because the cells in it have become flattened and compressed into clear glassy-looking plates (Fig. 101). Finally, there is the outer **horny** layer, composed of fully keratinized cells which ultimately flake off and are lost.

There are marked regional differences in the epidermis, and they determine two main types of skin. **Thick skin** is found in the palms of the hands and the soles of the feet; it has all the four layers just described, and in places like the side of the great toe there may be a layer of keratin 3 mm thick. Thick skin responds to repeated pressure and friction by increased keratinization, as witness the horny hands of labourers.

In contrast is the **thin skin** of the general body surface, in which only the germinative layer and a thin keratinized layer are always recognizable; the clear and granular layers are thin or absent. This skin merges into mucous membrane (p. 16), which has still less keratin, at the **muco-cutaneous junctions** in such places as the lips, the nostrils, and the anus. These junctions are moistened by mucus, and the horny protective layer is not so necessary.

The surface of the epidermis displays a fine network of creases, and in places like the finger tips there are ridges and furrows which are the basis of the individually unique fingerprints, palm prints, and footprints. These ridges facilitate the frictional grip of the hands and feet on outside objects.

Deep to the epidermis lies the **dermis,** a connective tissue layer. Its most superficial part forms a series of protruding **papillae** interlocking with the epidermis like the pieces of a jigsaw puzzle, and so helps to bind the two together when shearing stresses are applied to the surface. The papillae are very numerous and deep in places where there is much surface friction, such as the finger pads, and shallower and less frequent in more protected areas. The

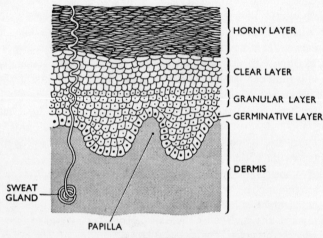

HORNY LAYER

CLEAR LAYER

GRANULAR LAYER

GERMINATIVE LAYER

DERMIS

SWEAT GLAND

PAPILLA

Fig. 101. Thick skin.

papillae contain loops of blood vessels which nourish the epidermis by diffusion, and those in the hand and foot may contain tactile corpuscles (p. 71). The cutaneous nerve fibres form a terminal plexus in the superficial part of the dermis (p. 71) containing both sensory and sympathetic fibres.

The deeper layers of the dermis merge imperceptibly into the superficial fascia (p. 21), and the combined depth of the two may be very considerable (up to 4 cm or more in the abdominal wall of a fat person); only about 0·5 cm or so of this total is usually described as dermis. The dermis contains the **sweat** and **sebaceous glands** (p. 204) and the roots of the hairs. In the dermis of the hairless skin of the palms, soles, and digits, there may be organized sensory receptors such as the lamellated corpuscles (p. 71). In the face and neck the dermis contains the somatic muscles of facial expression (p. 262). Visceral **arrector pili** muscles under sympathetic control are inserted into the hairs and erect them when the skin is cold. In the scrotum and the nipple there are other visceral muscles; the **dartos** muscle of the scrotum regulates the position of the testicle relative to the abdominal wall, and thus alters its temperature, so influencing the formation of spermatozoa (p. 192); the nipple muscle plays a part in erecting the nipple as a response to sucking.

Skin appendages

The nails

A nail consists of a **nail plate** of hard keratin scales fused together. This may be regarded as modified epidermis, and rests on a **nail bed** which is modified germinative layer plus dermis; there is no superficial fascia deep to a nail. At the root of the nail lies an area called the **matrix**, where the germinative layer is thicker than elsewhere, for here it is producing the keratinized cells which are added to the nail and push it outwards towards the tip of the digit. Under the free edge of the nail plate the epidermis of the finger tip is continuous with the nail bed, and the nail plate sends projections deeply between the germinative cells, so securing the bed and the plate together. The nail bed, and particularly the region of the root, is well supplied by sensory nerve fibres, and this perhaps accounts for the popularity of tearing out nails one by one as a method of torture.

Nails grow continuously throughout life, at a rate of a millimetre or more a week; if a nail is removed, it will be replaced by a new one, so long as the matrix is not destroyed.

The hairs

Hairs are also composed largely of keratin. Each hair has a shaft, and a root which lies in the dermis in an obliquely running cylindrical **hair follicle** (Fig. 102) containing one or several hairs. Into the swollen **hair bulb** at the end of the follicle protrudes a connective tissue papilla housing the **matrix** which gives rise to the hair and is responsible for its growth. Attached to every hair follicle is an arrector pili muscle, and opening into the follicle is the mouth of a sebaceous gland (p. 204), which lubricates the shaft of the hair and the surface of the skin.

The number of hairs per unit area varies considerably in different regions; there are about ten times as many on the top of the head as there are on the chin. The distribution of body and facial hair is one of the secondary sexual characteristics (p. 221).

Hairs grow for a time, then enter a resting phase, and are finally shed and replaced by new ones. Eyelashes have a life span of about 3 to 4 months, but scalp hairs have a span of 3 to 4 years; the rate of growth varies from 1 to 2 mm a week for the leg hairs of children to 3 mm a week for scalp hairs. Cutting does not affect the growth of hair.

The colour of hair depends on the amount of pigment incorporated in the shaft of the hair, and on the number and size of tiny air spaces within the shaft.

In other mammals, hair serves to keep the body warm and to reduce the risk of injury to underlying skin. In man, this is not a very important factor, though blows on the head may sometimes be cushioned by abundant hair. Some human hairs are profusely innervated, and serve as touch receptors.

Skin glands

Like mucous membranes, the skin is kept moist, but intermittently, and by a different mechanism. In the deeper layers of the dermis lie the **sweat glands**, tiny coiled tubes which send thin ducts up through the dermis and epidermis to open on the surface of the skin. The openings are easily seen with a lens, especially on the skin of the palm; they are most numerous on the palms and soles, and are

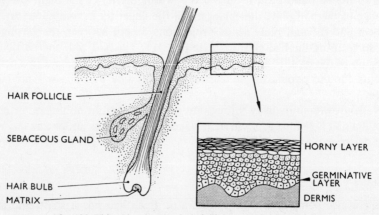

HAIR FOLLICLE

SEBACEOUS GLAND

HORNY LAYER

GERMINATIVE LAYER

HAIR BULB

MATRIX

DERMIS

Fig. 102. Thin skin, showing hair follicle and sebaceous gland.

absent from the margins of the lips and the beds of the nails. The majority are epicrine glands, but certain specialized glands found in the axilla and in the perineum secrete in an apocrine manner (p. 17).

The thin watery sweat rapidly evaporates from the surface of the skin, and provides an important means of adjusting the temperature of the body to external conditions (p. 216). The salt in the sweat is lost to the body, and must be replaced after excessive sweating (p. 216).

If the skin is soaked in water for a long time it becomes sodden and unhealthy, and the same sort of thing happens in prolonged sweating. In order to protect it the holocrine **sebaceous glands** opening into the hair follicles produce a thick greasy water-repellant material called **sebum**. In regions such as the nose, disturbed activity of the glands may give rise to blackheads. Sebum has a lubricant and a mild anti-bacterial action. There are no sebaceous glands in the palm or the sole.

In the external auditory meatus (p. 125) the specialized apocrine

ceruminous glands secrete the **wax** of the ear, which is yellow or dark in colour and protects the skin of this region.

The **mammary gland** is a modified and greatly enlarged apocrine sweat gland in the superficial fascia of the chest wall. It is a compound tubulo-alveolar gland opening by some fifteen ducts on the surface of the **nipple**, which lies in the centre of the round pigmented **areola**. Visceral muscle fibres (p. 50) run circularly round the ends of the ducts, which are so arranged that the gland consists of about 15 to 20 lobes, each separated from the others by septa of connective tissue, and positioned radially round the nipple. The breasts of babies of both sexes are often slightly active just after birth, for during intrauterine life they have been stimulated by the hormones circulating in the mother's blood and preparing her own breasts to secrete milk. Secretion of this kind in an infant is sometimes known as witch's milk, for it was a sign made use of by the witch hunters. After this the breast settles down until puberty, when the female breast begins to enlarge, mostly by the deposition in the surrounding areolar tissue of large quantities of fat, but also by some enlargement of the duct system and alveoli. After this the breast may swell a little and become tense with each successive menstrual period, but no real proliferation of the glandular tissues takes place unless pregnancy occurs.

After the child is born the mammary gland begins to secrete a watery fluid called **colostrum**, which only becomes true milk after a few days. Following weaning, the breast regresses in size, but nearly always remains somewhat larger than before pregnancy. At the menopause, the gland shrinks and the amount of fat surrounding it usually lessens also.

Milk secretion is controlled by hormones produced by the anterior lobe of the pituitary gland and the ovary. Breast milk alone will maintain satisfactory growth for some time, but it has too little iron in it for the proper development of bone marrow and blood, and soon requires to be supplemented by solid food (p. 213).

Part 2
The body as a whole

16 · Metabolism

The raw materials for the operations of cellular anabolism and catabolism (p. 8) are derived from the food which is eaten, and the oxygen which is inspired is essential for their performance. The term **metabolism** refers to all the changes which food and oxygen undergo within the body.

The fate of the absorbed food

There are three chief classes of foodstuff. The **carbohydrates** and **fats** are compounds of carbon, hydrogen and oxygen; they are used as sources of energy, and can be stored for future use. The **proteins** are compounds of carbon, hydrogen, oxygen and *nitrogen*; many of them contain also phosphorus and sulphur, and a few contain other elements, such as iron.

In the process of digestion **proteins** are broken down into aminoacids, which are the basic building blocks of the protein molecule. The ordinary diet contains many different kinds of protein, but they are all formed by different permutations and combinations of some twenty or thirty kinds of aminoacid, and these are brought to the liver by the portal vein from the villi of the small intestine (p. 177). Some of the aminoacids undergo **deamination**: the liver cells split off their nitrogen content in the form of ammonia, which is ultimately converted mainly into **urea**. This is then carried by the circulation to the kidney, which excretes it into the urine, so that it is lost to the body. (The kidney itself can deaminate certain aminoacids.) The deaminated material left behind in the liver is now utilized as a source of energy in much the same manner as the carbohydrates and fats.

But not all the aminoacids are thus deaminated; many are allowed to pass through the liver to the cells in other tissues, where they are utilized as building blocks with which to synthesize the characteristic proteins of the human body. In this way the aminoacids derived from the breakdown of (say) roast mutton can be reassembled in different groupings to form human fibrinogen or albumin. Both the liver and the tissue cells are capable within limits of converting one aminoacid into another by chemical means. However, some cannot be synthesized in this way, and are known as the **essential aminoacids**, since if they are not present in the diet the body will be starved of essential building material.

The **carbohydrates** are all absorbed in the form of simple sugars (p. 180), which are conveyed to the liver by the portal vein. Here four possible fates await them. The fructose and galactose and some of the glucose are converted into **glycogen** (p. 3), which is stored in the liver and in the muscles. This is a method of storing energy, for the glycogen can be broken down again into glucose on demand. Some of the glucose is oxidized immediately to supply energy for other chemical operations, and some is converted into fat and stored in the tissues. Finally, some remains in the blood stream as the **blood sugar**.

The blood sugar rises after a meal and gradually falls again over a period of a couple of hours to a steady level; this rise and fall, when charted, is called the blood-sugar curve (Fig. 103). The feeling of emptiness round about mid-morning and mid-afternoon which induces a desire for morning coffee or afternoon tea is due to the normal fall in the blood sugar after the preceding meal. The amount of sugar in the blood is the result of a balance between the breakdown of glycogen to form glucose, the absorption of glucose from the alimentary canal, and the synthesis of glycogen from glucose. These activities are affected by various hormones. For example, adrenaline stimulates the formation of glucose from glycogen, so that there is plenty of glucose circulating in the blood to meet any emergency (p. 88). On the other hand, insulin (p. 200) stimulates the formation of glycogen, and thus lowers the blood sugar. In diabetes mellitus, where insulin production is deficient, the blood sugar rises and sugar is excreted in the urine. Glycogen formation is reduced, and the inability to make proper use of the carbohydrate intake throws the strain of energy production on the fats and proteins of the diet. Fat metabolism cannot proceed normally if carbohydrate metabolism is defective, and so there is a general upset of the chemistry of the whole body which in severe cases results in loss of weight, weakness, coma and death. If a diabetic takes too much insulin, the blood sugar falls rapidly, the brain is unable to continue working, and a condition called insulin coma supervenes. In some diabetics the margin between diabetic coma on the one hand and insulin coma on the other is dangerously narrow.

The end-result of carbohydrate metabolism is always the production of carbon dioxide and water, derived from the ultimate breakdown of the simple sugars in the muscles or elsewhere. The carbon dioxide is excreted by the lungs and the water by the kidney.

Fats are absorbed into the lacteals and transported by them into the thoracic duct and so into the venous blood. Thus they do not have to pass through the liver before they reach the tissues, and the tiny droplets of fat absorbed may simply be stored in the **fat depots**, such as the mesentery and the anterior abdominal wall. The body will only store fats of certain kinds, and any which are unsuitable are allowed to pass back to the liver, where they are oxidized after being split up again into glycerol and fatty acids. Such fat as is required for storage is then manufactured by the liver from carbohydrate and sent round to the tissues again. Normally fat, like carbohydrate, is completely oxidized to carbon dioxide and water.

Energy requirements

If no food whatever is taken, the stores of fat and glycogen in the body are used to provide the energy necessary for the processes of living—for the heart beat,

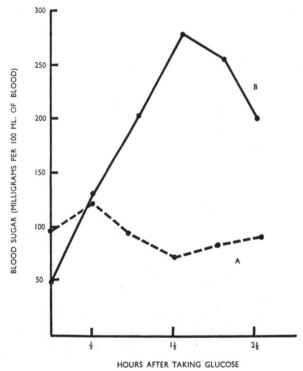

Fig. 103. Normal (A) and diabetic (B) blood-sugar curves after each patient had been given 50 grams of glucose.

the respiratory movements, and other involuntary activities, as well as such voluntary movements as the individual may undertake. Certain essential organs (such as the brain and heart) are well supplied with food material up to the last, while the stores in other tissues may be completely exhausted. A starving person deaminates the protein of his own body to help to supply his energy requirements, and therefore loses nitrogen in his urine. If the energy requirements are supplied by adequate fat and carbohydrate, but he is given no protein, the excretion of nitrogen falls, but never below the level corresponding to the consumption of 30–50 g of protein every day. This loss is due to the normal wear and tear of the body, and it must be made good by the intake of a corresponding amount of protein in the diet. (Actually, the minimum amount of protein given daily must be at least 100–150 g to ensure an adequate supply of essential aminoacids). After operations or injuries patients for some reason excrete more nitrogen, and have to be given extra protein in their diet.

Once the supply of nitrogen is assured, attention can be returned to the supply of energy. To find out how much food a patient needs to eat, we must know how much energy he expends, and how much energy the various classes of foodstuff he eats will give him. The second part of the problem is easy. One gram of either carbohydrate or protein, when burned in a calorimeter, yields 4·1 Calories of heat. (One Calorie is the amount of energy required to heat a litre of water through 1°C). Fat, on the other hand, yields 9·3 Calories per gram.

The energy used by the patient is much more difficult to determine. His **basal metabolic rate** is defined as his expenditure of energy measured after he has been lying down at rest for at least twelve hours after his last meal. Direct measurement in a large 'calorimeter' shows that the average male under these basal conditions uses up 70–80 Calories per hour. The exact figure depends mainly on the surface area of the body—obtained by consulting tables calculated from the height and weight of the patient—and the basal metabolic rate for a male adult is taken to be 40 Calories per square metre of body surface per hour (the average body has a surface area of rather less than 2 square metres). Children require more than this, and adults who are moving round require more still. The total amount of energy required per day obviously varies with the type of work the patient does; a lumberman uses up between 5000 and 6000 Calories, while an office worker may need no more than 2000 to 3000.

The **respiratory quotient** is defined as the ratio of the amount of carbon dioxide expired to the amount of oxygen taken up, and this can be directly measured by means of a **Douglas bag**. The patient breathes from the bag, and the carbon dioxide he produces is chemically absorbed, while the oxygen he has used is estimated from the analysis of the residual air in the bag. If he is utilizing *only* carbohydrate, the respiratory quotient will be unity, because the carbon dioxide produced by the oxidation of carbohydrate is exactly equivalent to the oxygen consumed. If he is utilizing *only* fat, more oxygen will be consumed than is represented by the carbon dioxide produced, and the respiratory quotient drops to 0·7. When protein *only* is being oxidized, the quotient is about 0·8. A determination of the respiratory quotient will thus indicate roughly in what *proportion* the three types of foodstuff are being used in the patient's body. If the total oxygen consumption per unit time is known, the Calorie output of the body can be calculated.

From information of this kind it is possible to construct a diet. All the energy expended must be taken from the food, or else the body will have to encroach on its own resources, and it cannot do this for very long. A diet must therefore contain sufficient foodstuffs to meet the energy requirements of the individual, and enough protein to balance the daily loss of nitrogen. It must, however, also supply sufficient quantities of essential aminoacids, inorganic salts and **vitamins**.

The vitamins

Vitamins are indispensable, but are needed only in minute quantities. Most of them are referred to by the letters of the alphabet—a system which has no relation to their chemical structure, but has the merit of convenience, since their chemical names are often rather fearsome.

Vitamin A is found in green plants and is necessary for the proper growth of bone, for the nutrition of the cornea, and for normal night vision (p. 120). Vitamin B is really a complicated group of vitamins, twelve of which have now been identified. Some of these are necessary for growth, and others for such things as the proper functioning of the nervous system and the proper formation of the erythrocytes. Vitamin C prevents scurvy and is found in fresh vegetables and in citrus fruits; in the eighteenth century a British naval surgeon named Lind prevented the hitherto inevitable outbreak of scurvy on board his ship by taking supplies of lime juice with him. This discovery allowed Captain Cook to make his voyages to Australia without a single case of scurvy occurring among his crew. Vitamin D prevents rickets (p. 36), vitamin E deficiency may lead to infertility, and so on. The already impressive list of vitamins is still increasing. A diet consisting largely of tea and toast, such as is taken by many old age pensioners, may be grossly deficient in vitamins and essential aminoacids.

Minerals

Iron is essential for the formation of haemoglobin, and the best sources are liver, kidneys, and cocoa. The average adult male needs about 12 mg a day, but women need more because of the recurrent losses of erythrocytes during menstruation. In pregnancy the demands of the fetus for iron are a drain on the maternal resources, and more iron must be supplied. Iron deficiency anaemia (p. 153) is thus more common in women than in men. Milk is a poor source of iron, and the new born baby contains a large store of iron which is gradually drawn upon until the time of weaning.

If the diet is deficient in **calcium**, bone formation and maintenance will be affected (p. 36). The best source of calcium in an ordinary diet is cheese. Calcium is also necessary for the formation and welfare of the teeth, the clotting of the blood, and the normal functioning of cardiac and skeletal muscle. The sources of **phosphorus** are much the same as those of calcium. Both are directly concerned in calcification, and the ratio of calcium to phosphorus is of some importance in this process. Phosphorus is needed for the proper functioning of the central nervous system and in the energy cycle of somatic muscle (p. 58).

Fluorine is another constituent of bone and teeth, and dental caries is commoner if it is deficient. Tea is a good source, but many places now add fluorides to their water supply. **Iodine** is needed by the thyroid gland, and a lack of iodine leads to goitre (p. 199). One of the best sources is fish.

Traces of many other elements, such as **zinc, manganese, copper,** and **cobalt** are necessary for specialized tissues or physiological processes, and their absence in the diet can cause deficiency diseases.

The calculation of a diet is a complicated business, and for ordinary healthy individuals it is comforting to know that 'a little of what you fancy does you good'. The trouble is that so many people all over the world have to attempt to exist on diets which are unsatisfactory both in quality and quantity.

Excretion

There are many pathways by which waste material can be discharged from the body. Some is excreted in the sweat, and some by the alimentary canal. Carbon dioxide is got rid of by the lungs, and a number of other volatile substances not normally produced by the body can be removed in this way. For example, in diabetes mellitus the disordered metabolism results in the production of ketones which are partly excreted in the breath, to which they give a characteristic smell.

The main excretory organs of the body are, however, the kidneys. Urine contains a complicated mixture of substances in solution, and is the main channel by which nitrogen is excreted, mostly as urea (p. 209). Many other urinary constituents, such as phosphates and urates, are also products of protein metabolism.

Fluid balance

The main component of protoplasm is water, and every cell in the body operates in a watery medium. The amount of water each cell contains is kept fairly constant by the operation of osmotic pressure. If the cell begins to run short of water its contents become more concentrated, and osmotic pressure causes more fluid to enter the cell from the tissue fluid which surrounds and bathes its walls. The tissue fluid is in turn derived from the blood plasma, with which it effects a free interchange as the blood travels slowly along the capillaries.

Water is also necessary for the formation of such fluids as the cerebrospinal fluid, the bile, the saliva and the intestinal juices. This water is abstracted from the blood plasma by various filtration mechanisms and there is always some device for recovering it—the lymph drains into the veins, the CSF is absorbed into the dural sinuses, the water of the digestive juices enters the veins again through the intestinal wall (p. 185). In such ways water is conserved as much as possible and used over again.

The water in the body is derived from two sources—that which is drunk and that which is constantly being formed by metabolic processes. On the other hand, water is constantly being lost by three main routes, the breath, the sweat, and the urine.

The amount lost in the breath naturally depends on the relative humidity of the inspired air. If this is already saturated with water vapour it cannot acquire any more water by passing over vascular mucous membranes, and thus the body loses no water by this route. If on the other hand the air is very dry, it readily picks up water and is expelled in a moister condition than when it entered. The water vapour in the expired air shows itself when it condenses on a cold surface such as a window pane.

Fluid is continually lost in the **insensible perspiration** which moistens the skin. In hot weather or during exercise the sweat glands are more active, to prevent the body temperature rising (p. 216), and this leads to a considerable loss of fluid.

About 1·3 litres of urine are excreted every day, though the amount varies enormously with the fluid intake, the loss by sweating, and many other factors; the total fluid loss must balance the total fluid intake. Certain substances stimulate the kidneys to produce large quantities of urine, and these are sometimes prescribed in an attempt to clear away an excess of fluid which has accumulated in the tissue spaces. This condition is known as **oedema**, and may be recognized by the swollen and puffy condition of the part, which will pit when pressed firmly by the finger. Oedema may result from various causes, ranging from a failing circulation (p. 150) to disease of the kidneys which renders them incapable of filtering an adequate amount of urine. If the kidneys stop work altogether, poisonous waste materials accumulate in the blood and tissue spaces, and death occurs in a week or so, the condition being called **uraemia**.

If much fluid is being lost by other channels, the kidney reduces its output, but it must always excrete some fluid in order to retain in solution the waste products which must be removed from the body. If the fluid intake is less than the total fluid loss, the tissue cells become dehydrated. This may occur in two sets of circumstances. Either the intake is restricted, as in the case of the ship-wrecked sailor, or the loss is abnormally increased, as in the disease diabetes insipidus (p. 199), in which very large amounts of extremely dilute urine are passed. Dehydration produces a number of symptoms, among which is the parched feeling of the mouth and throat which we describe as thirst; this sensation is due to stimulation of the nerves in these situations by drying up of their surrounding. The volume of the circulating blood is reduced and the blood becomes more viscous; it is consequently harder work for the heart to push it round the circulatory system. Children are particularly susceptible to the effects of dehydration, especially after a bout of vomiting, for a great deal of fluid can be lost by this means.

Salt balance

The concentrations of the inorganic salts in the body must be kept very constant if the cells are to function at their highest efficiency. The most important of these salts is sodium chloride, and a balance has to be struck between its intake and its output. Salt is excreted by the kidneys and also in the sweat. If there is much sweating the body becomes deprived of salt, and the normal dietary intake has to be supplemented. People in hot occupations, such as foundrymen, are in the habit of putting salt in their beer, and in this way they make good both fluid and salt losses. Severe exertion in a hot climate may produce a craving for salt, and large lumps of it may be eaten; salt is a common currency in parts of Africa.

Temperature regulation

The temperature of the body has to be kept fairly steady in order to keep constant the rate of the chemical reactions which take place within it. When the air temperature falls the bodily reactions of cold-blooded animals like snakes are slowed down and the creature becomes sluggish, but man, like other warm-blooded animals, has an internal thermostat and can work in a wide range of external temperatures.

The average temperature of the blood stream in a normal resting person is about 37·7°C, but this cannot be directly measured without inserting special apparatus into a large blood vessel, so for practical purposes we rely on the value obtained by putting a clinical thermometer under the tongue with the mouth shut. In this situation the thermometer is close to the blood vessels supplying the tongue, and in a closed cavity inside the body. An even better approximation to the blood temperature is obtained by putting the thermometer into the rectum; putting it in the armpit gives very inaccurate results.

The mouth temperature varies considerably; the value 37°C is engraved on most clinical thermometers as 'normal', but it is better to remember that the range of normal oral temperature in different people extends from about 36°C to over 37·2°C. The oral temperature is usually lowest in the early morning and highest in the evening; the daily variation may be as much as 0·5°C. The temperature rises after a meal, when additional heat is produced by the processing of food, and also during exercise; if heavy work is done under hot conditions the rise may be to 38·3°C or more. A rise of temperature is a feature of certain diseases, and is known as a **fever**. Recovery is unusual if an oral temperature of 41·5°C is exceeded. On the other hand, in certain surgical operations where the circulation has to be temporarily interrupted the patient may be refrigerated to reduce his oxygen demands during the crucial few minutes. In such patients the temperature can be reduced by 17°C or more without impairing their subsequent recovery.

Heat is produced in the body by metabolic processes and by muscular exercise. Heat is lost through the skin and the mucous membrane of the respiratory tract, and to a lesser extent by the voiding of urine. The processes involved are the familiar physical ones of radiation, conduction, and the vapourization of water. The temperature of the skin is always intermediate between the temperature of the blood and the temperature of the outside environment; the loss of heat is proportional to the gradient of temperature between the body surface and the environment. The surface of the skin is kept moistened by the activity of the sweat glands, and the evaporation of this sweat cools the body. Under cool climatic conditions there is no visible accumulation of sweat on the skin, for the 'insensible perspiration' is removed as it is formed. If the atmosphere is hot and dry, a great deal of evaporation may go on from the surface of the skin, and if it is hot and moist droplets of sweat may become visible if the rate of production becomes greater than the rate of evaporation.

In the respiratory tract the atmospheric air comes in contact with a very vascular mucous membrane and the heat of the blood becomes transferred to it in this way; the evaporation of moisture from the mucous membrane is also a means of losing heat.

Normally heat production balances heat loss. If the body is losing too much heat to its surroundings, more food may be eaten, voluntary exercise may be performed, or shivering may occur (p. 59). The rate of sweat production is reduced, and the blood supply to the skin is cut down. The temperature of the skin consequently falls, and the decrease in gradient between skin temperature and environmental temperature means that less heat is lost. If, on the contrary, the body is getting too hot, less food is eaten, more rest is taken, sweat production is increased, and the blood vessels in the skin dilate, allowing more heat to be lost to the environment. The autonomic nervous system controls the distribution of blood in the skin and the respiratory tract, as well as the activity of the sweat glands. For this reason it is suspected that the overall regulation of the body temperature is vested in the hypothalamus (p. 108), but direct evidence for this is not very satisfactory. At all events the various mechanisms concerned in the production and loss of heat are in some way correlated and integrated by the central nervous system. An important additional factor is the control of the basal metabolic rate by the thyroid gland (p. 199).

Adaptation and endurance

The world provides many environments which challenge the ability of the human body to survive. A man can subject himself to high or low atmospheric pressures; he can undergo severe cold or tropical heat; he can suffer lack of food or water; he can breathe or eat poisonous material.

The reactions of the body to changes in the external environment are designed to maintain as constant as possible the physical and chemical surroundings of the cells and tissues. Many compensatory mechanisms require time to develop, and if time is not allowed, death may ensue. The sudden reduction in atmospheric pressure due to failure of the pressurizing system of an aircraft flying at 20,000 feet may be fatal, but the gradual decompression achieved by a mountain climber ascending to the same altitude need not be. Rapid translation in an aircraft from a northern winter to a shade temperature of 49°C may cause death from heat exhaustion, but it the trip is made gradually by ship, the body has time to 'learn' how to deal with the situation. Time is also extremely important in other emergencies, such as the loss of blood (p. 154).

The restorative changes which preserve the internal equilibrium in response to external stress are called **adaptation**, and naturally vary with the nature of the stressing agent and its duration. They are thought to be initiated and controlled by the autonomic and endocrine systems, particularly the suprarenal cortex.

The ability to survive in adverse conditions is not solely a matter of functional compensation. The state of health, the physique, and the mental attitude of the individual are also important. Thus Channel swimmers who have a thick covering of adipose tissue can remain twenty hours or so in water at 15°C, while those without this advantage might well die after four or five hours. Again, determination drives Olympic distance runners beyond the apparent limits of their powers.

It is possible that there are some genetic adaptations to environment. Central Australian aborigines sleep comfortably in conditions which cause Europeans to shiver uncontrollably. The Eskimos have a high basal metabolic rate which helps them to withstand cold, but this is probably due to a diet rich in fat and protein.

Bodily rhythms

The daily variation in body temperature (p. 216) is only one of a large number of rhythmical bodily processes. The heart contracts rhythmically about 70 times per minute, the cycle of inspiration and expiration occurs about 12 to 16 times a minute, and peristaltic waves are initiated in the musculature of the stomach about three times a minute. The monthly cycles of the ovary and uterus are slower, and other processes have a longer period still. For example, there is a seasonal cycle of growth in height and weight (p. 223).

Many phenomena have a roughly daily periodicity, and sleep is an obvious example of such a **circadian** rhythm. Daily swings in mood and energy probably have a hormonal basis, for the suprarenal cortex has a daily rhythm of secretion; the amount of cortisone in the blood reaches a peak in the early morning. Somatotrophin (p. 198), on the other hand, reaches maximal levels during the night. There is therefore some truth in the old saying that you grow more in bed.

Several variations in excretory activity are circadian. The kidney excretes much less urine during sleep, and nitrogen as a source of energy is spared at night. The excretion of potassium and sodium in the urine rises and falls regularly every day, irrespective of times of meals, rest in bed, work, or diet; the peak for both is about mid-day. Following a plane trip from London to New York the excretion of potassium and sodium keeps Greenwich time for a few days, after which the peak sodium excretion adjusts itself to New York noon, followed after an interval of several days by the peak potassium excretion.

The nature of the internal 'clocks' which control such rhythms is so far unknown, but attempts have been made to incriminate the hypothalamus (p. 108), a region which is always fetched in to 'explain' something we do not understand. The characteristic rhythms of adult life are not present at birth, and develop slowly and at different times during the first four or five months of life; it is common, and bitter, experience that the sleep rhythms of a young infant do not necessarily correspond with those of his parents.

17 · Growth and ageing

Growth

The process of bodily growth begins at the moment of fertilization (p. 11) by the division of a cell about 100 μm in diameter. Nine months later, when the baby is born, it weighs on the average about 3·4 kg, and is about 50 cm long. In the next twenty years or so it may grow into a man standing perhaps 180 cm tall and weighing as much as 90 kg. All this tremendous increase in size and weight is derived from the food. A new born baby needs, per unit weight, much more food than an adult, since he has not only to provide for his own metabolism, but also for a surplus which will allow for his extremely rapid rate of growth.

Growth is, however, much more than a mere matter of putting on height and weight. It involves changes in the composition of the body, the progressive modification and re-shaping of its parts, the development of skills, and the acquisition of postures. Sometimes it involves the actual destruction of existing material which has served its purpose and must make way for new and different structures.

Growth does not cease when adult status is reached, but continues throughout life. In the early embryo, everything is subordinated to cell division. Later, differentiation (p. 11) begins, and a balance is struck between these two activities (if every cell differentiated, further growth in size would at once cease, and if none did so, the embryo would become an enormous mass of undifferentiated cells). In adult life everything is directed towards the maintenance of function. Cells which die or are lost to the body during the ordinary cycles of activity (p. 11) are replaced, and another balance is struck between death and replacement. Some tissues manage their affairs better than others; there is no repair in the central nervous system, but there are excellent replacement systems in the skin and in the walls of the gut. However, in time even these begin to fail, and in the fourth and last phase of growth the repair processes are inadequate to replace losses, and functional deterioration ensues.

Growth in height is determined largely by the behaviour of the bones (p. 31). In the first year of life the length of the baby increases by about 50 per cent; during the second year this figure is approximately halved, and after this growth slows down to a fairly uniform 5–6 cm a year until **puberty** is reached. This is the time when the secondary sexual characteristics begin to appear, and it is also the time (11 to 14 years in girls and 13 to 16 years in boys) that there occurs a considerable acceleration of growth known as the **adolescent spurt**. Following the spurt there is a rapid deceleration, and growth in height ceases with the closure of the epiphyses (p. 34) of the long bones, an event which occurs earlier in girls (16 to 18 years) than in boys (18 to 21 years); there is wide variation round these times. A certain amount of limited growth is still possible after this time; for example, the skull and the vertebral column may add a few millimetres to their vertical dimensions, and the time at which an individual attains his maximum height is difficult to predict.

Growth in weight is more difficult to predict than growth in height, since it depends to a greater extent on extraneous factors, such as intake of food, exercise, occupation, etc. In the first year the birth weight increases by about 200 per cent; the increase, like that of height, is approximately halved in the second year, and thereafter about 2·5 kg is added every year unil the time of the adolescent spurt. In the spurt growth in height tends to precede growth in weight; the child shoots up first of all, and subsequently 'fills out' and broadens.

If the diet contains more Calories then the body can use up, particularly if they are supplied by carbohydrate, the excess may be stored as fat. In middle age there is usually a decrease in physical and metabolic activity, but there may be no comparable diminution in appetite; the consequent accumulation of fat is a considerable health hazard.

Another cause of increase in weight is exercise of muscles, for this can cause hypertrophy (p. 12) and muscles account for about 40 per cent of the total weight of the body. Pregnancy may temporarily add as much as 13–14 kg to a woman's weight; some of this commonly persists after the birth has taken place.

During growth the composition of the body alters. For example, the relative amount of tissue fluid progressively declines throughout life. Again, a newborn baby has much subcutaneous fat, which helps to prevent him losing heat to his environment. This is important, because, like all small animals, he has a large surface area, through which heat can be lost, relative to his volume, which determines the amount of heat generated (p. 8). The fat content of the body increases until about the age of 1 year, and after this it actually decreases until the age of 6 to 8 years, when it begins to increase again. At puberty a great deal of subcutaneous fat is put on by girls, who as women have about double the quantity possessed by males. The distribution of this fat forms part of the secondary sexual changes which occur at this time.

Not all tissues or organs grow at the same rate at the same time. For example, epitheliolymphoid tissue is prominent throughout the body at birth, and grows rapidly in early childhood, but it begins to undergo a slow process of atrophy about adolescence, well before the growth of other tissues is complete.

The liver is relatively very large at birth, and so is the central nervous system, which develops rapidly in the first few years, and slows down later on, taking little part in the adolescent spurt. It contributes about 15 per cent to the weight of the new born baby, and only 3 per cent to the weight of the adult.

The suprarenal glands are large in the new born, decrease in size subsequently, and do not regain their birth weight until puberty. The genitalia and the organs of reproduction remain very small in early childhood, and only begin to grow in earnest with the advent of adolescence.

The proportions of the body also alter during growth. The large size of the brain at birth means that the head is relatively much bigger in the infant than in the adult. The shape of the skull undergoes progressive alterations from infancy to adult life, since the cranium enlarges but little, and the face, and particularly the jaws, grow a great deal (p. 36). It has been said that at puberty 'the face emerges from under the cranium'.

At birth the upper limbs are fairly well developed, but the lower limbs are poor affairs which make up only about one-third of the total body length; in the adult they account for about a half. The infant pelvis is similarly very small. At puberty, when the secondary sexual characteristics appear, the shape of the female pelvis becomes altered as a preparation for childbirth; the male pelvis merely continues to grow in size. The male muscles enlarge and the shoulder width increases; the female muscles remain relatively small and the shoulders stay narrow.

In the infant the combination of a small pelvis and a large liver means that there is inadequate room in the abdomen for its contents, and this causes the abdomen to bulge and forces the thoracic viscera up towards the neck. The heart is therefore higher and more horizontal in the infant, and the ribs are more or less horizontal, in the position adopted in full inspiration (p. 270). This means that further thoracic inspiration is virtually impossible, and the baby relies on abdominal breathing; an abdominal operation which causes pain on movement may therefore reduce the respiratory excursion to a level at which pneumonia becomes a risk.

The high centre of gravity of the infant means that he is top-heavy when he starts to walk, and the progressive descent of the centre of gravity with the development of the lower limbs makes walking progressively easier. At birth, the skull bones do not form a continuous protective covering for the brain (p. 35) which is therefore more vulnerable until after two years or so when the developing bones meet and close the gaps.

Some children grow fast and others more slowly. It is therefore difficult to compare the progress made by one child with that of another in terms of their chronological ages; a boy of fourteen may be almost an adult in size, weight, and sexual development, or he may scarcely have begun his adolescent spurt. It is often desirable, when considering a child suspected of being 'backward', to try to find out how far it has progressed along the road to anatomical and physiological, as opposed to legal, maturity. This can be done by estimating the **radiological age** (p. 33) from radiographs of the wrist region. Similar information can be obtained by studying the eruption and development of the teeth (**dental age**), the manipulative ability (**neural age**), or the development of the secondary sexual characteristics (**sexual age**). For this last purpose the menarche (p. 196) is an easily remembered landmark.

Growth is influenced by many factors, one of the chief of which is heredity; the best way of growing tall is to arrange to have tall parents. Faulty nutrition is perhaps the commonest cause of deficiencies in growth and ultimate stature. The diet should be properly balanced as well as ample, and should contain adequate supplies of vitamins and minerals.

Girls are ahead of boys in the rate of maturation of all tissues, and the adolescent spurt begins and ends considerably earlier in girls than in boys. There is thus a stage at which girls are taller and heavier than boys of the same age; in Britain this occurs round about 11 years old.

Growth in height proceeds faster in the Spring and growth in weight faster in the Autumn. Growth can be interfered with by disease, and on recovery proceeds faster than it would normally have done, indicating some control mechanism operating in the direction of making good the deficiency.

The endocrine glands play an important part in the control of growth. The growth hormone **somatotrophin** (p. 198) stimulates the body to form protein from aminoacids. The thyroid hormone affects growth, particularly of the skeleton and the nervous system, by stimulating metabolism (p. 199); the gonads and suprarenal cortex (p. 200) are important particularly at the time of the adolescent spurt.

Finally, there is a secular change in growth; children today are taller and heavier than children of the same age some years ago, and are reaching maturity earlier. It is not so clear that adult heights and weights are increasing also, but if they are, the increase is much less than is shown by the children; it is the rate of maturation which is altering so obviously, not the end result.

It is not certain what causes growth to stop; the secretions of the suprarenals and the gonads play an important part in determining the closure of the epiphyses, but there are other factors as well, of which the chief is probably a genetic one. Nor is very much known about the causes of abnormal growth, though a great deal of effort is currently being expended in cancer research.

Localized newgrowths (**neoplasms**) may occur at any stage of development from the embryo onwards, and their characteristic feature is that their cells multiply without regard to the needs of the body. The stimulus which initiates such growth can apparently be very diverse in its nature; thus, chemical irritation of the skin, infection by viruses, genetic factors, and hormonal influences can all produce tumours.

Ageing

No sooner has maturity been reached than a gradual decline in the efficiency of the body begins. As in growth, this decline becomes manifest in different organs and tissues at different times, the nervous system suffering first (or at least, most obviously) because of the lack of replacement of the nerve cells which are steadily lost (p. 11). Visual performance reaches its peak in adolescence, and thereafter deteriorates, although for ordinary purposes it functions satisfactorily until the onset of presbyopia (p. 119). About this time hearing becomes obviously less acute, although there is a progressive loss, particularly of higher notes, from adolescence onwards; children can hear high pitched noises, such as the squeaks of bats, much better than adults.

Mental efficiency declines progressively after about the age of 20, but this is usually masked (for a time, at any rate!) by the continuing acquisition of experience and skill. Athletic prowess—a combination of physical and mental skills—is maximal in young people, and in many sports nowadays a man is 'too old at thirty'. Postural co-ordination becomes less and less easy as we grow older, and the confident and effortless postures of youth are eventually replaced by the unsteadiness, tremor, and shuffling gait of old age. In old people thermal regulation fails, and there is danger from exposure to cold; autonomic control over smooth muscle decreases, and incontinence may follow.

The cause of these changes is the progressive loss of neurones, and a similar net loss of cells in other tissues and organs eventually shows itself as atrophy. The skin becomes thin, the gonads shrink, hair falls out and is not replaced, the breast is absorbed, the kidneys become smaller, and so on.

Families of cells appear to have a finite life; a sheet of fibroblasts cultured outside the body will double itself anything from 40 to 60 times, and will then die. Towards the end of this process, chromosomal anomalies appear (p. 9) and the process of mitotic duplication slows. If this happens in the intact body it would explain the gradual decline, not only in cell populations, but also in the capacity to make good losses and to repair injuries. The actual mechanisms of repair seem to be maintained in fairly good order until very late in life; it is the speed with which they come into action that becomes progressively slower.

Parallel with these cellular changes are profound changes in the intercellular matrix and the fibres which it contains (p. 19). The gradual loss of tissue fluid throughout life (p. 221), results in 'drying up' of the tissues, and there is a tendency for calcium salts to be deposited in the matrix.

The elastic fibres throughout the body deteriorate and stretch. Two important consequences of this are the stiffening of the walls of the elastic arteries (p. 133) and the less efficient recoil of the lungs in expiration. The first results in a loss of the elastic rebound which maintains the diastolic blood pressure (p. 134) and the second results in inadequate emptying of the lung. The loss of elasticity is well seen in the skin, where wrinkles appear; if the skin is pinched up, it does not return to its normal position for several seconds.

Collagen fibres, unlike elastic fibres, increase in number with age, and tend to replace other more specialized tissues, such as cartilage. Calcium salts are deposited around them, so that the rib cartilage, for example, become progressively calcified, leading to a diminution in the resiliency of the chest (p. 271) and an increased tendency to rib fractures. Articular cartilages exposed to strain may become partly calcified and also eroded by wear and tear. The fibrocartilaginous intervertebral discs may begin to break up and be partially absorbed in early adult life, and the bowed and shortened vertebral column of old age demonstrates their disintegration.

The typical stiffness of elderly joints may be in part due to the deposition of collagen round them, but it has also been suggested that there may be an increase in the viscosity of synovial fluid with age, and that the diminished resilience of the articular cartilage plays a part. At all events, the results are often dramatic—for example, there is a gross reduction in the ability to flex the spine in old age.

The replacement of one tissue by another less specialized tissue is known as **degeneration**, and other forms of degeneration take place in old age. The walls of the large arteries are partly replaced by fatty material in the condition known as **atherosclerosis**. This causes a weakening of the walls, and the arteries may burst when the pressure inside them is raised by emotion, straining, or other causes. Conversely, the protrusion of the invading patches of degeneration into the lumen of the vessel may lead to the blood clotting and blocking the channel (**thrombosis**). In either case the tissue supplied will lose its blood supply, and because the cells in the walls of the vessels involved have lost much of their enthusiasm for repair work and replacement, the collateral circulation which eventually develops may not do so in time (p. 136).

The heart muscle, like the somatic muscles, tends to degenerate in old age, and may be replaced by fibrous or by fatty material. The movements of the heart and the limbs are weakened, and a deficient circulation and general feebleness result. The loss of muscle tone, and even more, the loss of nervous co-ordination, contribute to the frequency with which old people sustain falls. The bones are more brittle in old age, because calcium has been withdrawn from them, and consequently they are more likely to break following a fall (p. 37). Having broken, they do not repair themselves so well as in youth, and many fractures remain ununited (p. 38).

The complicated hormonal changes which occur when the functions of the ageing ovary decline at the time of the menopause (p. 196) are still not completely understood. In the male there is no exactly comparable change, but rather a gradual deterioration of testicular function usually extending over a much longer period.

18 · Posture

The human body, like that of all animals, is engaged in a constant struggle against gravity. Man is specially affected by his own weight since he has reared himself erect and now stands insecurely balanced on a narrow base instead of firmly planted on four widely separated limbs. He moves about with his weight supported on one limb at a time, and his equilibrium is then thoroughly unstable. The bones which stiffen the various segments of his body prevent them from collapsing, but the joints which permit movement between these segments at once introduce weak points, and to protect these a complicated system of muscular defences has been developed.

Whatever position or **posture** the body takes up at a given instant the postural muscles (p. 53) have to maintain it against the pull of gravity. Some postures are more easily maintained than others—the recumbent posture, when the body lies horizontally, supported at many points, requires a minimum of protective mechanisms, and is therefore used for sleeping. It is possible to rest, and even sleep, while sitting supported by a chair, though it is common to wake up with a sore neck occasioned by the overstrain of one of the unsupported joints. The standing posture, on the other hand, requires constant adjustments of muscular tone to prevent overbalancing, and this means constant vigilance on the part of the nervous system.

Mechanisms of posture

The mere maintenance of position thus involves the tonic contraction of certain groups of muscles (p. 78), and it is against this background of **postural tone** that any voluntary movements are carried out.

A standing man is never able to remain perfectly still; his body moves slightly in one or other direction all the time, and is as constantly brought back to a central position by the postural muscles. In some people this **postural sway** is more pronounced than in others, and they are said to be 'unsteady on their feet'. Most patients after a period in bed exhibit well marked postural sway, for the maintenance of the upright posture is a complicated act, and needs constant practice. The difficulty of standing upright is greatly enhanced by the socially desirable, but physiologically indefensible practice of wearing high heeled shoes, in which the weight of the body is tipped forwards to a sometimes dangerous extent, so that the muscles at the back of the legs have to work hard to prevent the wearer falling forwards on to her nose. The high heel also increases the strain on the defences of the ankle region in response to lateral sway (p. 373).

The distribution of postural tone is controlled by the nervous system through the anterior horn cells of the spinal cord; the number of impulses reaching a given muscle depends on the number and pattern of the impulses reaching the central nervous system through a variety of different channels.

The most widespread of these are the afferents from the stretch receptors in the muscles themselves. Typical **stretch reflexes** (p. 78) are a feature of the postural or anti-gravity muscles only; when a postural muscle with an intact nerve supply is stretched, it responds by a rapidly developed steady response which may last for long periods. If, on the other hand, a muscle not directly concerned with posture is stretched, it responds only by a short-lived contraction. The stretch reflex is the basic factor in postural tone, and the impulses coming from the neck muscles have a particular importance, for they are responsible for maintaining the proper relationship between the head and the trunk. Not only this, but turning the head to one side alters the distribution of tone in the limb muscles. The way the head is carried is thus an extremely important factor in the posture of the body as a whole.

The stretch reflexes are modified and controlled by information coming up to the brain along the proprioceptive pathways (p. 114) from such sources as the joint capsules and the soles of the feet. These are reinforced by information from the vestibular apparatus (p. 129) and from the eyes. Movement of the head in space stimulates the semicircular canals and the retina; movements of the head relative to the trunk stimulate the neck muscles, and so the body posture is adjusted by a series of stages. In man the optical information is more important than the vestibular, and we depend very largely on our eyes to tell us where we are.

The cerebellum is the chief ganglion of the proprioceptive system; it is responsible for co-ordinating proprioceptive and vestibular information, and controls the distribution of tone throughout the body (p. 78). If a cerebellar hemisphere is damaged, there is diminished tone on the same side of the body and tremor of the hands, which cannot be held steady. The eyes are unable to look steadily at any point, but move slowly away from it and jerk back; this **nystagmus** is essentially a disturbance of the postural mechanisms of the eye muscles. On walking, the patient falls over to the affected side rather like a drunken man. Because of the disturbance of tone all movements become slow, clumsy and feeble.

The cerebral cortex is also involved in the maintenance of posture because it receives the important visual information. Just how the cerebrum and the cerebellum interact is not yet wholly clear, but both initiate impulses which ultimately find their way to the anterior horn cells of the spinal cord. These in turn send out to the muscles the integrated result of the operation of many reflex mechanisms.

Varieties of posture

People stand and walk differently, and we can often recognize a friend from a distance by his characteristic walk. In the 'correct' standing posture the head is held high and balanced easily on the neck, the abdomen is held in and the chest allowed to expand freely; the arms hang easily by the sides and the shoulders are not permitted to sag forwards. The ideal is not a stiff military 'standing to attention', but a graceful, relaxed position which will readily merge into any desired movement. Posture is not a static affair—a series of disconnected attitudes—but a constantly changing plastic background on which is superimposed the whole possible range of bodily movements. The criterion of good posture should thus be the grace and skill of the individual in motion rather than at rest. A great deal of physiological inefficiency is attributed to a failure to stand, sit and walk correctly, but the evidence for this is not always very satisfactory. Nevertheless, as every typist knows, faulty posture can cause discomfort, and can ultimately lead to deformity which may produce considerable disability.

The 'correct' standing posture is only gradually acquired. Small children tend to stand with the abdomen thrust out, while the rest of the body above the protuberance leans backwards to preserve the position of the centre of gravity. One of the basic factors which determines the posture of the whole body is the way in which the pelvis is tilted on the thighs and this question is discussed on p. 372.

The upright posture

The overwhelming advantage of the upright posture lies in the liberation of the forelimbs from the duties of locomotion. The hands are now free to grip and manipulate, and have eventually become the most versatile instruments at the command of any animal (p. 289). At the same time vision has been improved because the eyes have been brought further from the ground, and the skull in which they lie can now be poised more freely on top of the vertebral column. These are tremendous gains, but they have been bought at a considerable price.

Some of the drawbacks of standing upright have already been mentioned (p. 226): the precarious equilibrium, the demands on the organization of the nervous system, and the strain on the bones and joints of the lower limb need no further emphasis. The foot has had to become adapted to unassisted weight-bearing, and the prevalence of flat or painful feet shows that the task is sometimes too much for it (p. 373). The stresses in the human vertebral column are quite different from those imposed on the cantilever-like vertebral columns of four-footed animals. Adaptation of the human spine to the new system of weight-bearing is not yet perfect, and its lower part is a common source of aches and pains. The intervertebral discs are frequently damaged (p. 255) and the joints of the lower vertebrae tend to degenerate.

Other difficulties are perhaps less immediately obvious. The forelimbs, which in quadrupeds lie approximately parallel to the hindlimbs and at right angles to the body, now hang down at the side of the trunk; the resulting alterations in the structure and mechanics of the shoulder region have rendered some of the nerves and vessels entering the limb prone to compression and injury. The pelvis has had to become modified to bear weight in a different direction, and at the same time to continue to allow the passage of the baby's head during childbirth; the modifications are not always equal to the occasion, and such conditions as sacroiliac strain are unfortunately common in women who have borne many children (p. 279).

In the upright position the abdominal viscera tend to fall towards the pelvis, and the unaccustomed strains thus set up contribute to the production of herniae (p. 282). The musculature of the pelvic diaphragm now takes a heavy gravitational load, and when it is weakened the pelvic viscera may prolapse (p. 283).

The elevation of the brain above the heart means that gravity hinders the supply of blood to the organ most vulnerable to oxygen lack (p. 136). For the same reason the return of blood to the heart from the lower limbs is impeded and the thin-walled superficial veins in the leg and thigh frequently give way under the hydrostatic pressure of their contents and become varicose (p. 135). Gravitation has in fact necessitated a complicated system of regional vasomotor control to ensure that all parts of the body receive blood at an appropriate pressure, no matter what the posture adopted. If the autonomic system is not sufficiently alert, sudden changes in posture from lying or sitting to standing may result in the brain receiving blood at such a low pressure that fainting may occur.

In the four-footed position the ribs swing backwards and forwards relatively unimpeded by the pull of gravity. As soon as the chest is tipped on end, it requires much more muscular effort to pull the ribs upwards in inspiration (p. 270); expiration, on the other hand, becomes largely passive.

The recumbent posture

Clearly it is difficult, and sometimes even dangerous, to stand upright. But it is in some respects much more risky to take to one's bed. Although we spend rather more than one-third of our lives in bed, the body is adapted to an alternation of postures rather than to the prolonged continuance of any one posture, and confinement to bed, particularly in old people, may of itself produce considerable disability, and occasionally even death.

The maintenance of one position, the absence of exercise, and the diminished respiratory movements allow bronchial secretion to collect in the most dependent parts of the lungs, and pneumonia may result. The absence of normal movements also means that the venous blood is not so efficiently propelled towards the heart, and the flow of blood in the peripheral veins may be slowed down sufficiently for a clot to form, with pulmonary embolism as a possible sequel (p. 153). The heart itself adapts to the lessened demands of recumbency, and has difficulty in coping with the increased needs of the body when the time comes to get up.

The intricate system of sensory and motor control of willed movements, particularly of the lower limb, becomes disorganized through lack of practice, with the result that unsteadiness and lack of co-ordination are very marked when the patient first gets up. In diseases of the nervous system which interfere with co-ordination a period of rest in bed may produce a serious and possibly permanent deterioration.

The muscles undergo a disuse atrophy which makes them dangerously weak for the heavy job of supporting the body weight. Some muscle groups are more prone to melt away than others; the quadriceps femoris (p. 361) is particularly vulnerable, and has to be kept regularly exercised if the patient is to be able to walk satisfactorily when he gets up. The pressure of the bedclothes on the feet may stretch the tissues in front of the ankle and allow those behind it to contract; this condition of partial foot-drop is often seen in patients who have spent many years in bed, and must be guarded against by placing a wire cradle over the feet to take the weight of the bedclothes. Other joints may stiffen and allow contractures to occur simply through prolonged inactivity.

During a period of bed rest, calcium drains out of the skeleton into the blood stream, so that the bones become porous and less able to bear weight. It is partly for this reason that in old people fractures may fail to unite if the patient is in bed, but succeed in doing so if they are ambulant. The calcium is excreted by the kidney, and may contribute to the formation of 'stones' in the urinary tract. The control of micturition may become disturbed, partly because of the difficulty of using the bed-pan or bottle, and constipation is common for similar reasons. The skin, from being continually pressed upon, may become eroded and allow the formation of bed-sores (p. 63).

For these reasons rest in bed is no longer prescribed so freely as it once was, and patients in hospital are encouraged to become ambulant as soon as they are fit to get up.

19 · Muscular activity

Effects of muscular exercise

The chemical changes in muscular contraction have already been considered (p. 58). The end products are carbon dioxide and water, and the immediate sources of energy are the phosphates and glycogen the muscle fibre itself contains. When the glycogen is exhausted, more has to be formed from the glucose in the blood, and the drain on the blood sugar is met by mobilizing glycogen from the reserves in the liver. These processes require additional energy, and this may be provided by the destructive breakdown of all three classes of foodstuff. In fact, the respiratory quotient during exercise is much the same as it is at rest, suggesting that the body uses up protein, carbohydrate and fat in similar proportions whether it is exercising or not.

As exercise progresses, there is a slight rise in the concentration of carbon dioxide in the blood, which consequently becomes slightly less alkaline. In severe exercise some lactic acid may accumulate in the muscle as a result of incomplete oxidation; it is rapidly dealt with after the exercise stops.

Muscular exercise has a profound effect on the cardiovascular system. As each muscle contracts it compresses the blood which it contains, and this results in a greatly increased venous return to the heart. This is augmented by the increase in the depth and frequency of the respiratory movements, which help to pump the blood into the thorax (p. 135). Secondly, the rise in the amount of carbon dioxide in the blood stimulates the sympathetic innervation of the heart. The heat produced by the muscles (p. 59) raises the blood temperature slightly, and this directly stimulates the sinu-atrial node and also affects the vagus nucleus.

The net effect of all these factors is that the heart rate may increase to 150 beats per minute or more in severe exercise. Further, the stretching of its chambers by the augmented venous return produces an increased contraction force, so that more blood is expelled at each beat. The total amount of blood discharged by the heart in one minute may therefore increase up to five or sixfold.

At the same time the blood in the peripheral circulation is redistributed. The secretion of adrenaline, which occurs in any emergency, constricts the vessels in the skin, gut, and other relatively inactive tissues, and dilates the vessels in the active muscles. In this it is assisted by the local accumulation of carbon dioxide and other substances produced by the muscular contractions, for these relax the walls of the capillaries and arterioles in the working muscles. The dilatation of the vessels in the active muscles outweighs the constriction in the non-essential areas, and the total peripheral resistance of the circulation is decreased. Accordingly, the blood pressure rises only moderately in spite of the increased cardiac activity.

After a time, the demand for blood by the muscles conflicts with the need to lose heat through the skin, and the blood vessels in the skin again dilate (p. 217). The sweat glands are also set in full action to cool the body by evaporation, and are helped by the concomitant increase in the respiratory rate and depth, which allows more heat to be lost through the respiratory tract.

The respiratory centre is stimulated by the excess carbon dioxide in the blood and by the rise in body temperature; afferent impulses from the distended veins and from the active muscles themselves may also affect the centre directly. The consequent increase in the rate and depth of the respiratory efforts leads to an increased ventilation of the lungs. This increase may take some time to settle down after exercise is terminated because of the oxygen debt incurred (p. 58). Thus, after a fast quarter-mile run it may take more than half an hour for breathing to return to normal.

During exercise and recovery, therefore, more oxygen gets into the lungs per minute, and the increased quantity is also more efficiently absorbed, for the blood arriving at the lungs is more than usually deficient in oxygen, the active muscles having used up more than their normal quota. Consequently the pressure gradient between the oxygen in the air and the oxygen in the blood is increased, and more is able to pass through the alveolar wall into the blood. Exactly similar considerations apply to the excretion of carbon dioxide.

Oxygen absorption through the lungs is also greatly increased because of the increased volume of blood which is being sent every minute through the pulmonary circulation. The increase in oxygen consumption which these two methods have to keep pace with is sometimes prodigious. At rest the average consumption of oxygen is of the order of a quarter of a litre per minute, but during running at 10 miles per hour it rises to nearly four litres per minute.

A puzzling phenomenon connected with the respiratory effects of exercise is the occurrence of 'second wind'. Just when respiratory distress and exhaustion are at their height there is a sudden relief; breathing becomes easier and distress is alleviated. The reason for this is quite unknown.

Another aspect of muscular activity that has no satisfactory explanation is the appearance of **fatigue**. The nerve fibres which supply the muscles are virtually unfatiguable, and will continue to conduct impulses for as long as they are required to do so. If a muscle is removed from the body and stimulated through its nerve supply for a sufficient time of power of the contraction gradually begins to fail and finally no contraction can be elicited. If now the muscle be stimulated directly it will once more respond quite well, though after a further time this form of stimulation will also become ineffective. From such experiments it has been suggested that when the nerve supply is intact fatigue may be due to something occurring at the motor end-plate (p. 76), but this does not satisfactorily explain the fatigue which occurs in the intact body after exercise. In this situation the central nervous system is believed to give way first, as a result of the accumulation of carbon dioxide and the reduction of oxygen in the blood which supplies it. Stiffness and discomfort in the muscles which have been active are possibly due to afferent impulses set up by the chemical changes within the muscle. The onset of fatigue depends on many factors, among which is the nature of the work done. If this provides a variety of movements fatigue is delayed, but if the same few movements are repeated over and over again it comes quickly. Environmental conditions are important, and fatigue occurs rapidly if the ventilation is poor or if the temperature and humidity are high. Endurance depends also on psychological factors, in particular on motivation and the interest taken in the task. Encouragement or competition may postpone for a time the onset of fatigue.

Each kind of task has an optimum rate of performance; below this rate the output of finished work suffers, and above it the quality of the work done is impaired. Similarly each individual has his own optimum rate of working at a given task, and also a maximum rate of working which can be sustained only for a short period. If an individual works at his maximum rate, fatigue is early, and frequent rests are necessary to allow the muscular apparatus to recover. Work which requires an energy expenditure of over 700 Calories per hour cannot be continued for more than about one hour by an untrained man.

At the optimum rate there is a steady output of energy which can be maintained for much longer periods. The novice mountain climber dashes off at high speed on the lower slopes while the guide walks steadily behind; some hours later the beginner is completely exhausted but the guide is still keeping up the same steady pace.

Experiments on the metabolic cost of climbing steps show that the optimum rate is about one step per second. Similar results have been obtained for bicycle pedalling. The optimum rate depends partly on the viscosity of the muscle itself. When a movement is made, time and energy have to be expended in overcoming this viscosity, and a contraction which lasts only a short time may not be able to achieve any external work at all. The muscles flexing the elbow produce a maximum external effect when the flexion lasts at least two seconds, and it is calculated that for muscles in general the optimum duration of a contraction is about one second. The viscosity of muscle is actually a valuable safety device, for it means that the force is gradually rather than suddenly applied, so reducing the chance of breaking bones (p. 37) or tearing tendons.

The acquisition of skill is also important in determining the level of energy output and the onset of fatigue. A skilled performance always makes the task look easy, and in fact it *is* easier, because the muscles concerned are working to the best advantage all the time, with no waste effort. Skating is very hard work when we begin, but becomes almost 'effortless' when we have mastered it.

Training

The repeated performance of a given task or exercise has far-reaching effects. The number of movements required is reduced and so less energy is expended; the way to achieve this may have to be pointed out by an expert in 'work study'. After training the task requires less concentration and so causes less mental fatigue. Training leads to an increased pulmonary ventilation and increased oxygen utilization, a postponement of fatigue and a more rapid recovery of circulatory and respiratory disturbance after the end of exercise. Perhaps surprisingly, it does not greatly increase the mechanical efficiency of the muscles concerned. In untrained people the muscles convert 20–30 per cent of the total energy they produce into useful work, and training can only improve this figure to about 25–35 per cent. The heart rate at rest may be lowered in the trained subject, and the output per beat may be increased.

In specialized exercises the muscles concerned may increase in size and in strength; the increase in size is partly due to a true hypertrophy of the muscle fibres, but there is also an increase in the amount of connective tissue and in the number of blood vessels within the muscle. The way to develop strength and muscle size is to exercise the muscle against increasing resistance, either by using it to pull against increasing weights or springs, or by utilizing the weight of the body. Repeated exercise without increasing the resistance may actually slightly decrease the size of the muscle. The increase in endurance of a trained muscle is much greater than its increase in size might suggest, and is possibly associated with the increased intramuscular blood supply.

The energy cost of exercise

The amount of energy expended in various forms of exercise can be measured by enclosing the subject in a giant calorimeter (p. 212). This is expensive and rather impractical, and more usually the indirect method of measuring oxygen consumption (p. 212) is adopted. This is done at rest, during the specified activity, and during the period of recovery. (It is necessary to measure the oxygen consumption during recovery because of the oxygen debt; the true cost of the exercise is given by the excess of oxygen consumed during both the activity and the period of recovery.) The respiratory quotient does not alter much in exercise, so it can be assumed that it remains at the resting value of about 0·85. With this value of R.Q. one litre of oxygen produces 4·86 Calories when it is used in metabolism, and from this figure the Calories produced by the exercise can be calculated.

Many investigations have used a treadmill running at different speeds. This device allows the subject to remain stationary and so facilitates the collection of samples, and it also frees the subject from having to concentrate on maintaining an even pace, since this is done for him. A similar device is the bicycle ergometer, in which the subject pedals a fixed bicycle, the work done being measured either mechanically or electrically. It is found that standing takes about 10 per cent more energy than lying down, but there is little difference between lying down and sitting in a comfortable chair. Sleeping costs about 70 Calories per hour, and at the other extreme running at the speed of 19 miles per hour is paid for at the rate of over 9000 Calories per hour. One table gives the average expenditure of Calories on 'housekeeping' as 115 per hour, and 'dancing, vigorous' as 340 per hour. One pound of body fat yields about 3500 Calories, and a simple calculation will show that slimming by exercise is an uphill task.

Part 3
Functional topography

20 · Topographical information

Topographical methods

So far we have studied the component parts of the body, and some of the ways in which they co-operate. We must now consider the relationships of tissues, systems and organs in different regions.

The traditional way of studying regional topography is by dissecting a dead body. This has the advantage that leisurely observation is possible, and that structures deep in the body can be examined just as easily as more superficial ones. But there are also disadvantages. The bodies used are usually bequeathed to the anatomy department by elderly people, so that their muscles are small and often wasted, and old age and possibly disease have altered the normal relationships of their organs and distorted their structure. The process of embalming, which is essential if the body is to be preserved, itself shrinks and hardens the soft tissues.

Dissection has provided the bulk or our static knowledge of topographical anatomy, but anatomy is not a dead but a living subject, and it is the dynamic functional anatomy which matters. Several methods are available for the study of the living body.

The simple methods of clinical examination—inspection, palpation, percussion, and auscultation—can discover a great deal. **Inspection** reveals the swelling of muscles and the play of their tendons when a movement is made, the pulsation of arteries, and the position of superficial bony prominences. The superficial veins are often visible through the skin. **Palpation**, or examination by touch, gives more detail about superficial structures, and allows deeper structures to be detected. The structures felt are naturally those which are resistant to the examining fingers, and the method is useful for mapping the position of bone, nerve trunks, dense connective tissue, and solid viscera, as well as for examining the contraction of the deeper-lying muscles. **Manipulation** will demonstrate the types and ranges of movement of which the various joints are capable, and will give some idea of how they are constructed. **Percussion** involves laying the finger of one hand on the body and striking it firmly with a finger of the other hand. Vibrations are set up in the finger in contact with the skin, and a dull or resonant note is produced according to whether the finger lies over a solid or an air-containing organ. The method can be used to map out the outlines of such organs as the heart, the lungs, the full bladder, and the stomach. **Auscultation** (listening) with a stethoscope gives some idea of the position and size of such organs as the heart and lungs, for their normal functioning creates noises.

Special instruments increase the range of information obtainable from the living body. Electrodes placed on or in muscles can record the part they play in specified movements. Instruments can be introduced through the appropriate orifices to inspect viscera such as the oesophagus, the stomach, the nose, larynx and bronchial tree, the urinary bladder, the rectum and lower part of the colon, the vagina and uterus. The ophthalmoscope permits examination of the retina, and the auriscope allows the drum of the ear to be seen. Surgical operations provide further anatomical information.

Finally, **radiography** is of the greatest value. Plain radiographs demonstrate the relationships and structural details of the bones, and also the shape, size, and position of the heart and lungs in different phases of activity and in different postures. The hollow organs can be made visible if substances opaque to X-rays are swallowed or injected so as to coat their lining membranes. Air can be introduced to outline various body cavities. Cineradiography can show the mechanical functioning of the joints, the heart, lungs, and alimentary canal.

By such means we have been able to learn a great deal about the factors which influence the position of a given organ in the living body. If, for example, the patient stands up, radiography shows us that gravity pulls his heart down substantially below the position it occupies in relation to the chest wall when the patient is lying down. The mobility and range of position of the internal organs were not appreciated until well within living memory, and in many books the position of the viscera is still given with reference to the dead body. It must be clearly recognized that the cadaveric positions and relationships are grossly modified by such simple acts as breathing, sitting, and bending down.

Biological variation

In topographical anatomy **biological variation** is an important factor. Just as some of us are taller and heavier than others, so the dimensions and weight of our component parts may vary. One healthy adult may have a liver which is only three-quarters as heavy as that of another healthy adult. There is nothing 'abnormal' about such people—they are merely exemplifying the principle of individual variation. Nevertheless most textbooks only give average figures for size, weight, and performance, and such figures tell us nothing about the next patient we meet. Because the average person can bend his wrist through a range of 80° to 90° it does not follow that a patient who can only accomplish 70° of bending has anything wrong with his wrist. In such cases both sides of the body should always be examined—the healthy side as well as the side under suspicion—to get an idea of the normal performance for the particular patient.

Another form of variation is not so much quantitative as qualitative. Let us suppose that an artery divides into three branches (Fig. 104). This may happen in several ways, of which one is perhaps slightly more common than the others. The net result for the tissues is exactly the same; they receive the same amount of blood by the same eventual channels; the only thing is that the local plumbing arrangements differ in different people. In a textbook it would be inordinately tedious to give all the possible ways in which every artery has been known to divide in normal people, and so the usual procedure is to give only the commonest arrangement. In the Figure this is arrangement (A), but from the percentage frequency figures there is clearly nothing unusual about the people who have arrangement (B) or arrangement (C). Nor is there anything 'abnormal' about arrangement (D), though one might say that it was uncommon. If we were all constructed to exactly the same pattern it would be a very dull world.

Surface anatomy

Nevertheless, the general pattern in all of us is basically similar, and we can lay down a set of rough rules for finding any given structure from its relationships to other structures. Since we are nearly always concerned to find the position of a structure in the living body we must make our approach through the intact skin, and the name given to the set of instructions we devise is the appropriate one of **surface anatomy**. This is naturally the most important form of anatomy for the majority of people, who are neither anatomists nor surgeons.

Of all the structures in the body, the bones exhibit the least normal variation, and we therefore use them to take our bearings from whenever possible. We say, for example, that a given nerve is to be found exactly halfway between two easily felt and unmistakable bony prominences. Using a rule of this kind makes it unnecessary to worry about the fact that different patients are of different sizes. The proportion of the bony structures to each other remains the same during growth, and will also be the same in fat and thin people. For this reason rules which express relationships as a proportion sum are greatly to be preferred to rules which express them in terms of centimetres. If, for example, we say that the apex of the heart lies 9 cm from the midline, we are faced with a complicated calculation when we try to discover how far it would be from the midline in a child of five years of age.

Frequently it is impracticable to use proportional methods; we cannot say that an artery lies two-thirds of the way along a line joining two bony points when only one suitable bony point is to be found in the vicinity. The position of structures is then referred to the nearest bony landmark, the measurement from it being made *in units derived from the patient's own body*. We may say, for example, that the artery is three fingersbreadths away from the landmark selected. The distance between the bony point and the artery will remain relatively the same throughout life if expressed in this way, for the patient's fingers will also grow with him. Other measurement units used are the breadth of the thumb and the breadth of the hand.

The limbs are extremely mobile, and as they move, the relationships of the soft parts within them alter. In the forearm, for example, the bones are parallel when the back of the hand is laid on the table, but cross over each other when the palm is downwards. In consequence of such movements the soft parts twist over and round each other. To avoid distortion of this kind measurements must always be made with the limb in a specified position.

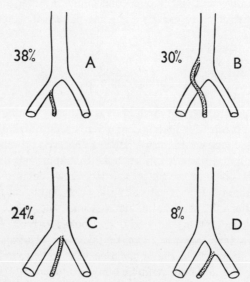

Fig. 104. Variations in the way in which a given artery may arise. The percentages indicate the frequency of each arrangement.

In the thorax and abdomen, where bony landmarks are scanty, a 'grid' of artificial lines is superimposed as a substitute (Fig. 105). Perpendiculars dropped from the midpoints of the clavicles (collar bones) are called the midclavicular lines. Similar lines are dropped from the anterior and posterior folds of the armpit (the anterior and posterior axillary lines), and from a point midway between the two (midaxillary line). Horizontal lines are drawn at the level of the lowest point reached by the chest wall (subcostal plane) and the highest points reached by the bony pelvis (intercristal plane). The points of intersection of these lines may be used as substitutes for bony landmarks, and the lines themselves divide the abdomen into zones which are used for the purposes of description.

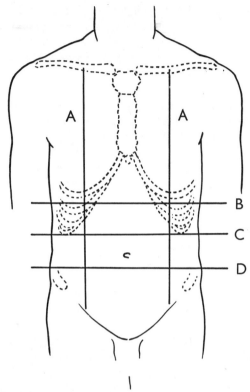

Fig. 105. Some lines used as artificial landmarks in surface anatomy. A: midclavicular line; B: transpyloric plane (through 1st lumbar vertebra); C: subcostal plane; D: intertubercular plane (through the tubercles of the iliac crests).

Essential information

When studying deep fascia the essential things to know are its attachments and its arrangement into partitions and investing planes, because muscles are often attached to deep fascia, and because fluid may be conveyed from one place to another along these planes (p. 22).

In regard to **bone**, the description is vital, and the attachment of muscles and ligaments, the mechanics of the bone, and its surface anatomy are all important, since the bones are the framework on which the body is built up, and from which all surface features are located.

When dealing with a **joint**, the primary thing is the type and range of the various movements permitted; to understand the mechanics of a joint, it is necessary to know something of its capsule, its intrinsic and extrinsic ligaments and the muscles which operate on it. It is also advisable to think about the factors which render the joint stable or unstable, and to know its main relations, for some of these could be injured if the joint should dislocate.

In regard to a **somatic muscle**, the basic facts include its origin, insertion, nerve supply, actions and functions (p. 59)—particularly the last two, for the most important thing about a muscle is what happens when it contracts. In most parts of the body the *exact* attachments of muscles are not of much functional significance, but an appropriate knowledge of them is essential if the action of the muscles is to be understood rather than merely memorized.

For a **peripheral nerve** it is necessary to know how it originates, its course and surface anatomy, its main relations, and its distribution. Nerves are exposed to injury as they travel through the tissues, and a knowledge of their course and surface anatomy enables one to predict whether or not they are likely to have been injured by a given accident; it also enables intramuscular injections and incisions to be made at a safe distance from nerves which might be injured by the procedure. The distribution of the nerve—the structures which it innervates—determines what happens when the nerve is damaged, for the structures it supplies will be wholly or partly paralysed (p. 91).

Similar information is required in the case of **arteries**, for similar reasons, though the distribution is not of such great importance since most peripheral arteries can be blocked or tied off without a catastrophe ensuing (p. 136). Most **veins** accompany arteries, and if the relations of the artery are known there is no point in memorizing the relations of the vein as well. However, some veins, particularly the superficial veins of the limbs and neck, are not accompanied by arteries, and their course and relations are just as important as those of a major artery. The most necessary things to remember about the **lymphatic system** are the position of the various groups of lymph nodes and the territories which they drain, since this enables an enlarged node to point the way to the discovery of the source of the infection or malignant growth which has caused it to swell.

Finally, regarding a **viscus**, the approximate range of normal size, weight, and capacity should be known, together with something about its position, surface anatomy, blood supply, nerve supply, and mode of emptying.

Most of these points will become clearer in the course of the succeeding chapters, which are not intended to give a complete coverage of the regional topography of the body, but rather to emphasize certain principles of functional topography.

21 · The vertebral column

Description

The vertebral column is the skeletal support of the neck and back, and is formed from 33 bony **vertebrae**, separated and connected to each other by wads of fibrocartilage known as **intervertebral discs** (Fig. 106). This construction gives the column mobility, for each vertebra can move slightly on its neighbours, and resilience, for the discs absorb shocks which might otherwise break the bones.

Each vertebra is formed from spongy bone enclosed in a thin coating of compact bone. In the adult the vertebrae are one of the chief sites of formation of erythrocytes. Seven **cervical vertebrae** lie in the neck, and the next twelve are the **thoracic vertebrae**, which have ribs articulating with them. Below these the five **lumbar vertebrae** form the bony support of the abdomen. The five **sacral vertebrae** in the pelvis are separate in the child, but later fuse together into a single bone, the **sacrum**. Articulating with the lower end of the sacrum is the **coccyx**, formed by the fusion of four **coccygeal vertebrae**.

A 'typical' vertebra (Fig. 107) has a relatively massive **body** anteriorly, and a **vertebral arch** behind. The vertebral arch forms a section of the vertebral canal which surrounds the spinal cord and cauda equina (p. 79). Each arch consists of two **pedicles** which spring from the posterolateral aspect of the body, and two **laminae**—flat plates of bone which stem from the pedicles and meet each other in the midline posteriorly. At the point where the pedicle becomes the lamina a large **transverse process** juts out on each side, and where the laminae come together the **spine** of the vertebra projects posteriorly. Each pedicle is rounded above and below so that when the bodies of the vertebrae are separated by the intervertebral discs there is a hole betweeen adjacent pedicles. Through this **intervertebral foramen** the spinal nerve emerges from the vertebral canal (p. 79). Finally, two pairs of **articular processes** (superior and inferior) arise from the junctions of pedicles and laminae, and articulate with the corresponding processes of the vertebrae above and below by plane joints.

The body of the vertebra is convex in front and slightly concave posteriorly, where it forms the anterior wall of the vertebral canal: its upper and lower surfaces are flattened to receive the attachments of the discs. Each disc has a central soft and elastic portion, the **nucleus pulposus**, encased in an outer

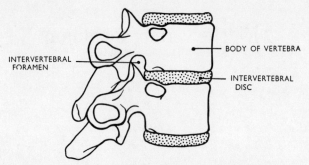

Fig. 106. Diagram showing the relationship of the intervertebral discs to the vertebrae. Notice the intervertebral foramen, through which the spinal nerve emerges; if the disc disintegrates the nerve is likely to be pressed upon by the adjacent bone.

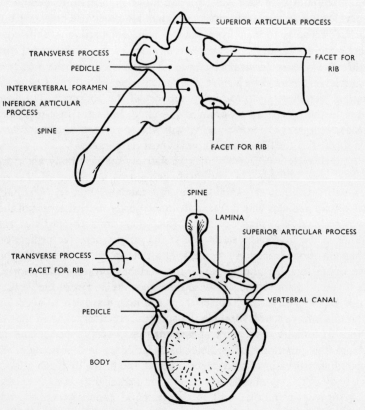

Fig. 107. A 'typical' (thoracic) vertebra from the side and from above; to show the constituent parts of a vertebra.

fibrous shell called the **anulus fibrosus**. Approximately a quarter of the total length of the column is made up by the discs, and their gradual disintegration after middle age contributes to a progressive diminution in stature. Because of compression of the discs the height of the body is approximately 1·5 cm greater in the morning than it is at night, unless the day has been spent lying down. If a sudden strain is taken by the vertebral column, as in lifting a heavy weight, the anulus fibrosus may tear, allowing the nucleus pulposus to protrude as a semi-fluid mass into the vertebral canal. The lower lumbar discs are most prone to behave in this way, for they are subject to the greatest mechanical stresses. A 'prolapse' of a nucleus pulposus in this region may press on one of the roots in the cauda equina. Because of the great mobility of the cervical region of the column prolapses occur in this region also, and these affect the roots of the brachial plexus (p. 328).

Regional differences

The first cervical vertebra, the **atlas**, is unique in having no body and no spine, being merely a ring of bone (Fig. 108). Its transverse processes are very wide apart. The upper articular facets are concave, and support the condyles of the occipital bone.

The body of the atlas is incorporated with that of the second cervical vertebra, the **axis**, and sticks up from it like a rounded pillar occupying the ring of the atlas. Round this pillar, which is known as the dens, the atlas and the skull which it bears rotate when the head is turned from side to side.

The atlas and the axis are specialized vertebrae, but from the third cervical down to the fifth lumbar there is only a gradual transition of anatomical features. The **bodies** of the vertebrae are wider from side to side than from front to back in the cervical and lumbar regions, but the converse is true of the midthoracic vertebrae. Their size increases steadily from the cervical to the lumbar region; the thoracic bodies have facets for articulation with the heads of the ribs. The **intervertebral discs** are thickest in the lumbar region and thinnest in the upper thoracic part of the column.

The **spines** of the vertebrae all incline downwards, but their obliquity varies greatly. It is most marked in the middle thoracic region, where the tip of the spine of one vertebra is opposite the middle of the body of the vertebra below. The spines of the lower cervical vertebrae (except for that of the seventh) are bifid, and the spines of the lumbar vertebrae (except for the fifth) are shaped like a hatchet or tomahawk. The spines can be felt, and in the lumbar region seen, in the midline of the back.

The **transverse processes** are largest in the atlas, and then diminish down to the end of the thoracic region; they become large and strong again in the lumbar region. All the cervical vertebrae have a foramen in the transverse process, and in the upper six this transmits the vertebral artery as it climbs into the skull to supply the brain.

The **laminae** tend to overlap each other in the thoracic region, but elsewhere leave gaps which are largest in the lumbar part of the spine.

The **intervertebral foramina** increase in size from above downwards.

In the cervical region the **superior articular facets** face backwards and upwards; in the thoracic region they face backwards, upwards and sideways, and in the lumbar region they are nearly vertical, facing medially and slightly backwards. (The behaviour of the inferior facets naturally mirrors that of the superior ones.)

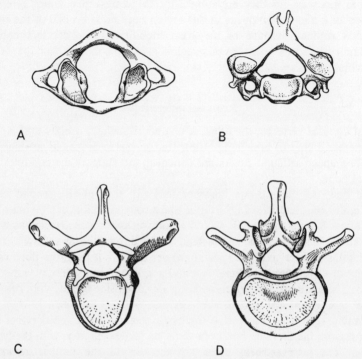

A   B

C   D

Fig. 108. Individual vertebrae. A: the atlas. B: a cervical vertebra. C: a thoracic vertebra. D: a lumbar vertebra.

Typical vertebrae from each region of the moveable part of the column are therefore easily recognizable (Fig. 108). Occasionally, however, there is some trouble where two regions meet. It is not uncommon for the first sacral vertebra to be wholly or partly separate from the others. This is called **lumbarization**, and if the separation is complete there are said to be six lumbar vertebrae and only four sacral ones. Similarly the fifth lumbar vertebra may be completely or partially fused with the first sacral—a process of **sacralization**—and the twelfth thoracic vertebra may be lumbarized, having no rib attached to it; it then has a form intermediate between a thoracic and a lumbar vertebra. Such conditions usually cause little or no disability, but if there is an **cervical rib** related to the seventh cervical vertebra, the lower roots of the brachial plexus (p. 328) may be stretched over it, causing tingling or even paralysis in their distribution (p. 336).

The **sacrum** differs from the moveable part of the vertebral column in that its five component vertebrae are usually firmly welded together, The bodies are fused in the midline, and the other elements of the vertebrae also fuse to form the **lateral mass**. Between the lateral mass and the body are the **anterior** and **posterior sacral foramina**, through which the ventral and dorsal rami respectively of the sacral nerves emerge. In the midline at the back the remnants of the sacral spines reveal the composite nature of the bone. The **sacral canal** in the middle of the bone is the lowest part of the vertebral canal, and contains the sacral nerve roots and the filum terminale (p. 96).

The **coccyx** is a small irregular bone which articulates with the sacrum at the sacrococcygeal joint, which may become fused in later life. It represents the tail, but its chief importance is that it may become fractured by a fall and give rise to a great deal of pain.

Ligaments

The vertebrae are bound to each other strongly by ligaments. In front of the bodies, stretched like a strip of sticking-plaster from the coccyx to the skull, is the **anterior longitudinal ligament**, and corresponding to this on the backs of the bodies is the **posterior longitudinal ligament**. These ligaments bind together bodies and discs. The spines are joined by a series of **supraspinous** and **interspinous** ligaments; in the neck the spines are sunk well below the surface of the body, but the ligaments joining them extend posteriorly in the mid-line to form an intermuscular septum which is attached to the back of the skull. In four-legged animals this **ligamentum nuchae** helps to support the weight of the head, but in man this is chiefly done by muscles. The laminae are joined together by a series of ligaments called the **ligamenta flava** from their yellowish colour. These are the only human ligaments which have a considerable admixture of elastic tissue, and they help the muscles to maintain the upright posture of the vertebral column. They are mainly of use when the column is flexed, and they are probably the only ligaments which normally take a steady rather than an intermittent strain.

Curvatures

In the infant there are two **primary curvatures** of the column, both concave forwards. The lower one is formed by the sacrum and coccyx, and the upper one by the whole of the rest of the column. The point where the two meet is called the **sacrovertebral angle** (Fig. 109). The adult vertebral column has two other

secondary curvatures, which compensate for the primary ones by being concave backwards. The **cervical** curvature is formed when the infant begins to lift his head up from his chest, and the **lumbar** curvature about the time when the child starts to walk. The two primary curvatures are formed by the shape of the bodies of the vertebrae; the two secondary ones are due largely to the shape of

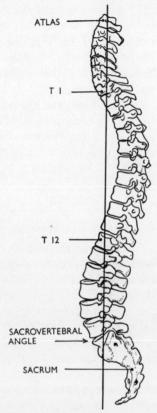

ATLAS

T I

T 12

SACROVERTEBRAL
ANGLE

SACRUM

Fig. 109. Diagram showing the normal curvatures of the vertebral column; the line of gravity is shown.

the intervertebral discs. If these discs collapse or atrophy, as in old age, the effect is to diminish the compensatory secondary curves, and so the spine bows forwards.

In addition to the flexion–extension curvatures, there is usually a **lateral** curvature in the thoracic region, and if this is pronounced secondary curves will develop above and below it. The lateral curvature is said to be due to the greater use of one or other arm, and is usually convex to the right; this is not, however, the only explanation. Sometimes the curvatures of the vertebral column get out of hand and produce a deformity. An abnormal increase in the flexion of the thoracic spine is called **kyphosis,** an increase in the extension curve of the lumbar region is called **lordosis,** and an exaggerated lateral curvature is known as a **scoliosis.** A contribution to any of these deformities may be made by poor postural habits of standing, sitting or walking (pp. 371, 373), but there is still a good deal of mystery surrounding their causation, particularly in the case of scoliosis. Postural scoliosis is common in adolescence, and can be corrected by voluntary effort; it disappears if the spine is flexed. In contrast, structural scoliosis cannot be corrected voluntarily, and does not disappear on flexion. It may be due to developmental causes such as a failure of one side of the body of a vertebra to ossify, or it may be due to disease, such as paralysis of the muscles supporting the column. But many facts require explanation; for example, scoliosis is four times commoner in females, and in paralytic scoliosis there is little correlation between the severity of the deformity and the severity of the muscular weakness. In 80 per cent of cases the curvature is convex to the right.

Scoliosis usually involves a rotatory deformity as well, the spine becoming twisted on its own axis, and the combination of lateral curvature and rotation may have disastrous effects on the function of the chest. As the deformity increases the vital capacity falls, and breathing may become difficult. The heart is not usually displaced, and cardiac function remains more or less normal.

In normal subjects a vertical line through the centre of gravity usually passes through the dens, the front of the body of the second thoracic vertebra, the middle of the twelfth thoracic body, and the back of the fifth lumbar body (Fig. 109). The sacrovertebral angle is a point of danger in the vertebral column, for here the weight of the upper part of the body tends to drive the fifth lumbar vertebra downwards and forwards off the first sacral vertebra (p. 279).

Joints and movements

The movements of the column are flexion and extension, rotation, and lateral bending. The amount of movement between adjacent vertebrae is small, but the sum total is quite impressive. Movement occurs both at the plane joints between the articular processes and at the symphyseal joints between the vertebrae and the discs, and the most important factors influencing the movements of different parts of the column are thus the relative thicknesses of the discs and the set of the articular processes.

When considering flexion and extension of the column, it is useful to think of the vertebra as rocking on the nucleus pulposus of the invertebral disc (Fig. 110). The spines protruding backwards from the vertebrae afford a considerable mechanical advantage to the muscles attached to them; the length of this lever compared to the distance to the front of the vertebra is about 3.25 to 1. This lever action is responsible for the fact that in full flexion of the vertebral column a tape measure held over the spines shows an increase of 10 cm in the distance between the spines of the seventh cervical and the first sacral vertebra.

Movements of all kinds are most free in the neck. The skull flexes and extends on the atlas, but cannot rotate upon it because of the curvature of the articular facets; the **atlanto-occipital** joint is thus a pure hinge joint. The **atlanto-axial** joint between the atlas and the axis is a pure pivot joint in which the atlas, bearing the skull, rotates round the pivot of the dens. At the joints between the other cervical vertebrae small amounts of flexion, extension, lateral bending, and rotation can take place. The result of all the cervical rotatory movements is that the head, carrying with it the eyes, has a range of movement of some 90° on the trunk. This is sufficient to enable the full horizon to be observed without moving the rest of the body. Lateral bending in the neck is always accompanied by some rotation, owing to the set of the joints.

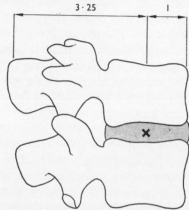

Fig. 110. Leverage on lumbar spine. X, the position of the nucleus pulposus, can be considered the fulcrum.

In the thoracic region rotation is much hampered by the presence of ribs and sternum. Flexion and extension are limited by the articular processes, but lateral bending is fairly free and important; it causes the ribs on the bending side to be bunched up together, and those on the the the other to be splayed apart.

In the lumbar region the discs are thick, and there is no thoracic cage to impede movement, but rotation is limited to some 30° by the set of the articular processes. Flexion, extension and lateral bending are free. This region of the spine can be trained to perform considerable feats of extension, as when the head is thrust forwards between the legs in order to pick up with the teeth a handkerchief placed on the floor. (This trick also demands a greatly increased range of movement at the hip joints.)

Muscles

The movements of the spine are performed by a series of longitudinally running muscles supplied by the dorsal or ventral rami of the spinal nerves. The most important of these muscles are those which help to maintain the secondary (and only precariously stable) curvatures. Both these are concave backwards, and the main muscle mass is therefore posterior to the vertebrae. The powerful **erector spinae** (Fig. 111) is made of long superficial and short deep fasciculi which take origin from the dorsum of the sacrum and from the dense fascia which covers the muscle itself. From this mass three thin columns of muscle run up the back. The most lateral is attached to the ribs, the intermediate one—the largest—to the ribs and transverse processes, and the medial one to the spines. It is a waste of time to commit the exact attachments of this complicated muscle to memory, yet it is one of the most important postural muscles in the body.

The other important muscles on the back are the **semispinalis capitis** and the **semispinalis cervicis**. They do for the cervical curve what the erector spinae does for the lumbar curve, and the semispinalis capitis, which is inserted into the occipital bone, is the chief factor in balancing the head on the neck. The centre of gravity of the skull lies in front of the atlanto-occipital joint because of the attachment of the heavy jaw to the front of the skull, and the semispinalis capitis counteracts this by pulling the back of the skull down, the atlanto-occipital joint acting as the fulcrum (Fig. 24). If the muscle relaxes its hold, as for example during an anatomy lecture, the head falls forwards on the chest.

Deep to the erector spinae and the semispinalis muscles, in the angle between the transverse processes and the spines of the vertebrae, lies a series of longitudinal muscles which assist their larger neighbours, and can also help in rotation of the spine. In the cervical region the important movement of rotation is largely controlled by the **sternocleidomastoid** muscle (p. 263).

Lateral bending of the vertebral column is effected by muscles lying on each side of the column; the most important are the **sternocleidomastoid** (p. 263), the **psoas major** (p. 282) and the **quadratus lumborum**. The last is a strong muscle which passes from the poerior part of the crest of the ilium (p. 277) to the transverse processes of the lumbar vertebrae and to the twelfth rib; by contracting it pulls the vertebral column over to its own side and depresses the twelfth rib. If the weight of the body is supported on the other leg it helps to take the weight of the free leg by pulling the pelvis up to the ribs.

Flexion of the vertebral column is mainly performed by muscles which lie at some distance in front of it. The chief flexor of the lumbar spine is the **rectus abdominis** (p. 279) in the anterior abdominal wall, and the chief flexors of the neck are the sternocleidomastoids (p. 263). But in the common situation, where gravity is the prime mover, the main control of flexion is vested in the erector spinae, which gradually 'pays out', so permitting the spine to bend. In the position of full flexion the erector spinae relaxes, and the tension is wholly taken by the ligaments of the vertebral column.

Extending the back from the position of full flexion is a difficult and potentially dangerous movement. In the first phase of the movement the lumbar spine is compressed down on the sacrum by strong contraction of the abdominal and back muscles and the straightening takes place at the hip joints, the pelvis being rotated backwards on the heads of the femora (p. 357).

After the half-way point has been passed, the compression of the lumbar spine is released, and the erector spinae now pulls the lumbar spine into its normal curvature, so straightening the back as a whole. There are thus two distinct phases in regaining the erect posture, and a fault in the mechanism is

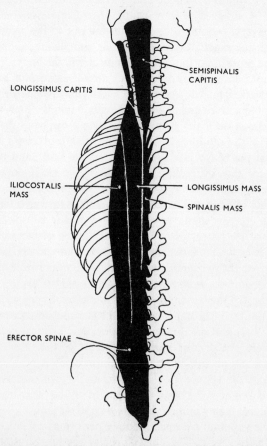

SEMISPINALIS CAPITIS

LONGISSIMUS CAPITIS

ILIOCOSTALIS MASS

LONGISSIMUS MASS

SPINALIS MASS

ERECTOR SPINAE

Fig. 111. Diagram showing the positions of the erector spinae and the semispinalis capitis muscles; all other back muscles have been suppressed for the sake of clarity. The three columns of the erector spinae—the iliocostalis, the longissimus, and the spinalis muscles—are artificially demarcated. Note that only one small component (the longissimus capitis) of the erector spinae reaches the skull.

often made manifest at the change-over stage, when the patient may adopt a trick movement, usually a rotation, which gets him past the difficult part of the movement; if this fails, he may use his hands to 'climb up' his thighs. The maximum power of the erector spinae is required at the change-over stage, and after this further extension becomes progressively easier.

The act of picking up a heavy weight can be dangerous if incorrectly performed. It is best to hold the back as straight as possible and to do the lifting by flexing and straightening the knees; in this way the strain is transferred from the arms and neck (p. 306) through the chain of vertebral bodies and discs, and the spinal muscles are not called upon. But if the lift is made from the stooping position the erector spinae is brought into full play to oppose the flexing action of the load, and the abdominal muscles contract, so raising the intra-abdominal pressure. Lifting a heavy weight may thus cause not only spinal injuries such as prolapse of an intervertebral disc (p. 247), but also hernia or prolapse of the viscera (p. 283).

Surface anatomy

In the neck the spines of the second to the sixth cervical vertebrae lie deeply, and cannot be distinctly felt. On either side of a slight median depression are the longitudinal muscular swellings of the semispinales capitis, covered superficially by the thin trapezius muscles. In the midline between these lies the ligamentum nuchae, which stands out as a median ridge when the head is bent forwards. This ridge leads from the external occipital protuberance on the back of the skull (p. 256) to the spine of the seventh cervical vertebra, which is easily felt, and is therefore called the **vertebra prominens**; it is the landmark from which all other vertebral spines are identified. From this level downwards all the vertebral spines can be palpated, and are often visible as surface protuberances.

In the anatomical position, the base of the spine of the scapula lies opposite the third thoracic spine, and the inferior angle of the scapula opposite the seventh thoracic spine. A line drawn through the highest points of the iliac crests, felt from behind, cuts the spine of the fourth lumbar vertebra, and this is used as a guide to the site of a proposed lumbar puncture (p. 101). The first three sacral spines are usually readily felt under the skin, but the others are often poor affairs and may be absent. The coccyx is found in the depression between the buttocks some distance behind the anus. On either side of the lumbar spines is the mass of the erector spinae, covering over the deeper layer of spinal muscles, and itself clothed by the superficial layer of the thoracolumbar fascia (Fig. 126). The lumbar spines thus lie at the bottom of a median groove, which becomes a ridge when the trunk is flexed. In the sacral region the erector spinae is covered over by the gluteus maximus (p. 357) which forms the fleshy bulge of the buttock.

22 · The head and neck

The skull

The skull is in two parts, for the lower jaw, or **mandible**, is separate from the rest and articulates with it at the **temporomandibular joint**. The larger portion of the skull contains the brain and the organs of special sense, and incorporates the upper jaw. The **base** supports the brain, and the **cranial vault** roofs over it, like the lid of an Easter egg; in front of these lie the bones of the **face**.

The vault of the cranium (Fig. 112) is made up of several bones. The **coronal suture** runs across the skull from side to side and separates the single **frontal bone** from the paired **parietal bones** behind. In the infant there are two frontal bones with a suture between, and where the four bones meet there is a diamond-shaped unossified patch called the **anterior fontanelle** (Fig. 112). Here the vault is formed of membranous tissue only, and when the baby cries the rise in intracranial blood pressure renders the soft area tense. By the age of 2 years the fontanelle has usually become ossified, and later the suture between the two frontal bones is absorbed, though it may occasionally persist into adult life. The frontal bones are named from their association with the region of the forehead, and the bulges at their points of maximum convexity are called the **frontal eminences**. The **parietal eminences** are similar bulges on the parietal bones (Fig. 112).

The two parietal bones are separated by the **sagittal suture**, which runs backwards in the midline until it bifurcates into two diverging sutural lines which together called called the **lambdoidal suture**. Between the limbs of this suture, and articulating with both parietal bones, is the unpaired **occipital bone**, which extends also on to the base of the skull. The most prominent portion of the occipital bone behind is called the **external occipital protuberance**, and passing laterally from it on each side is a ridge called the **superior nuchal line**.

Seen from in front, the main features of the skull are the two **orbits**, the bony aperture of the nose, and the upper and lower jaws (Fig. 113). The orbits are pyramidal cavities whose medial walls are approximately parallel, but whose lateral walls slope inwards (Fig. 57). They are roofed by a process of the frontal bone on each side, but their floors and medial and lateral walls are formed by contributions from several different bones. At the back of the orbit the

256

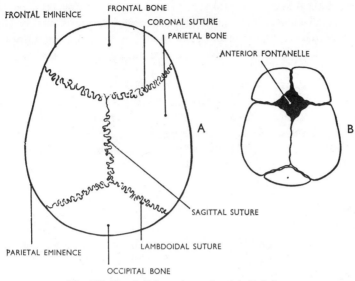

Fig. 112. The skull from above, A: adult. B: infant.

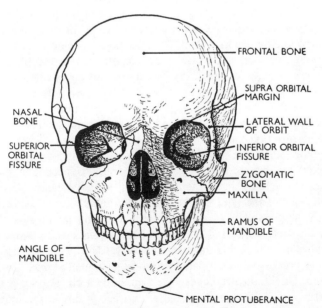

Fig. 113. The skull from in front.

superior orbital fissure transmits vessels and nerves running between the orbit and the cranial cavity; the smaller **inferior orbital fissure** is largely closed up by fascia. Medial to the superior orbital fissure is the small round **optic canal** for the optic nerve (p. 120). The supraorbital margin is sharp and can easily be felt in the living body; the inferior margin is more rounded, but the lateral margin is again sharp, and is used as a landmark.

The aperture of the nose lies between the two bones of the upper jaw— the **maxillae**—and is roofed in by the small **nasal bones**; the front of the bony part of the **nasal septum** can be seen sunken below the surface in the midline. To this irregular bony margin are attached the various nasal and septal cartilages which stiffen the pliable part of the nose; the junction between bone and cartilage is readily felt in the living subject.

The two maxillae carry the upper teeth, and each is hollowed out by a cavity called the **maxillary sinus**, which is the largest of the **accessory air sinuses**. These sinuses lighten the facial bones, and so take some of the strain off the posterior neck muscles (p. 253), but they are an insufferable nuisance, for they constantly become infected from the nose, and their drainage outlets are sited in a way which might be splendid in a four-footed animal, but which in the upright posture is well suited to spread infection from one sinus to another.

Lateral to the maxillae the two **zygomatic bones** form the prominences of the cheeks and also most of the lateral walls and floors of the orbits. Lastly there is the mandible, carrying the lower teeth and shaped like a horseshoe. The portion which bears the teeth is called the **body** of the mandible; the two halves of the body unite in the **mental protuberance** which forms the point of the chin.

There are 32 teeth in the jaws of an adult; each maxilla has two **incisors**, one **canine**, two **premolars** and three **molars**, working from the midline laterally, and the teeth in the mandible correspond. The last molars (wisdom teeth) may never appear, or may make an embarrassing and painful appearance in old age, but usually the dentition is complete by about the age of twenty. The adult or **permanent teeth** are preceded by a set of **deciduous** or **milk** teeth, which are shed seriatim as the haw becomes large enough to accommodate the permanent ones. In the course of evolution the mandible has become smaller, and it is now comparatively rare that the lower teeth exactly meet their opposite numbers in the upper jaw when the jaws are closed. If the disparity is too great (malocclusion) the bite is seriously affected, and this accounts for the wire entanglements to be found in the mouths of many school children.

From the side of the skull can be seen the cranial bones already examined, and two more. A portion of the **sphenoid bone** forms the lateral side of the vault just behind the orbit (Fig. 114), and behind this again is the **squamous part** of the **temporal bone**. From this there thrusts forwards a process which, together with a similar process stretching back from the zygomatic bone, forms a flying buttress on the side of the head called the **zygomatic arch**. The space above and medial to this arch is the **temporal fossa**, and below this is the **infratemporal fossa**. The posterior end of the zygomatic arch leads to the bony **external**

auditory meatus (p. 125), which also lies in the temporal bone. When the skull is in the anatomical position the upper margin of this meatus is in the same horizontal plane as the lower margin of the orbit. Medial to the meatus a thin bony spike of variable length called the **styloid process** juts down and forwards from the temporal bone, and behind the meatus is the solid chunky **mastoid process** (Fig. 114). In front of the meatus lies the socket for the **head** of the mandible, a spindle-shaped affair separated by a narrow **neck** from a wide

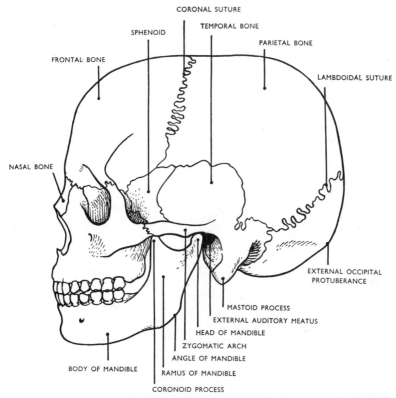

Fig. 114. The skull from the side.

quadrilateral of bone called the **ramus** of the mandible. The ramus carries on its anterosuperior angle a wide sharp protrusion called the **coronoid process**, which lies between the zygomatic arch and the sphenoid. The point where the body joins the ramus is called the **angle** of the mandible.

On the case of the skull the main feature is the great **foramen magnum** (Fig. 115) through which the spinal cord passes to become the medulla (p. 102). The foramen magnum is wholly enclosed by the occipital bone, and on each side of it lie the **occipital condyles** which articulate with the atlas. The line of the atlanto-occipital joint lies just in front of the mastoid process. Blocking in the horseshoe of the upper teeth is the **hard palate**, formed by processes from the maxillae and the **palatine bones**, and above the posterior edge of the hard palate may be seen the posterior apertures of the nose, superior to which is the floor of the cranial vault. Hanging downwards from each end of the maxillary horseshoe are a pair of processes which belong to the sphenoid bone and are called the **medial** and **lateral pterygoid plates**; they form the medial boundaries of the infratemporal fossa, which is bounded laterally by the zygomatic arch.

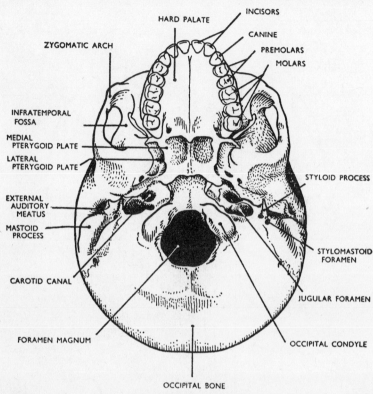

Fig. 115. The skull from below.

There are numerous foramina all over the skull; some of these will be mentioned later in connexion with the cranial nerves.

The hyoid bone

The small **hyoid bone** (Fig. 77) is suspended by muscles and ligaments in the front of the neck at the level of the axis vertebra. It is shaped like a 'U' with the open end pointing backwards, and the larynx is attached to it by both ligaments and muscles.

The temporomandibular joint

This is a synovial joint between the socket in the temporal bone and the head of the mandible, which is spindle-shaped; a complete fibrous articular disc intervenes between them (p. 41). Hinge movements take place between the mandible and the disc, but whenever the jaw opens more than a few degrees the head of the mandible and the disc together leave the concave posterior part of the socket and glide forwards on to a convex portion of bone just in front of it. When the jaw is closed, both retreat backwards again. In protrusion (protraction) of the jaw the mandible moves forwards in the same way; indeed protraction is often accompanied by a degree of opening. Retraction is the converse of protrusion. Side to side chewing movements are possible with the jaw either in its protracted or its retracted position; the mechanism is that of a condylar joint (p. 43), the two condyles being separated by the width of the skull. The joint has a strong capsule which is reinforced laterally by the **temporomandibular ligament**. Dislocation of the jaw can occur if it is too widely opened, as in yawning; the head of the mandible slips forwards off its rather precarious perch on the temporal bone and enters the infratemporal fossa. Here the tension of the strong muscles which close the jaw will not allow it to escape downwards again, and the jaw is held wide open until it can be put back by manipulation.

Important muscles

The jaw is closed by three main muscles, the **temporalis**, the **masseter** and the **medial pterygoid**. The **temporalis** arises from the medial wall of the temporal fossa, and its fibres converge like a fan on its handle to be inserted into the coronoid process of the mandible. The anterior fibres are almost vertical, and close the jaw strongly; the posterior ones become more and more horizontal, and retract the jaw.

The **masseter** (Fig. 116) arises from the zygomatic arch and is inserted into the outer aspect of the ramus of the mandible.

The **medial pterygoid** comes from the medial side of the lateral pterygoid plate and is inserted into the medial surface of the ramus of the mandible. The direction of its fibres is such that it closes the jaw and pulls it to the opposite side, producing a chewing movement.

The other main muscle acting on the joint is the **lateral pterygoid**, which arises from the lateral side of the lateral pterygoid plate and runs backwards to the head of the mandible and the articular disc. It protracts the jaw, and pulls it towards the other side, so assisting the medial pterygoid in the movement of chewing.

All these muscles are classified as **muscles of mastication**, and they are all supplied by the mandibular division of the trigeminal nerve (p. 264). Closely associated with them in the process of mastication are the muscles which move the tongue, all of which are supplied by the hypoglossal nerve (p. 264).

Two sets of muscles act on the tongue; the **intrinsic** ones form the substance of the tongue and alter its shape, and the **extrinsic** ones anchor the tongue to surrounding bones and move it bodily about. In this way the tongue is rendered extremely mobile, and can be used to compress food into a bolus for swallowing, to explore the mouth, or to perform the extremely rapid and fine movements demanded by speech. The tongue rests upon the **mylohyoid** muscle, which forms a sling or diaphragm stretching between the mandible and the hyoid bone. When the muscle contracts, the floor of the mouth is pulled tight as a sheet is stretched for someone to jump into during rescue operations. The tongue is thus pressed against the roof of the mouth, and at the same time the hyoid, and with it the larynx, are drawn upwards. This movement is an essential part of the process of swallowing (p. 172).

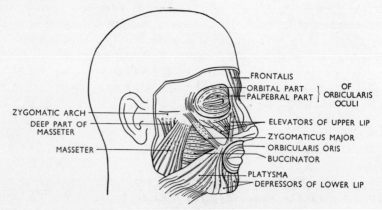

Fig. 116. The muscles of facial expression and the masseter.

The **muscles of facial expression** form a complex pattern on the face (Fig. 116), the details of which it is unnecessary to remember. The pattern centres round two large sphincter muscles surrounding the eye and the mouth. The **orbicularis oculi** is attached to bone medially, but laterally its fibres are attached merely to skin. Some of them run round the orbit, and screw up the eyes against a bright glare, while others enter the eyelids, and close the eyes tightly. Both movements are accompanied by a medially directed pull on the skin at the lateral side of the orbit. This assists the flow of tears from the lacrimal gland across the eye to the lacrimal sac (p. 125), but it furrows the skin at the lateral side of the eye and thus accounts for the 'crows feet' of sailors and the popularity of sunglasses among young women.

Many muscles are inserted into the circular sphincter fibres of the **orbicularis oris**, and they give rise to a variety of facial expressions. Those which lift the corner of the mouth are used in laughing, those which elevate the upper lip in puzzlement or sneering, and so on. Most of them have some origin from bone, and if they all contract, the mouth is widened. The orbicularis oris itself purses up the mouth as in whistling, and the general system is similar to the constriction and dilatation of the pupil of the eye (p. 118).

All the muscles of facial expression are supplied by the facial nerve (p. 264), which also supplies the **buccinator** muscles forming the bulk of the cheeks. These are normally used to compress the space between the gums and the cheeks during eating, and so to force the food which accumulates there back into circulation.

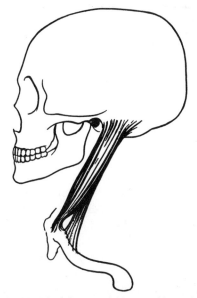

Fig. 117. The left sternocleidomastoid muscle.

In the neck lie many muscles which operate on the cervical spine, the atlanto-occipital and the atlanto-axial joints (p. 252). Posteriorly the **semispinalis capitis** is partly concealed by the upper fibres of the **trapezius** (pp. 253, 307). Anteriorly the **infrahyoid** and **suprahyoid** muscles join the hyoid bone and the larynx to the sternum and to the jaw. These muscles pull up or down the larynx in addition to bending the neck forwards.

But the main muscular feature of the neck is the **sternocleidomastoid**, which arises by two heads from the front of the manubrium of the sternum, the sternoclavicular joint, and the medial third of the clavicle, and passes upwards and backwards on the side of the neck to be inserted into the mastoid process (Fig. 117). It is supplied by the spinal part of the accessory nerve, and it flexes the neck and turns the point of the chin over towards the opposite shoulder—a pose much favoured by society photographers. The sternocleidomastoid is pathologically contracted in the condition of wry neck, and the fibres of the muscle may have to be divided in order to restore symmetry.

Important nerves

Some of the cranial nerves (p. 84) need further mention. The large **trigeminal** nerve has three sensory divisions which supply the skin of the face and most of the scalp. The **ophthalmic** division runs to the eye, orbit, forehead and scalp, the **maxillary** division supplies the skin over the upper jaw, and the **mandibular** division that over the lower jaw. Each of these divisions supplies deeper structures also. Thus the ophthalmic division sends sensory fibres to the eyeball; the maxillary supplies the mucous membrane of the nose, the air sinuses, the upper teeth and the hard palate; the mandibular gives branches to the lower teeth, the tongue, and the mucous membrane of the mouth and cheeks. The sensory root of the nerve springs from the large **semilunar ganglion** lying just inside the skull and runs backwards to enter the pons (p. 102). All three divisions of the nerve converge on this ganglion, and the much smaller motor root of the nerve, which lies under it, is entirely distributed with the mandibular division to the muscles of mastication.

The **facial** nerve is the motor nerve to the muscles of expression. Inside the skull it travels close to the inner ear to reach the medial wall of the middle ear, and eventually leaves the skull by passing down a long tight canal to the **stylomastoid** foramen lying between the styloid process and the mastoid process. Here the nerve enters the substance of the parotid gland (p. 171) and divides into several branches which radiate to supply the facial muscles, as well as a superficial muscle of the neck called the **platysma** (Fig. 116). The facial nerve is subject to a number of hazards in its course and can be damaged by infection in the middle ear, disease of the parotid, or wounds of the face. But the commonest cause of facial paralysis (known as **Bell's palsy**) is compression of the nerve in the tight canal in the skull. This is usually a temporary affair lasting a few months at most. In such patients the side of the face affected is smooth and expressionless, and the healthy muscles pull the mouth over to the opposite side. The patient cannot whistle or show his teeth, nor can he shut his eye tightly. The loss of tone in the lower eyelid may lead to it sagging away from the eyeball, so that tears spill over on to the cheek.

In the neck, the **vagus** nerve, which leaves the skull through the jugular foramen (Fig. 115) sends motor and sensory branches to supply the pharynx, the larynx, the oesophagus and the trachea, before it continues on its way to the thorax and abdomen. The **hypoglossal** nerve runs forwards under the angle of the mandible to supply all the intrinsic and extrinsic muscles of the tongue. If either of these nerves is cut speech and swallowing are grossly impaired.

The **spinal part of the accessory** nerve passes downwards from the jugular foramen to enter the sternocleidomastoid, which it supplies. The nerve leaves the muscle half-way down its posterior border, and runs laterally and posteriorly across the deeper muscles to enter and supply the **trapezius** (p. 307), which, among other duties, supports much of the weight of the upper limb. The nerve is vulnerable as it lies superficially, and if it is injured the affected shoulder falls and the contour of the side of the neck is altered.

The ventral rami of the upper four cervical nerves take part in the **cervical plexus**, a simple loop plexus with several named branches. Among these are the **great auricular** nerve, which helps to supply the skin of the external ear and the side of the face, and the **supraclavicular** nerves, which run to the skin over the clavicle and shoulder. Other cutaneous branches of the plexus are the **lesser occipital** and the **transverse cutaneous nerve of the neck**, both names being more or less self-explanatory. The motor branches of the plexus supply the deep muscles of the neck and many of the hyoid muscles.

The upward prolongation of the **sympathetic trunk** (p. 87) lies behind the carotid sheath (p. 265), and has three ganglia, of which the first is much the largest. This **superior cervical ganglion** gives grey rami communicantes to the first four cervical nerves, and also provides the sympathetic supply to the eye. This runs into the skull along the internal carotid artery and is distributed along its ophthalmic branch. The **middle cervical ganglion** corresponds to and feeds the fifth and sixth cervical nerves, while the **inferior cervical ganglion**, which is often fused with the first thoracic ganglion, supplies the grey rami communicantes to the last two cervical nerves. Many of these fibres run to the upper limb, and the combination of inferior cervical and first thoracic ganglia represents a bottleneck through which the whole of the sympathetic supply to the upper limb must pass. Both ganglia are thus attacked in the operation of cervical sympathectomy (p. 90). All three cervical ganglia give branches which run down into the thorax to supply the heart.

Important vessels

The main arteries of the head and neck are the **common carotids**, each of which runs up vertically along a line drawn from the sternoclavicular joint to the angle of the mandible. At this point the artery divides into an **internal** and an **external carotid** artery; at the point of division all three arteries are slightly dilated to form the **carotid sinus** (p. 146), and close to this lies the **carotid body** (p. 169).

The **internal carotid** runs into the skull through the carotid canal (Fig. 112) and ends by taking part in the **circle of Willis** (p. 266) and giving off the **anterior** and **middle cerebral** arteries. It also gives rise to the **ophthalmic** artery which runs through the optic canal with the optic nerve to supply the eye and orbit.

The **external carotid** has a large number of branches which supply more superficial structures. Some of these are used by the anaesthetist when he wants to feel the pulse: for example the **facial** artery, which crosses the mandible just in front of the masseter on its way to the face, the **superficial temporal**, immediately in front of the ear as it runs upwards to the scalp, and the **occipital**, between the mastoid process and the external occipital protuberance.

The common and internal carotid arteries are enveloped in a strong binding of deep fascia which forms the **carotid sheath**, and this encloses also the vagus nerve and the **internal jugular vein**. This is the largest vein of the head and neck, and begins at the jugular foramen (Fig. 115) as a continuation of one of the venous sinuses of the skull. It runs down along the side of the internal and common carotid arteries to end by joining the **subclavian** vein behind the sternoclavicular joint, so forming the **brachiocephalic vein** (p. 273). At this point on the left side of the neck the **thoracic duct** enters the venous system (p. 147).

The main artery of the upper limb, the **subclavian**, puts in a brief appearance in the neck on its way to the arm, which it enters opposite the midpoint of the clavicle. It gives off branches to structures in the neck, and also the **vertebral** artery which runs up the foramina in the transverse processes of the cervical vertebrae to enter the skull at the foramen magnum. The vertebral ends

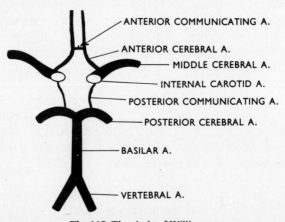

Fig. 118. The circle of Willis.

by joining its fellow to form the **basilar** artery inside the skull ; this runs forward on the under surface of the pons and divides into two **posterior cerebral** arteries which diverge outwards on the under surfaces of the cerebral hemispheres. Communications between the posterior and middle cerebral arteries complete the circle of Willis posteriorly (Fig. 118).

Surface anatomy

The whole of the cranial vault and the facial skeleton lie very close to the skin, and all the bony landmarks in these regions are easy to make out. Even the sutures can often be felt, and in a bald person sometimes actually seen, for the skin tends to become tacked down to them.

On the side of the skull the zygomatic arch is easily felt in its whole length. The space above it—the temporal fossa—is filled in by the temporalis, and the masseter blocks in the space below it. When the jaw is clenched the hard line of the anterior border of the masseter is readily felt. Running horizontally across it is the duct of the parotid gland, which can be rolled against the muscle. The temporalis can be felt contracting about a hand's breadth above the arch.

The body, angle and posterior border of the ramus of the mandible can be made out clearly, and the head of the bone can be felt moving forwards, as the jaw is opened, from a position immediately in front of the external auditory meatus. The coronoid process of the mandible and the tendon of the temporalis can be felt from inside the mouth.

Just behind the ear is the easily palpated mastoid process, and running medially and backwards from this to the external occipital protuberance is the superior nuchal line. The transverse process of the atlas can be felt as a deep resistance between the mastoid process and the ramus of the mandible.

In the front of the neck several structures can be made out in the midline. Under the chin the resistance of the mylohyoid muscles forms the floor of the mouth, and at the level of the second cervical vertebra lies the hyoid bone (Fig. 77). The upper border of the thyroid cartilage has a distinctive notch, and inferior to the thyroid lies the cricoid cartilage at the level of the sixth vertebral body. Inferior to this again is the softer resistance of the isthmus of the thyroid gland (p. 199) as it crosses the upper rings of the trachea. Finally there is the upper border of the manubrium of the sternum (p. 268), which is known as the **suprasternal notch.** On each side of this notch can be felt the tendinous sternal attachment of the sternocleidomastoid muscle; more laterally is its fleshy attachment to the clavicle. The muscle may be made to stand out by pushing the head forwards against resistance applied to the forehead. Halfway down its posterior border the spinal accessory, great auricular, lesser occipital and transverse cutaneous nerves leave the shelter of the muscle to run to their destinations. The upper trunk of the brachial plexus is sometimes palpable as it runs from under the muscle at the level of the cricoid cartilage to leave the neck at the midpoint of the clavicle. The lateral border of the trapezius can be felt running down in a curve from the superior nuchal line to the lateral third of the clavicle.

The surface markings of the main arteries have already been given.

23 · The thorax

The thoracic cage

The thorax contains the lungs, the heart, and the great vessels; it is protected by an extensive bony cage. This cage is not rigid, but expansible, so that air can be taken in and out of it during breathing.

The vertical support of the cage is given by the thoracic vertebrae behind and the flat **sternum** in front. Between these stretch the semicircular hoops of the **ribs**. The sternum is much shorter than the thoracic part of the vertebral column (Fig. 119). Its main part, or **body**, articulates with two smaller portions. The **manubrium** joins the body at a slight angle known as the **sternal angle**, and the **xiphoid process** may not be fully ossified. In young people the joints between the three pieces of the sternum allow some movement, but after middle age they fuse together into one rigid bone.

There are twelve pairs of ribs, each of which has a **head** and a **tubercle** at its posterior end (the last three pairs of ribs have very rudimentary tubercles). The heads of the ribs articulate with the intervertebral discs and the bodies of the thoracic vertebrae posteriorly, and the tubercles articulate with the transverse processes of the vertebrae. These joints are freely movable synovial articulations. The upper **costotransverse** joints are ellipsoid, and allow the rib to rotate but not to slide about on the transverse process. The lower joints, on the other hand, permit slight vertical gliding movements as well as rotation.

Each rib is continuous anteriorly with a hyaline **costal cartilage**. The first of these cartilages is attached to the manubrium, the second to the region of the sternal angle, and the third to the sixth inclusive to the body of the sternum. The seventh is attached to the sixth, the eighth to the seventh, and so on until we reach the eleventh and twelfth cartilages, which are unattached, thus earning for their ribs the title of 'floating ribs'. The continuous border formed by the costal cartilages is the **costal margin**.

The thoracic cage is flattened from before backwards, and its anteroposterior diameter is further reduced by the protrusion forwards of the bodies of the thoracic vertebrae into the thoracic cavity (Fig. 121). On each side, however, there is a deep bay posteriorly. The point of maximum convexity of the ribs occurs here, and in each rib it is known as the **angle**. The **inlet** of the

thorax is an imaginary plane drawn through the first pair of ribs, the manubrium, and the superior border of the first thoracic vertebra; the **outlet** is the
irregular line formed by the costal margin and xiphoid process in front and the
lower border of the twelfth thoracic vertebra behind. The narrow opening of the
inlet is packed tight by the trachea and oesophagus and the great vessels and
nerves. The outlet is blocked up by the **diaphragm**, the great muscular partition
between thorax and abdomen (Figs. 122, 129).

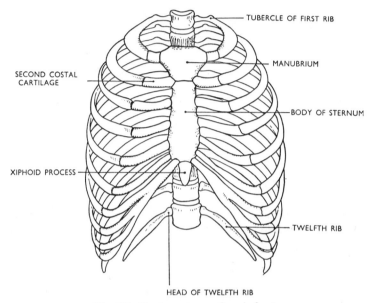

Fig. 119. The thoracic cage from in front.

The mechanics of the ribs

Each rib can rotate upwards around a line drawn through its head and its
tubercle (Fig. 120), and this movement would, if the rib were free at its anterior
end, force the costal cartilage upwards and laterally. In the upper six ribs this
movement is limited by the attachment of the relatively short cartilages to the
sternum, but the lower ribs are not so hampered, for their longer and thus more
pliable cartilages are only indirectly fixed to bone; in consequence their anterior
ends can be splayed out laterally and upwards to a greater extent. As the ribs
rise, the sternum is pushed forwards and rises slightly; the body tilts on the
manubrium, so allowing an increase in the range of movement; this of course
cannot occur after the joint at the sternal angle is obliterated. The net result of
these movements is an increase in the anteroposterior and oblique diameters of
the chest (Fig. 120).

The lower ribs also move in another way which has been compared to the movement of bucket handles. The head of the rib and the anterior end of its costal cartilage may be taken as the attachments of the handle to the bucket, and as the handle is lifted up the tubercle of the rib glides upwards on the transverse process of the vertebra concerned, and the costal cartilage becomes twisted on itself. The lower costotransverse articulations allow this to occur more readily than those in the upper part of the thorax, and the movement increases the transverse diameter of the chest.

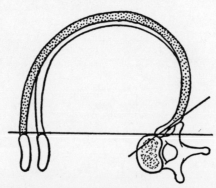

Fig. 120. Movement of a rib. The lines indicate the axes round which movements take place; the position of the rib in inspiration is shaded.

Respiratory movements

The act of breathing is very complicated, and what follows is a very much simplified account.

In the movement of **inspiration** the first rib is steadied by the **scalene** muscles, which come from the transverse processes of the cervical vertebrae and are inserted into the first and second ribs. The succeeding ribs are then pulled up to the first by the contraction of the **intercostal** muscles which run from the lower margin of the rib above to the upper margin of the rib below. The excursion of the third rib is greater than that of the second, and so on down, till the maximum movement is found in the region of the seventh and eighth ribs. The ribs lower than this move less because the twelfth rib is tethered down to the iliac crest by the quadratus lumborum (p. 253).

This movement of **thoracic** inspiration increases the anteroposterior and lateral diameters of the chest because of the way in which the ribs move. In the upper part of the apparatus anteroposterior expansion is greater than lateral expansion, but lower down, because of the bucket-handle movement, lateral expansion predominates. As the capacity of the thorax is increased the atmospheric pressure forces air into the lungs (p. 165).

Abdominal inspiration occurs simultaneously with thoracic inspiration, and cannot be divorced from it. The **diaphragm** is the chief muscle of inspiration. It is attached to the xiphoid process, to the bodies of the first three lumbar verte-brae, and between these two points to the inner surfaces of the lower six ribs on each side. The muscle fibres rise up from this origin and curve over like an inverted pudding-basin to be inserted into a central tendon which lies at the level of the sixth costal cartilage. The right side of the dome, which lies above the liver, is slightly higher than the left side, which underlies the heart. The diaphragm is supplied by the phrenic nerves (p. 272). When the muscle contracts, the dome descends, compressing the abdominal viscera and increasing the intra-abdominal pressure. This descent increases the vertical diameter of the thorax. After a time the dome is brought to a halt by the increasing resistance of the abdominal contents, and the diaphragm now acts from its insertion on its origin (p. 59) and pulls the lower ribs up towards the central tendon, thereby contributing to thoracic inspiration.

If the whole lung is to be adequately ventilated, both types of respiration are necessary, and the prevention or impairment of abdominal respiration by tight lacing in the Victorian era probably contributed to the prevalence of respiratory tuberculosis. If there is an obstruction to inspiration, as when the bronchioles are constricted in asthma, muscles not normally connected with respiration may be brought in to help. Thus the patient can grip the bed rail to fix the insertion of the pectoralis major (p. 310), which can then act from its insertion on its origin from the upper six costal cartilages. In this way it raises up the front of the chest in an effort to allow still more air to rush in; it is then said to act as an accessory muscle of inspiration. Similarly, if the skull is fixed by other muscles, the sterno-cleidomastoid (p. 263) can pull the sternum up towards the mastoid process and aid in inspiration. Another accessory muscle of inspiration is the pectoralis minor (p. 308).

Expiration is not such an active process as inspiration. Thoracic expiration results from the relaxation of the intercostals, so that the ribs fall back to their original position and air is forced out of the chest. At the same time the diaphragm relaxes, and the rebound of the compressed abdominal viscera forces it upwards again, so decreasing the vertical dimensions of the thoracic cavity. If there is an obstruction to expiration the muscles of the abdominal wall contract and by compressing the viscera drive the diaphragm upwards, at the same time pulling down the lower ribs, to which they are attached (p. 279). The chest can also be compressed by the contraction of the latissimus dorsi (p. 310), which squeezes the lower ribs during sharp expiratory efforts, such as coughing.

In old people calcification of the costal cartilages and the fusion of the pieces of the sternum make pulmonary ventilation less efficient than in the young; at the same time the tone of the abdominal muscles decreases and the abdominal viscera sag, so that expiration becomes more difficult. Old people therefore tend to develop infections in the stagnant backwaters of their lungs.

Important nerves

The **intercostal nerves** run forwards in each intercostal space and supply the skin of the chest and the intercostal muscles. At the anterior end of the intercostal spaces the lower members of the series escape from the chest wall behind the costal margin, and end by supplying the muscles and skin of the anterior abdominal wall (p. 285).

The **phrenic nerves** (Fig. 121), which supply the diaphragm, are derived on each side from branches of the ventral rami of the third, fourth and fifth cervical nerves. Each phrenic nerve runs downwards along the great vessels and the fibrous pericardium. The nerve contains sensory as well as motor fibres, and if the pleura over the diaphragm is inflamed the pain may well be felt in the shoulder, which is the region supplied by the cutaneous branches of the segments from which the phrenic nerve is derived (p. 95).

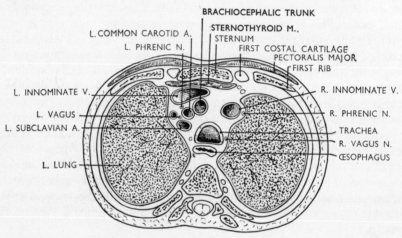

Fig. 121. Transverse section through the chest at the level of the third thoracic vertebra.

The **vagus** nerves (Fig. 121) also enter the chest along the great vessels, and during their passage through the chest on their way to the abdomen they give branches to the heart, lungs, oesophagus and trachea. The two **sympathetic trunks** (Fig. 122) lie on the necks of the ribs at each side of the vertebrae, and supply the various thoracic organs, as well as giving off **splanchnic nerves** which pass down into the abdomen (p. 285). The autonomic fibres in the chest form **cardiac** and **pulmonary plexuses** from which the heart and lungs are innervated.

Important vessels

The **superior vena cava** and the **inferior vena cava,** the largest veins in the body, feed the heart with venous blood. The superior vena cava is formed behind the manubrium of the sternum by the junction of the two brachiocephalic veins (p. 266, and the inferior vena cava pierces the central tendon of the diaphragm; it has only a very short course in the thorax. Both veins empty into the right atrium. The **pulmonary arteries** take the venous blood from the right ventricle to the lungs; there is a large common pulmonary trunk, which then divides into right and left branches which lie in the same straight line, making a sort of 'T-junction'. The blood is returned from the lungs by the four short pulmonary veins which immediately enter the left atrium. Finally, the blood which has passed from the left atrium to the left ventricle (p. 143) leaves it again by the **ascending aorta,** which rises upwards in front of the pulmonary trunk to about the level of the middle of the manubrium. Behind the manubrium the aorta curves backwards and slightly to the left, forming the **aortic arch,** and from the convexity of the arch spring the **brachiocephalic trunk,** the **left common carotid,** and the **left subclavian** arteries (Figs. 121, 128). The brachiocephalic trunk splits into **right common carotid** and **right subclavian** arteries behind the right sternoclavicular joint. The arch of the aorta has within its concavity the left bronchus and the left pulmonary artery. At the side of the fourth thoracic vertebra the **descending aorta** begins its run down the posterior wall of the thorax. This part of the aorta gives off the bronchial arteries (p. 168) and the **posterior intercostal arteries,** which pass forward in the intercostal spaces till they meet and anastomose with the **anterior intercostals** given off by the **internal thoracic** artery. This branch of the subclavian runs downwards by the margin of the sternum, deep to the costal cartilages on each side. The lower anterior intercostals are given off by a terminal branch of the internal thoracic called the **musculophrenic** artery. This runs laterally behind the costal margin. In the adult female, the intercostal arteries of the second, third and fourth intercostal spaces are large, for they form the chief blood supply of the breast.

Important viscera

Between the lungs is a wide median septum called the **mediastinum,** in which lie the heart and the great vessels, the trachea and the oesophagus, the phrenic and vagus nerves. The **trachea** travels down the midline as far as the level of the fourth thoracic vertebra, where it divides into the **right** and **left bronchi.** The right bronchus diverges gradually to the right to enter the right lung obliquely, while the left bronchus runs more horizontally across to enter the left lung inferior to the arch of the aorta.

From above downwards the **oesophagus** (p. 172) is posterior to the trachea, the left bronchus and the heart. It inclines gradually to the left, and leaves the chest by piercing the muscular part of the left dome of the diaphragm. The oesophagus is at first medial to the arch of the aorta, then anterior to the descending aorta, and finally lateral to it (Fig. 123).

The **heart** occupies the part of the mediastinum in front of the oesophagus

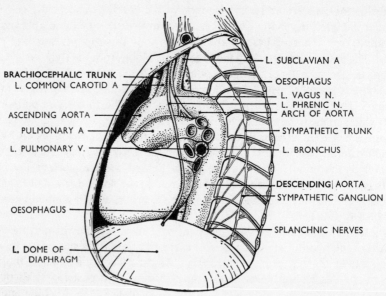

Fig. 122. The left side of the heart and aorta; the left lung and pleura have been removed.

and the trachea, and it rests inferiorly on the central tendon of the diaphragm, to which the pericardium is attached. The heart is rotated to the left (Fig. 67), so that the front of it is formed by the right ventricle and the back by the left atrium; the right atrium forms its right margin, and the left side is formed by the left ventricle. To the right and left the pericardium is in direct relationship with the pleura and the lungs, which overlap the heart anteriorly on both sides.

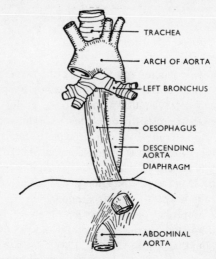

Fig. 123. The relationship of the trachea, oesophagus and aorta; seen from in front.

The **lungs** lie on each side of the mediastinum in their pleural sacs (p. 165). Each lung has an **apex** and a **base.** The apex rises up above the first rib and the clavicle into the neck; expansion of the lung is least at this point, and it is a favourite site for invasion by tuberculosis. The base lies in relation to the diaphragm.

Surface anatomy

The whole of the anterior surface of the sternum can be felt just under the skin; the lateral portions of it are only slightly obscured by the origins of the pectoralis major. The sternal angle can be located by inspection or palpation; it is an important landmark since it enables the second rib to be accurately identified (Fig. 119). Other ribs are always located by counting down from the second; the twelfth rib may be too small to protrude beyond the erector spinae, and for this reason it is not safe to identify ribs by counting from below up. The costal margin is easy to feel anteriorly but not so definite laterally.

The scapula and its attached muscles obscure the posterior aspect of the chest, and the pectoral muscles cover much of its anterior aspect. The female breast extends from the second to the sixth ribs, and from the lateral border of the sternum to the mid-axillary line (p. 242); the nipple usually lies at the level of the fourth interspace.

The lower border of the pleural sac can be marked out by a line passing from the xiphisternal joint to the twelfth rib at the lateral border of the erector spinae, cutting the tenth rib in the mid-axillary line. The lower border of the lung can be represented by a line from the xiphisternal joint to the tenth thoracic spine, passing through the eighth rib in the mid-axillary line. In full inspiration the lung may move downwards as much as 6–7cm, and its marking will then almost correspond with that of the pleura.

The apex of the lung is opposite the transverse process of the seventh cervical vertebra, about three fingers' breadths above the medial third of the clavicle. The oblique fissure of both lungs begins near the apex and runs downwards and forwards towards the xiphisternal joint; its position is roughly indicated by the medial border of the scapula when the patient places his hand on the back of his head. The horizontal fissure of the right lung passes laterally to meet this line at the level of the fourth right costal cartilage.

The transverse diameter of the heart is approximately half the transverse diameter of the thorax at the same level, and its **apex beat** (produced by the thudding of the left ventricle against the chest wall) is usually in the fifth or sixth left interspace, depending on such factors as the position of the patient, and the state of inflation of his lungs (p. 240). The right margin of the heart protrudes about a finger's breadth lateral to the right hand margin of the sternum, and the apex lies just medial to the midclavicular line (Fig. 105). Because of the movement of the left dome of the diaphragm during breathing, the apex of the heart moves upwards and downwards during respiration more than the part of the heart resting on the central tendon, so that the heart alters its shape considerably during deep breathing.

24 · The abdomen and pelvis

The skeletal framework of the abdomen is the lumbar part of the vertebral column (p. 245). Unlike the thorax, the abdomen has no bony support anteriorly, and is only imperfectly protected by bone laterally. Because the viscera have to be able to expand to accommodate a surfeit of suet pudding, the anterior wall is made of muscle which can bulge as required. Support is given to the abdominal viscera from below by the bony pelvis, which is the incomplete rigid casing for the pelvic organs, and protection is afforded by the overhanging thoracic cage to those organs which lie under the diaphragm.

The bony pelvis

The pelvis consists of two **hip** bones which articulate with each other in front at the **pubic symphysis**, and behind are separated by the sacrum (Fig. 124). The hip bone, which is more fully described on p. 344, is itself a composite of three bones, all of which meet at the socket of the hip joint. The large irregular plate of bone which stretches up towards the ribs is the **ilium,** and its upper margin is the **iliac crest**, which ends in **anterior** and **posterior superior iliac spines.** The ilium has an outer surface which looks towards the buttock, and an inner one which forms with its fellow a shallow basin called the **false pelvis.** This basin is very incomplete, and ends centrally in a fairly abrupt rim of bone called the **pelvic brim** (Fig. 124); behind and below this brim lies the **true pelvis**, formed by the sacrum and the other two components of the hip bone. The **ischium** is a thick block of bone whose **tuberosity** is the part of the skeleton on which we sit; it has also a **spine** which projects medially and backwards towards the pelvic cavity. The third part of the hip bone is the **pubis**, which lies anteriorly. It has a body, which articulates with its fellow of the opposite side at the symphysis; on the body there is a small **tubercle** directed upwards and laterally. The pubis has also two **rami**, which form the upper and medial margins of the large oval **obturator foramen** (Fig. 124), which is almost completely closed in life by fibrous tissue constituting the **obturator membrane** (p. 345). The inferior ramus of the pubis meets a similar projection forwards from the ischium, and the bar of bone so formed is called the **conjoined ramus** of the ischium and pubis.

Between the two conjoined rami is another gap which extends backwards from the **subpubic angle** to the ischial tuberosities; this gap is filled up in life by the **perineal membrane,** which is pierced by the urethra in the male and by the urethra and vagina in the female.

The sacrum and coccyx have already been described (p. 249). The sacrum is held between the two ilia at the **sacroiliac joint** and is tilted backwards at a considerable angle (to put the pelvis in the anatomical position, the anterior superior spines must be in the same vertical plane as the pubic tubercles). Only the first two or three pieces of the sacrum enter into the joint; the remaining pieces, with the coccyx, project in a curve below and between the ilia, and are tied to the ischium on each side by two very strong ligaments, the **sacrotuberous,** which runs to the tuberosity, and the **sacrospinous** ligament, which runs to the ischial spine. These ligaments fill in much of the lateral gap between the sacrococcygeal wall of the pelvis and the hip bones.

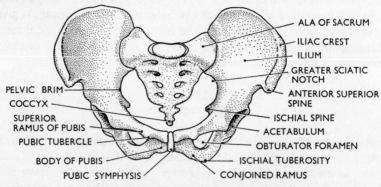

Fig. 124. Female bony pelvis from in front.

The pelvis is the part of the skeleton which shows the most obvious differences between the two sexes. In childbirth the head of the baby has to traverse the cavity of the true pelvis, and in consequence this has to be as roomy as possible. In the female pelvis the diameters of the pelvic brim are proportionally larger and the cavity of the true pelvis is wider and shallower. The subpubic angle is more obtuse, and the hip joints are thrust wider apart; the articulation of the sacrum with the ilia is shallower, and the curve between the ischial spine and the articular surface, which is called the **greater sciatic notch** (Fig. 170, p. 345), has a much less acute bend.

The joints

The **pubic symphysis** is a secondary cartilaginous joint (p. 39). During pregnancy fluid accumulates in the substance of the cartilage, and so allows a degree of extra movement during childbirth. The same applies to the ligaments of the **sacroiliac** joints, at which very little movement normally takes place. The articular surfaces are irregularly roughened, and the protuberances on the sacrum fit into the depressions on the ilium, and vice versa. The bones are joined by one of the strongest (albeit one of the shortest) ligaments in the body—the **interosseous sacroiliac ligament**, which lies behind the articular surfaces. Further, there are strong ligaments on the posterior surface of the joint called the **short** and **long posterior sacroiliac** ligaments; the long one becomes continuous with the sacrotuberous ligament, and so is attached to the ischium. The sacrotuberous and sacrospinous ligaments are extrinsic ligaments of the joint, binding the sacrum to the innominate bone. These extensive and strong ligamentous supports are necessary because the weight of the body is trying to drive the front part of the sacrum downwards between the ilia, and there is a consequent tendency for the posterior part of the sacrum and coccyx to tilt upwards like the other end of a seesaw; the ligaments tether it down. In contrast to the strength of the joint posteriorly is the thin and weak anterior part of the capsule. Repeated pregnancies impose a strain on this joint which it may be unable to survive unscathed, and one of the causes of low back pain in women who have borne children is strain of the sacroiliac ligaments.

The **lumbosacral** joint also takes an enormous amount of punishment. In the upright position the weight of the trunk tends to drive the fifth lumbar vertebra off the obliquely tilted upper surface of the sacrum (Fig. 109). To prevent this, the body of the fifth lumbar vertebra is markedly wedge shaped, and so is the lumbosacral intervertebral disc; the immensely strong **iliolumbar ligament** tethers the transverse process of the fifth lumbar vertebra to the posterior end of the crest of the ilium and so prevents the vertebra being pushed forwards and downwards. Lastly, the articular processes of the first sacral vertebra form a strong shelf on which much of the weight is taken.

Important muscles

The anterior abdominal wall depends entirely on its muscles, and when these get a little tired about middle age the pressure of the abdominal organs may cause it to sag. On either side of the midline the **rectus abdominis** muscles run upwards from the anterior surface of the body of the pubis, spreading laterally as they go, to be inserted into the front of the xiphisternum and the anterior surfaces of the fifth, sixth and seventh costal cartilages (Fig. 125). They are divided into segments by horizontal bands of fibrous tissue.

Lateral to the rectus muscles the abdominal wall is formed by three layers of muscle. Outermost is the **external oblique**, which runs downwards and forwards from the outer surfaces of the lower eight ribs to a complex insertion which includes the outer lip of the front part of the iliac crest and an aponeurotic sheath which encloses the rectus muscle. The lower border of the aponeurosis is strung like a bowstring across from the anterior superior spine to the pubic tubercle, and forms the **inguinal ligament,** behind which vessels, nerves and muscles enter and leave the thigh. The fleshy posterior border of the muscle, stretching from the ribs to the iliac crest, is overlapped by the inferior border of the latissimus dorsi (Fig. 146).

The muscle in the middle layer is the **internal oblique**, which runs upwards and forwards from the front part of the iliac crest and the lateral part of the inguinal ligament to be inserted into the lower borders of the cartilages forming the costal margin and into an aponeurosis which splits to form a sheath round the rectus abdominis muscle.

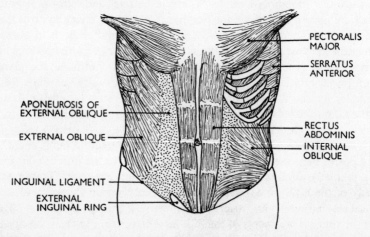

Fig. 125. The anterior abdominal wall. On the left side the external oblique has been removed to display the internal oblique muscle. The anterior wall of the rectus sheath has been removed to show the rectus abdominis. The external inguinal ring is the outer end of the inguinal canal; notice how the lower border of the internal oblique arches over above the canal.

The innermost muscle in general runs transversely, and is therefore called the **transversus abdominis.** It takes origin from the front part of the iliac crest and the lateral part of the inguinal ligament; other fibres come from the inner surfaces of the lower six ribs, and the whole muscle joins the posterior layer of the rectus sheath.

The **rectus sheath** (Fig. 126) is formed mainly by the splitting of the internal oblique aponeurosis; the front layer is reinforced by the aponeurosis of the external oblique, and the posterior layer by the aponeurosis of the transversus. The two sheaths fuse with each other in the midline to form a dense and strong fibrous band called the **linea alba,** which runs from the symphysis pubis to the xiphoid process.

All four muscles are supplied by the ventral rami of the lower six thoracic nerves. Their functions are numerous. By contracting, they lower the ribs and raise the intra-abdominal pressure, and are thus used in forced expiration and in coughing or sneezing (p. 161). They may be used to empty the abdominal organs by compressing them, as in vomiting, defaecation and childbirth, or they may contract to protect these organs against a blow. In patients who have inflammation inside the abdominal cavity the muscles may be of a reflex rigidity so intense that it is described as 'boarding'; this is to prevent movement and keep the part at rest. By their tone the abdominal muscles prevent the abdominal viscera from pulling away from their attachments to the posterior wall of the abdomen. They also steady the pelvis by holding it firmly to the thorax, so that

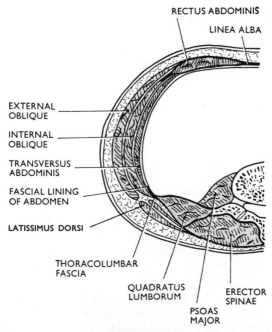

RECTUS ABDOMINIS

LINEA ALBA

EXTERNAL OBLIQUE

INTERNAL OBLIQUE

TRANSVERSUS ABDOMINIS

FASCIAL LINING OF ABDOMEN

LATISSIMUS DORSI

THORACOLUMBAR FASCIA

QUADRATUS LUMBORUM

PSOAS MAJOR

ERECTOR SPINAE

Fig. 126. Diagrammatic cross-section through the abdominal wall, showing the formation of the rectus sheath and the attachments of the muscles to the thoracolumbar fascia. The aponeurosis of the internal oblique splits round the rectus muscle; the aponeurosis of the external oblique passes in front, and that of the transversus passes behind the rectus. The erector spinae and the quadratus lumborum are enveloped in septa of the thoracolumbar fascia which are attached to the spines and transverse processes.

the thigh muscles have a firm platform from which to pull the lower limb about. In walking, for example, the rectus abdominis is in constant use, and it may be felt contracting strongly when the patient raises his legs off the bed without using his hands or bending his knees. In this instance the muscles are resisting the tendency for the weight of the lower limbs to tilt the pelvis round. The rectus abdominis is one of the chief flexors of the lumbar spine, and the internal oblique of one side combines with external oblique of the other to rotate the chest on the pelvis or vice versa.

The anterior abdominal wall is not uniformly strong, and it is possible for some of the more mobile viscera, particularly the small intestine (p. 181) to force a passage through some of the weak spots and protrude as a **hernia.** The commonest site for a hernia is just above the inguinal ligament, where there is an oblique tunnel through the muscles of the abdominal wall (Fig. 125). This is the **inguinal canal**; it is nearly 5 cm long, and in the male represents the passage along which the testicle travelled on its way from the abdominal cavity to the scrotum (p. 192). The male inguinal canal still transmits the **spermatic cord**, a bundle composed of the vessels and nerves going to and from the testicle, together with the **vas deferens** (p. 192); in the female the canal contains the **round ligament of the uterus.** A protrusion of a loop of gut along this canal is called an indirect inguinal hernia. Other points where herniae may occur are the **umbilicus** (navel), where there is a weak spot in the linea alba, or the place behind the inguinal ligament where the femoral vein enters the abdomen. Herniae in these positions are called umbilical and femoral herniae respectively.

The posterior wall of the abdomen is much more solid and unyielding than the anterior wall. On each side the **psoas major** muscle (Figs. 126, 129) originates from the transverse processes, the sides of the bodies and the intervertebral discs of the five lumbar vertebrae, and runs downwards and laterally behind the inguinal ligament, being joined on the way by the fibres of the **iliacus**, which comes from the inner surface of the ilium, and forms the floor of the false pelvis. Both muscles (forming the **iliopsoas**) are inserted into the lesser trochanter of the femur. The psoas is the chief flexor of the hip joint (p. 359) but it has also a powerful action in lateral bending of the lumbar spine; it is supplied by the ventral rami of the lumbar nerves, while the iliacus is supplied by the femoral nerve (p. 375).

Lateral to the psoas is the quadratus lumborum (p. 253), and lateral to this again lie the posterior margins of the transversus abdominis and the two obliques, all being attached to the layers of the **thoracolumbar fascia** (Fig. 126).

From the obturator membrane and the inner surface of the bone surrounding the obturator foramen spring the fibres of the **obturator internus**. The tendon of the muscle leaves the pelvis through the lesser sciatic notch to reach the greater trochanter of the femur (p. 345). The action of the muscle is dealt with in connexion with the hip joint (p. 357), but it is mentioned here because it forms a resilient lining to the anterolateral aspect of the true pelvis. Posterior to the obturator internus the gap between the spine of the ischium and the sacrum and coccyx is partly filled in by the sacrospinous ligament, covered internally by the **coccygeus** muscle, which also acts as a soft lining.

The **levator ani** is a very important muscle, for it forms a strong but flexible floor called the **pelvic diaphragm** which supports the pelvic viscera. It extends from the symphysis pubis along the fascia covering the inner surface of the obturator internus to the spine of the ischium, and from this origin the fibres pass downwards and medially to form a hammock for the bladder and the rectum, as well as the uterus in the female. The fibres meet each other in the midline, and resist the pressure of the pelvic viscera every time the abdominal muscles come into play. The pelvic diaphragm is pierced in front by the vagina and the urethra, and behind by the rectum, over which its fibres continue downwards to form a sling for the anus and give the levator ani its name. During defaecation it pulls the anus up over the descending faeces. In childbirth it is stretched, and it may not fully recover its tone after being subjected to repeated insults of this nature, so that it sags, and allows the viscera to sag too. When this happens, the uterus and the rectum may descend into the vagina—a condition of **prolapse**. The levator ani is supplied by the ventral rami of the sacral nerves.

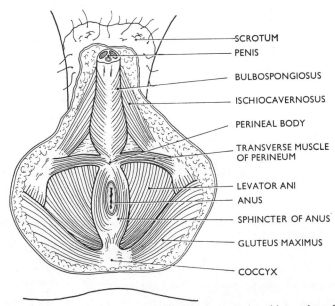

SCROTUM
PENIS
BULBOSPONGIOSUS
ISCHIOCAVERNOSUS
PERINEAL BODY
TRANSVERSE MUSCLE OF PERINEUM
LEVATOR ANI
ANUS
SPHINCTER OF ANUS
GLUTEUS MAXIMUS
COCCYX

Fig. 127. Muscles of the male perineum. Notice the central position and mechanical importance of the perineal body.

The **perineum** is the region bounded on either side by the conjoined rami of the ischium and pubis, anteriorly by the pubic symphysis, and posteriorly by the two sacrotuberous ligaments. It is thus a diamond-shaped area (Fig. 127) divided into an anterior **urogenital triangle,** through which pass the urethra in both sexes and the vagina in the female, and a posterior **anal triangle,** through which the anal canal descends to open on the exterior at the anus.

In the centre of the perineum, and forming a staging point for most of the muscles of the region, lies a fibromuscular mass called the **perineal body** (Fig. 127) and to this are attached the deep and superficial fascia. It is a most important structural support, and if it is torn in childbirth, the resulting weakening encourages prolapse.

In the anal triangle the anus is surrounded by the three layers of the **external anal sphincter** (p. 185), some of whose fibres surround the canal circularly, while others extend backwards round the canal from the perineal body to the coccyx. On either side of the anal canal lies a pyramidal cavity called the **ischio-rectal fossa,** filled with fat. This fat allows the anal canal to distend during defaecation (Fig. 93).

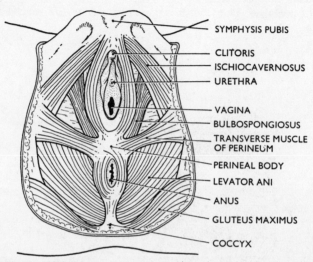

SYMPHYSIS PUBIS

CLITORIS
ISCHIOCAVERNOSUS
URETHRA

VAGINA
BULBOSPONGIOSUS
TRANSVERSE MUSCLE
OF PERINEUM
PERINEAL BODY
LEVATOR ANI
ANUS
GLUTEUS MAXIMUS
COCCYX

Fig. 128. Muscles of the female perineum. Notice the central position and mechanical importance of the perineal body.

In the male urogenital triangle, the muscles connected with the penis gain attachment from the conjoined rami and from the under surface of the perineal membrane (p. 278). Between them and these rigid surfaces lie the components of the penis, which they can therefore compress. The **ischiocavernosus** muscles (Fig. 127) compress the **crura** of the penis, which form the main body of the organ, and which are filled with cavernous vascular tissue (p. 138), while the **bulbospongiosus** muscle compresses the **corpus spongiosum** which surrounds the uretha; it is concerned with the expulsion of urine or semen along the urethra. The whole of this apparatus is covered and concealed by the scrotum.

In the female, the urogenital triangle contains the outlet of the **birth canal** along which the baby must pass during childbirth. This is the **vulva** (p. 194). The bulbospongiosus muscle splits to encircle this opening, and the ischiocavernosus muscles surround the crura of the **clitoris**, a small homologue of the penis which overhangs the anterior end of the vulva. On either side of the vulva lie two folds of fibro-fatty tissue corresponding to the scrotum and the floor of the male urethra; these are the **labia majora** and the **labia minora**.

Important nerves

The terminations of the lower five **intercostal** nerves and the **subcostal** (twelfth thoracic) nerve run forward in the anterior abdominal wall to pierce the rectus sheath and supply the rectus abdominis. They supply all the muscles and most of the skin of the anterior abdominal wall.

In the posterior wall of the abdomen lies the **lumbar plexus**, formed by the ventral rami of the first four lumbar nerves and embedded in the substance of the psoas major muscle. The main branches of this plexus emerge from the muscle, the **femoral** nerve on its lateral side and the **obturator** nerve on its medial side (p. 376). Other, smaller branches are of less importance, but the first lumbar nerve sends the **iliohypogastric** and the **ilio-inguinal** nerves round to supply the anterior abdominal wall (Fig. 129).

Part of the fourth lumbar ventral ramus forks off to join the fifth lumbar, and this is the beginning of the **sacral plexus**, which involves the ventral rami of the fourth and fifth lumbar, all the sacral, and the coccygeal nerves. This plexus supplies most of the lower limb; its main branch is the **sciatic** nerve (p. 378). Branches of the sacral plexus supply the pelvic diaphragm, and the **pudendal** nerve supplies the external genitalia and perineum. Still other branches pass out of the pelvis into the buttock (p. 377).

The **vagus** nerves enter the abdomen along with the oesophagus, and supply the stomach, the intestines, and their associated glands with parasympathetic fibres which reach distally as far as the middle of the large intestine. The gut below this level, and all the pelvic organs, receive their parasympathetic supply from the pelvic outflow (p. 89). The **sympathetic trunks** and their thoracic **splanchnic** branches (p. 272) enter the abdomen behind and through the diaphragm respectively, and feed a number of visceral plexuses of which the largest is the **coeliac** plexus (p. 88). These plexuses, and the continuation downwards of the sympathetic trunk, supply all the abdominal and pelvic viscera with sympathetic fibres which reach their destinations by running alongside the blood vessels.

Important vessels

The **descending aorta** passes behind the diaphragm to become the **abdominal aorta**. Opposite the fourth lumbar vertebra it divides into two **common iliac** arteries which diverge from each other along the line of the pelvic brim. At the sacroiliac joints each divides again into an **internal** and an **external iliac** artery. The internal iliac artery supplies the pelvic viscera, the perineum and the external genitalia, as well as providing branches which escape into the thigh and buttock. The external iliac artery runs along the pelvic brim and emerges under the inguinal ligament into the thigh, where it changes its name and becomes the **femoral artery** (p. 381).

The main visceral branches of the abdominal aorta are the paired **renal** arteries which run to the kidneys, and the unpaired **coeliac, superior mesenteric** and **inferior mesenteric** arteries. The coeliac artery has three large branches which supply the spleen, liver, gall bladder, and stomach, as well as the pancreas and the first part of the duodenum. The superior mesenteric supplies

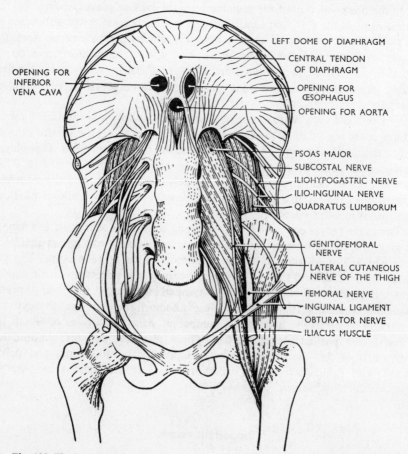

Fig. 129. The lumbar plexus and the muscles of the posterior abdominal wall. On one side the psoas major has been removed to expose the manner in which the branches of the plexus are formed within its substance.

the small intestine and the proximal half of the large intestine, and the inferior mesenteric supplies the rest of the gut down as far as the anus (Fig. 130).

The external and the internal iliac arteries are accompanied by corresponding veins, which join together to form two **common iliac** veins; these in turn unite as the **inferior vena cava**, which receives the veins draining the posterior abdominal wall, the suprarenals and the kidneys. The blood coming from the alimentary canal drains into the **portal vein** (p. 138). The inferior vena cava lies posterior to the liver and receives from it a number of **hepatic veins** before passing through the central tendon of the diaphragm.

Important viscera

The organs comprising the digestive and genito-urinary systems have already been briefly described (pp. 173–197).

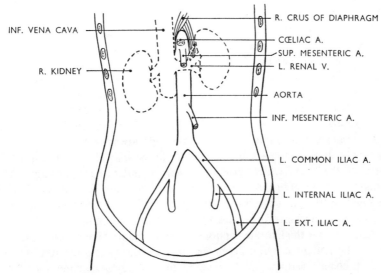

INF. VENA CAVA

R. KIDNEY

R. CRUS OF DIAPHRAGM

CŒLIAC A.

SUP. MESENTERIC A.

L. RENAL V.

AORTA

INF. MESENTERIC A.

L. COMMON ILIAC A.

L. INTERNAL ILIAC A.

L. EXT. ILIAC A.

Fig. 130. The main branches of the abdominal aorta. The renal arteries lie behind the renal veins and concealed by them.

Surface anatomy

The anterior and posterior superior spines of the ilium are easy to feel; the first usually makes a visible prominence, and the other lies at the bottom of a dimple in the skin. Between them the whole length of the iliac crest can be made out. The body and symphysis of the pubis are well defined, and between the pubic tubercle and the anterior superior spine the inguinal ligament is felt as a taut resistance.

If the abdominal muscles are well developed and the subject is thin, the rectus sheath stands out clearly, as do the fibrous bands which intersect the rectus muscle.

The umbilicus lies halfway between the xiphoid process and the upper border of the symphysis pubis, at the level of the disc between the third and fourth lumbar vertebrae.

The only structures normally palpable in the abdomen are the pelvic colon (p. 183), the lower pole of the right kidney (on full inspiration), and possibly the descending aorta in a thin person. An over-distended bladder can be felt between the symphysis pubis and the umbilicus, and the pregnant uterus forms a palpable swelling which in the later stages may reach up to the costal margin.

The oesophagus joins the stomach about 10 cm deep to the tip of the seventh left costal cartilage, opposite the tenth thoracic vertebra. The pylorus only lies near the misleadingly named **transpyloric plane** (Fig. 102) when the body is recumbent. This plane cuts the first lumbar vertebra and lies halfway between the suprasternal notch and the upper border of the pubic symphysis; when the body is upright and the stomach is full the pylorus may be as much as two vertebrae lower down. Between the cardia and the pylorus the stomach is extremely variable in position.

The mesentery of the small intestine stretches from the left side of the second lumbar vertebra down to the junction of the right midclavicular plane and the intertubercular plane (Fig. 102); this is also approximately the marking for the opening of the appendix into the caecum. The jejunum lies in the upper left part of the abdomen, and the ileum in the right lower part.

The right flexure of the colon is usually lower than the left because of the presence of the liver, which extends upwards under the right dome of the diaphragm to the level of the fifth rib in the midclavicular line. Most of the liver is protected by the rib cage, but a portion of the left lobe extends across the upper part of the abdomen just deep to the anterior abdominal wall (Fig. 92); in the midaxillary line the liver lies opposite the seventh to the eleventh ribs.

When the patient is lying down the gall bladder is opposite the tip of the right ninth costal cartilage, at the level of the transpyloric plane, but when he stands up it may descend as far as the fifth lumbar vertebra.

In the recumbent patient, the hila of both kidneys are approximately opposite the second lumbar vertebra, but they may move downwards by anything up to a hand's breadth if the patient stands up and breathes in.

This mobility of the abdominal viscera with changes in respiration and posture is even more marked than that of the thoracic viscera (p. 240).

25 · The upper limb

Introduction

The human thumb is opposable—that is, it can be brought to lie opposite each of the four fingers. This power of opposition has made of the hand a gripping instrument of great strength, delicacy and suppleness which allows man to make and use his own tools, to explore and alter his environment, to prepare and cultivate his food, to record his thoughts in writing, and in general to earn his living. The whole upper limb is a system of joints and levers designed to bring this gripping instrument to bear at any desired point, and to maintain it there steadily and accurately. The slightest impediment in this beautifully organized machinery may be reflected in a gross drop in the patient's capacity for everyday existence. No artificial limb has yet been designed which possesses more than a very small fraction of the efficiency of the real article, and for this reason any injury to the upper limb is always treated with the greatest caution; any portion of tissue which seems at all capable of surviving is preserved. In contrast, artificial lower limbs, because of the simpler tasks the lower limb has to perform, are often very serviceable, and it is possible for a man with an amputation at the hip to play golf with a plus handicap.

The skeleton

The arm is attached to the trunk by the **shoulder girdle,** which consists of two bones, the **clavicle** and the **scapula**.

The **clavicle** (Figs. 131, 132, 133) is a thin strut of bone, 12-15 cm long in the adult, which thrusts the shoulder joint away from the trunk so that the arm can move freely without being obstructed by the thorax. When the clavicle is broken, the shoulder falls inwards towards the midline. The medial end of the clavicle articulates with the manubrium sterni at the sternoclavicular joint; it is relatively thick and square in cross-section, having four surfaces (superior, inferior, anterior, and posterior). The lateral half of the bone is flattened from above downwards, so that it has only two surfaces (superior and inferior), separated by anterior and posterior borders. The clavicle is curved like a very open 'S' and is convex forwards in its medial half and concave forwards in its lateral half. The flattened lateral end articulates with the acromion process of the scapula in the acromioclavicular joint. On the under surface of the clavicle, at the junction of the outer and middle thirds of the bone, is a small roughened area to which the important coracoclavicular ligament is attached. The clavicle

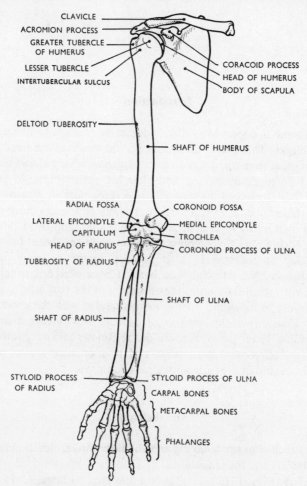

CLAVICLE
ACROMION PROCESS
GREATER TUBERCLE OF HUMERUS
LESSER TUBERCLE
INTERTUBERCULAR SULCUS
CORACOID PROCESS
HEAD OF HUMERUS
BODY OF SCAPULA
DELTOID TUBEROSITY
SHAFT OF HUMERUS
RADIAL FOSSA
CORONOID FOSSA
LATERAL EPICONDYLE
MEDIAL EPICONDYLE
CAPITULUM
TROCHLEA
HEAD OF RADIUS
CORONOID PROCESS OF ULNA
TUBEROSITY OF RADIUS
SHAFT OF ULNA
SHAFT OF RADIUS
STYLOID PROCESS OF RADIUS
STYLOID PROCESS OF ULNA
CARPAL BONES
METACARPAL BONES
PHALANGES

Fig. 131. The bones of the upper limb from in front.

is superficial, and is easily felt along its whole length. Just above the bone there is a depression in the neck called the supraclavicular fossa, and in thin people there may be a similar infraclavicular fossa below the bone, so that the clavicle stands out as a prominent bony ridge directed laterally and slightly upwards.

The **scapula** (Figs. 131, 132, 133) has a thin flattened triangular **body**, with superior, inferior and lateral angles and superior, medial and lateral borders.

The medial border runs vertically upwards from the inferior to the superior angle, and the superior border runs downwards and outwards to the lateral angle, which is greatly thickened and expanded to provide the **glenoid socket** for the shoulder joint. The portion of the body immediately surrounding this thickened block is the **neck** of the scapula. From the lateral angle the lateral

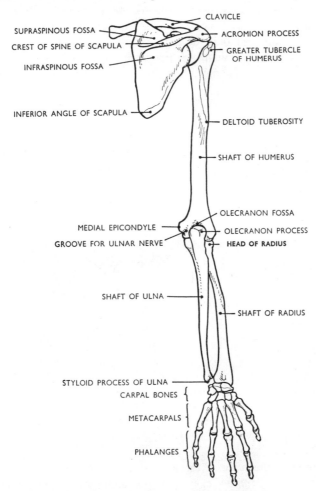

Fig. 132. The bones of the upper limb from behind.

border extends downwards to the strongly built inferior angle. The **costal** surface of the body is very slightly concave, and the **dorsal** surface is slightly convex. From the neck of the scapula the strong angulated **coracoid process** juts forwards and laterally. The **spine** of the scapula divides the dorsal surface into a **supraspinous** and an **infraspinous fossa**. The spine itself thrusts backwards to end in a free margin called the **crest** of the spine. This meets the medial border of the body in a shallow triangular slope, and gradually runs backwards and laterally away from the body until it spreads out to form the **acromion process**, a flattened quadrangular block of bone posterosuperior to the glenoid socket (Figs. 131, 132, 133). The whole of the spine is subcutaneous and can be felt and sometimes seen; the tip of the acromion process is the landmark from which measurements of the arm and of the upper limb are made. The inferior angle can be seen moving when the arm is abducted, but the other angles and borders can only be felt through a clothing of muscles. In the anatomical position the scapula extends from the second rib to the seventh rib, and the root of the spine is opposite the spine of the third thoracic vertebra.

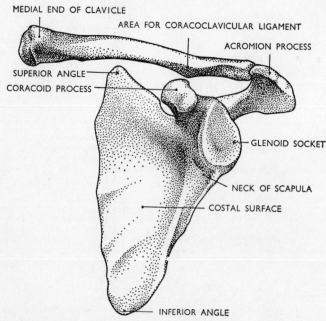

Fig. 133. The left clavicle and scapula from in front, showing the glenoid socket. Note that the socket faces forwards as well as outwards.

The **humerus**, the bone of the arm, is a typical long bone, and its **head** articulates with the glenoid socket of the scapula in the shoulder joint. The head forms rather less than a hemisphere, and its curvature is similar to that of a golf ball; round its margins runs a groove called the **anatomical neck**. The **lesser tubercle** (Fig. 131) lies anterior to the articular surface, and is separated by a vertical furrow—the **intertubercular sulcus**—from the **greater tubercle** which lies to the lateral side and can be felt as an indefinite mass distal to the acromion process. The upper half of the shaft is cylindrical, but the lower half is flattened from before backwards and has sharp **medial** and **lateral borders.** The region of the shaft just distal to the tubercles is called the **surgical neck**, because fractures often occur here. Below the surgical neck a shallow **radial groove** winds obliquely forwards and downwards round the lateral side of the humerus. Half way down the shaft, on the lateral side, is an irregular roughened area called the **deltoid tuberosity** because the deltoid muscle is inserted into it. The lower end of the humerus articulates with the **ulna** at a pulley-shaped articular surface called the **trochlea**, and with the head of the **radius** at a rounded **capitulum** lying lateral to the trochlea and continuous with it (Fig. 131). Above the trochlea in front lies the **coronoid fossa**, into which the coronoid process of the ulna pushes when the joint is flexed, and posteriorly the **olecranon fossa** accommodates the olecranon process when the joint is extended. The bone between these two depressions is very thin and sometimes translucent. Another much smaller depression above the capitulum in front is called the **radial fossa**, because the head of the radius lies in it when the elbow is fully flexed. The lateral border of the humerus broadens out inferiorly into a small **lateral epicondyle,** and the medial border leads to the much larger **medial epicondyle,** which may be picked up between finger and thumb in the living elbow. On the posterior aspect, between the medial epicondyle and the trochlea, is a deep groove for the ulnar nerve on its way to the forearm. The epicondyles are both used in the living body as landmarks for making measurements. The shaft of the humerus is thickly covered in front and behind by muscles, but its lateral and medial aspects can be felt fairly clearly all the way down.

The longer of the two forearm bones is the **ulna,** which lies on the medial side, and has a massive upper end shaped like the claw of a spanner (Fig. 134) because of two projections, the **olecranon** and the **coronoid processes.** Between them is the **trochlear notch**, which articulates with the trochlea of the humerus. When the elbow is fully extended, the two epicondyles and the tip of the olecranon lie in the same straight line; when the elbow is fully flexed they form an equilateral triangle. On the lateral side of the coronoid process is a small articular depression for the head of the radius called the **radial notch**, and just distal to the coronoid process is a rough area called the **tuberosity** of the ulna. The shaft of the bone is triangular in cross section: the **interosseous** border facing the radius is sharp, but the anterior and posterior borders on the medial side are more rounded. The borders mark out three surfaces—anterior, posterior and medial (Figs. 134, 155). The shaft narrows as it approaches the distal end, where the small rounded **head** of the ulna articulates with the lower end of the radius: medial to the head a small cylindrical **styloid process** projects downwards. The posterior border is subcutaneous in its whole length, from the prominent olecranon process, which forms the tip of the elbow, to the styloid process, which can be felt by pressure on the medial side of the wrist. The head of the ulna shows as a rounded knob on the back of the wrist, most evident when the hand is turned with the palm downwards.

The **radius** is shorter than the ulna by the width of the trochlear notch. The **head** of the bone, at its proximal end, is a cylindrical disc, concave on its upper surface to accommodate the rounded capitulum of the humerus, and fitting medially into the radial notch of the ulna. Distal to the head the bone narrows sharply to form the **neck,** and just below this on the medial side is the well marked **tuberosity** of the radius (Fig. 134). The shaft of the radius, like that of the ulna, has three borders; the **interosseous** border faces the ulna and is sharp, while the anterior and posterior borders (Fig. 155) are rounded and indistinct. The anterior, lateral, and posterior surfaces which these borders demarcate are

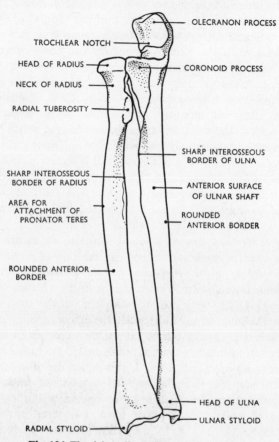

OLECRANON PROCESS

TROCHLEAR NOTCH

HEAD OF RADIUS

NECK OF RADIUS

CORONOID PROCESS

RADIAL TUBEROSITY

SHARP INTEROSSEOUS
BORDER OF ULNA

SHARP INTEROSSEOUS
BORDER OF RADIUS

ANTERIOR SURFACE
OF ULNAR SHAFT

AREA FOR
ATTACHMENT OF
PRONATOR TERES

ROUNDED
ANTERIOR BORDER

ROUNDED ANTERIOR
BORDER

HEAD OF ULNA

ULNAR STYLOID

RADIAL STYLOID

Fig. 134. The right radius and ulna from in front.

therefore more or less continuous with each other laterally. Half way down the lateral surface a roughened area marks the attachment of a muscle called the pronator teres (Fig. 134). Just as the shaft of the ulna narrows from above downwards, so the shaft of the radius thickens. The lower end is expanded, and carries on its lateral side a **styloid process** thicker and stronger than that of the ulna. The distal surface of the lower end is covered with articular cartilage for the radiocarpal joint, and the medial edge of it is slightly cupped to receive the head of the ulna at the distal radioulnar joint. In front, the lower end of the radius is slightly concave, and posteriorly it is much grooved by the passage of tendons over it on their way to the hand. One of the prominences so formed is the **radial tubercle,** which can be felt on the back of the wrist. The styloid process can be felt on the lateral side of the wrist, lying a finger's breadth distal to the level of the styloid process of the ulna. The whole of the distal end of the radius and the lower part of the shaft are easily felt, but the upper part of the

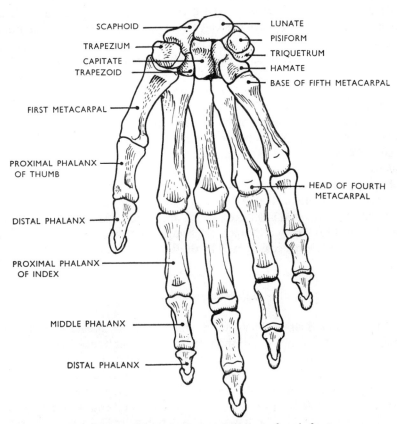

Fig. 135. The carpus, metacarpus and phalanges from in front.

shaft is covered by muscles. The head may be felt distal to the lateral epicondyle, particularly if the forearm is rotated to make it move.

The eight small bones in the wrist are collectively known as the **carpus** (Fig. 135). Seven of them are tightly bound together into two rows. The proximal row contains—from lateral to medial—the **scaphoid,** the **lunate,** and the **triquetrum:** the distal row contains—again from lateral to medial—the **trapezium,** the **trapezoid,** the **capitate,** and the **hamate.** The eighth bone, the **pisiform,** lies in the tendon of the flexor carpi ulnaris (p. 315), and is sometimes regarded as a sesamoid bone. It forms a small knob on the medial side of the front of the wrist, and can be moved about slightly on the anterior surface of the triquetrum if the wrist is held slack. The **hook of the hamate** is felt on the medial side of the palm, distal to the pisiform and about a thumb's breadth from it. The **tubercle of the scaphoid** is felt on the front of the wrist immediately above the ball of the thumb; it makes a visible projection if the wrist is bent backwards. Just distal to this it is often possible to identify the **tubercle of the trapezium**. The carpus is concave anteriorly, and it forms the posterior wall of the **carpal tunnel** (p. 324) which conveys the long flexor tendons of the fingers and thumb into the hand.

Articulating with the carpal bones are the **bases** of the five **metacarpals**, the shortest and stoutest of which is the metacarpal of the thumb. Each is a miniature long bone, with a shaft and two ends. The shafts of the metacarpals can be felt on the back of the hand, and are slightly concave forwards.

The distal end, or **head**, of each metacarpal articulates with a **proximal phalanx**, a miniature long bone with a slightly hollowed proximal end for the rounded metacarpal head, and a saddle-shaped distal end for the smaller **middle phalanx**. The middle phalanx has saddle surfaces at each end, and the **distal phalanges** have only one articular surface, and end distally in a roughened projection. The shafts of the phalanges are flat in front and convex posteriorly. The thumb has only two phalanges, which are shorter and stouter than those of a finger (Fig. 135). When the fingers are bent, the rounded projections known as the knuckles are in each case formed by the distal end of the proximal bone.

Joints and movements

The clavicle and scapula make up a unit called the **shoulder girdle** which amplifies the movements at the shoulder joint, and prevents them from being hindered by the proximity of the thorax. The two bones are bound together by the **coracoclavicular ligament**, which runs from the coracoid process to the roughened inferior surface of the clavicle directly above it. It has two parts, the **conoid** and **trapezoid** ligaments, which twist on each other as the scapula moves relative to the clavicle. The coracoclavicular ligament suspends the scapula from the clavicle as a ham is hung from a hook in a beam; but most of the weight of the upper limb is probably taken by muscles (p. 306). The second connexion between the clavicle and the scapula is at the **acromioclavicular joint**, a small plane joint which contains an incomplete disc. Here gliding and twisting movements take place between the two components of the shoulder girdle, which are attached to each other by the weak capsule of the joint.

The **sternoclavicular joint** is the only joint between the shoulder girdle and the trunk. The socket for the medial end of the clavicle is formed by the manubrium of the sternum and the first costal cartilage. A complete articular disc divides the joint into two separate synovial cavities, and this enables the incongruous surfaces to function more or less as a ball and socket joint. The disc also tethers the medial end of the clavicle to the sternum, and so prevents it being driven upwards and medially by a blow on the shoulder. The capsule is reinforced by strong anterior and posterior ligaments which are helped by the fibres of muscles arising from the joint capsule—particularly the sternocleidomastoid muscle (p. 263). Above the joint the **interclavicular** ligament joins the medial ends of the two clavicles and attaches them to the upper border of the manubrium (Fig. 136). Inferior to the joint is the very strong **costoclavicular** ligament, which connects the first costal cartilage to the surface of the clavicle.

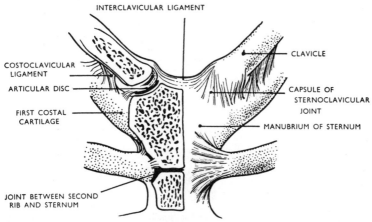

Fig. 136. The sternoclavicular joint. Both clavicles have been elevated, and on one side the capsule has been removed to show the articular disc.

This important joint allows a considerable range of movement: the lateral end of the clavicle can be elevated and depressed through some 45°, and its anteroposterior swing is nearly as extensive. In the movement of circumduction (p. 44) the outer end of the clavicle describes a circle. True rotation of 30°–40° occurs in relation to twisting movements of the scapula.

The scapula, attached to the trunk by muscles, moves over the surface of the thorax in all directions. The contour of the thorax is not quite the same as the portion of a sphere traced out by the lateral end of the clavicle, and so the acromioclavicular joint is inserted in the mechanism to allow the necessary twisting of the scapula on the clavicle. The scapula moves forwards round the chest wall in such movements as pushing or thrusting, and backwards towards the midline in pulling or climbing. It can also be made to rotate round a point about the middle of its spine, so that the glenoid cavity looks upwards instead of outwards (Fig. 147). This rotation, necessarily accompanied by rotation and elevation of the clavicle, approximately doubles the range of abduction of the limb which the shoulder joint alone could achieve.

The **shoulder** joint is the ball and socket joint between the head of the humerus and the glenoid socket of the scapula. The two surfaces have very different radii of curvature, and there is a great discrepancy in size; the socket is built up by a circular lip of cartilage called the **labrum glenoidale** to make a better fit. Even so, the humerus can readily be dislodged from the scapular articular surface when the muscles surrounding the joint are relaxed. This arrangement makes for immense mobility, but at the same time renders the shoulder the most frequently dislocated joint in the body. It is, in fact, the classical example of a joint which has sacrificed stability for mobility.

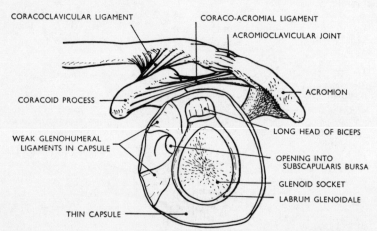

Fig. 137. Side view of the shoulder joint after removal of the humerus. The capsule and the intrinsic ligaments are weak and lax, but the extrinsic coracoacromial ligament is very strong.

The joint capsule is attached to the margins of the glenoid socket and labrum medially, and laterally to the anatomical neck of the humerus, except on its medial side, where the attachment dips down on to the surgical neck. The capsule (Fig. 137) is weak and lax, and the joint depends for protection entirely on the strong muscles which closely surround it and serve as 'extensile ligaments' (Fig. 138). The joint is well protected by such muscles in front, above, and behind, but not inferiorly. There are no important ligaments in the joint capsule, but the strong **coracoacromial** ligament superior to the joint forms a kind of supernumerary socket which prevents the humerus being dislocated upwards by a fall on the outstretched hand (Figs. 137, 138).

Movements of the humerus on the scapula are difficult to dissociate from movements of the shoulder girdle, for normally both always occur together. The glenoid socket looks forwards as well as outwards, and abduction and adduction are defined as taking place in the plane of the body of the scapula. Flexion and extension are at right angles to this plane, so that flexion is a movement forwards and medially across the front of the body. Extension moves the arm away from the body as well as backwards (p. 44). When the arm is moved above the head, whether the path followed is one of flexion or of abduction, fully half of the total movement is due to the shoulder girdle; even so, the shoulder joint is the most mobile joint in the body. The most limited of its movements (taking the zero point as the anatomical position) is that of extension (about 60°), and the least limited is flexion (about 160°). Adduction is limited by the chest wall, and abduction takes place to about 90°. If the arm is moved away from the body in the coronal plane, the humerus rotates laterally through 90° because of the pull of the scapulohumeral muscles; if lateral rotation is deliberately prevented, by medially rotating the arm, shoulder joint

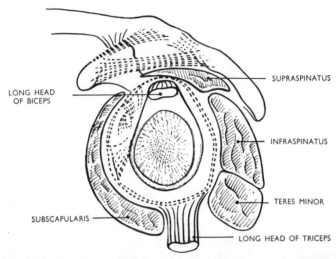

Fig. 138. Side view of the shoulder joint after removal of the humerus, showing the 'rotator cuff' of muscles (p. 111) which surrounds the joint and act as extensile ligaments. The capsule and ligaments are indicated by dotted lines (compare with Fig. 137).

movement is so limited that it is impossible to raise the arm above the head. With the arm at the side, about 70° of medial and 70° of lateral rotation of the humerus are possible.

If a patient falls on his elbow with the arm abducted, the greater tubercle of the humerus may impinge on the acromion and lever the head of the humerus downwards out of its socket. Nearly all dislocations of the humerus take this path because it is inferiorly that the capsule is weakest and least protected by muscles. Dislocations are liable to recur with increasing frequency, and eventually a state may be reached where the joint will dislocate under such trivial stress as raising the arm to brush the hair.

The **elbow** joint is a pure hinge joint between the trochlea of the humerus and the trochlear notch of the ulna, but its simplicity is complicated by the intrusion into it of the **superior radioulnar** joint—a pure pivot joint. The radius, bound to the ulna by ligaments and muscles, travels with it during flexion and extension, articulating with the capitulum of the humerus. The only movements at the elbow joint are flexion and extension; extension is limited by the ligaments of the joint, and flexion by the degree of development of the muscles on the front of the arm and forearm. The humerus and the ulna are not in the same straight line (Fig. 142): the angle between them, which is called the 'carrying angle', brings the hand over towards the mouth when the elbow is flexed, and it is only by laterally rotating the shoulder that it is possible to touch the shoulder with the hand of the same side. When the palm faces medially the bones of the forearm are more nearly in line with the humerus; this minimizes the shearing strain at the elbow when weights are carried.

The joint capsule is common to both elbow and superior radioulnar joints. It is attached round the articular surfaces on the humerus, both epicondyles being excluded, and to the coronoid process of the ulna. The lateral part of the capsule blends distally with the **anular** ligament of the superior radioulnar joint, which is attached to the sides of the radial notch of the ulna, and forms with it a ring in which the neck of the radius rotates. The capsule of the elbow, as in all hinge joints, is weak in front and behind, but laterally and medially strong ligaments are developed in it. The **radial collateral** ligament runs from the lateral epicondyle down to the annular ligament, and the triangular **ulnar collateral** ligament connects the medial epicondyle, the olecranon, and the coronoid process (Figs. 139, 140).

The **inferior radioulnar** joint is another pivot joint. The medial margin of the lower end of the radius is attached to the base of the styloid process of the ulna by a triangular plate of fibrocartilage. So tethered by the articular 'disc', the lower end of the radius moves in an arc of a circle round the ulnar styloid. The inferior aspect of the disc forms part of the socket of the radiocarpal joint, and the superior aspect moves over the head of the ulna during rotation of the radius. The cavity of the inferior radioulnar joint is thus L-shaped (Fig. 141).

Many structures other than the capsules of the two radioulnar joints connect the radius to the ulna. The chief of these is the **interosseous membrane**, a fibrous sheet attached to the interosseous borders of the two bones. The fibres in it run distally and medially from radius to ulna.

Movement at the radioulnar joints takes place round an axis passing through the head of the radius and the styloid process of the ulna (Fig. 142). Starting from the anatomical position, movement so that the thumb turns towards the midline of the body and the palm faces backwards is called **pronation**; movement in the reverse direction is **supination**. When the hand is fully supinated, as in the anatomical position, the bones of the forearm are parallel, but in full pronation the radius crosses over the ulna (Fig. 142). The interosseous membrane is tight in the position midway between full pronation and full supination. The ulna moves backwards and forwards slightly during these movements; the radius alone rotates, carrying the wrist and hand with it. Pronation and supination are most effective with the elbow at a right angle, but

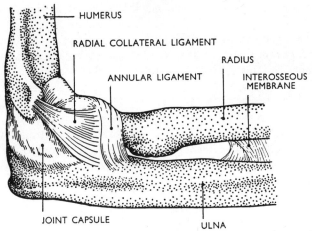

Fig. 139. The right elbow joint from the lateral side. The forearm is turned so that the thumb points upwards.

rotation of the hand relative to the trunk is greatest when the elbow is extended, for then the rotation of the forearm can be supplemented by rotation of the arm at the shoulder joint. The two movements are always associated, and are difficult to dissociate voluntarily. It used to be said that supination was a stronger movement than pronation, and that this was the reason why screws were usually made with a right hand thread, so that they could be screwed in by a movement of supination. Actual measurements, however, show little to choose

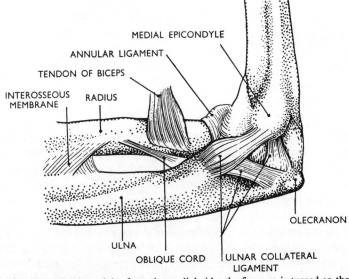

Fig. 140. The right elbow joint from the medial side; the forearm is turned so that the thumb points upwards.

between the power of the two movements, the verdict being slightly in favour of pronation.

The **radiocarpal** joint is an ellipsoid joint between the socket formed by the lower end of the radius and the articular disc and the convex ovoid mass formed by the scaphoid, lunate, and triquetrum. In the anatomical position, the scaphoid and lunate make contact with the socket, but the triquetrum is related to the medial ligament of the joint; it is only when the joint is adducted that the triquetrum moves round to come in contact with the articular disc (Fig. 141). The capsule is relatively weak in front and behind, and strong at each side, where an **ulnar collateral** ligament runs from the ulnar styloid to the triquetrum, and a **radial collateral** ligament from the radial styloid to the scaphoid.

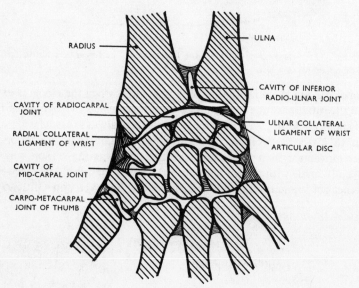

Fig. 141. Diagram of a section through the inferior radioulnar joint, the radiocarpal joint, and the midcarpal joint.

The carpal bones are closely united by strong ligaments, but the plane joints between adjacent bones allow of gliding movements. The important **midcarpal** joint between the proximal and distal rows of the carpus is also an ellipsoid joint, and a great deal of the movement of the hand relative to the forearm takes place here. It is impossible to separate radiocarpal from intercarpal movements, and both must be considered together as 'movements of the wrist'. In flexion of the hand on the forearm the midcarpal joint plays a very large part; extension is mostly due to radiocarpal movement. Normally about 80–90° of flexion are possible, but extension is rather less free. Adduction of the hand is also called ulnar deviation, and is much more extensive (about 45°) than abduction (radial deviation), which has a range of only 10–15° from the anatomical position (p.

47). In adduction movement occurs mostly at the radiocarpal joint, but abduction takes place almost entirely at the midcarpal joint. As in all condyloid joints, clumsy circumduction is possible, but there is no rotation.

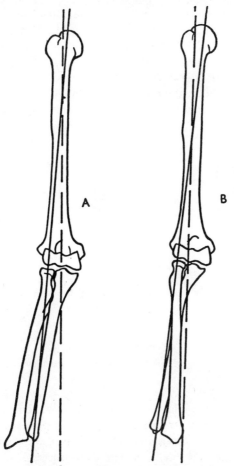

Fig. 142. Supination (A) and pronation (B). The solid line is the axis round which the movement takes place; the dashed line is the axis of the humerus. Note that the 'carrying angle' at the elbow is largely undone in pronation, when the lower end of the radius rotates round to the medial side of the ulna.

In the anatomical position, from which the figures for adduction and abduction are measured, the middle finger is in line with the bones of the forearm. This is not, however, the position of rest of the wrist. If the hand is allowed to 'flop' it will be found that in the true position of rest the middle finger is in line with the shaft of the humerus, and that there is thus a compensatory angulation at the wrist which exactly neutralizes the carrying angle at the elbow. Abduction and adduction from this position are more or less equal in range.

The **carpometacarpal** joint of the thumb is the best example in the human body of a saddle-shaped joint. It allows flexion and extension in the plane of the palm, adduction towards the plane of the palm, and abduction away from it (Fig. 143). Notice that movements of the thumb are at right angles to the correspondingly named movements of the fingers; the dorsal surface of the thumb faces laterally, and the front faces medially. The surfaces of the first carpometacarpal joint also allow about 30° of medial rotation, and this makes it possible for the pad of the thumb to be brought into contact with the pads of the other fingers in the extremely important movement of **opposition**.

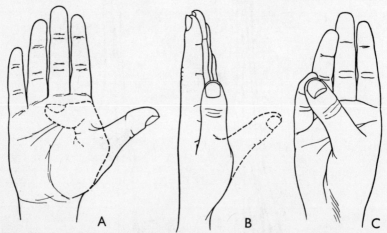

Fig. 143. The movements of the thumb. A: flexion and extension. B: adduction and abduction. C: opposition.

The carpometacarpal joints of the index, middle and ring fingers permit very little movement, but the metacarpal of the little finger can be slightly flexed and rotated across the palm towards the metacarpal of the thumb. This movement 'cups' the medial side of the hand as objects are grasped in the palm.

The ellipsoidal **metacarpophalangeal** joints are mainly concerned with flexion and extension, but allow a considerable range of abduction and adduction, circumduction, and also rotation, particularly in the thumb and index finger. There are strong **collateral** ligaments on each side of the joints, and anteriorly a hard fibrous plate is fixed firmly to the base of the phalanx, and weakly to the neck of the metacarpal; it moves proximally and distally as the joint is flexed and extended. Dorsally the capsule of the joint is replaced by the extensor tendon of the finger (p. 319). In the fingers the movements of abduction and adduction are considered to take place away from and towards an imaginary line drawn through the neutral position of the middle finger (p. 44). Adduction of the little finger thus occurs in the opposite sense to adduction of the index; the middle finger can be abducted away from its imaginary axis in either direction, and adducted back to it again. When the metacarpophalangeal joints are fully flexed, the collateral ligaments tighten up so that abduction and adduction become impossible.

The **interphalangeal** joints are pure hinge joints allowing flexion and extension only. The lengths of the phalanges are such that when the fingers are flexed all the finger tips touch the palm simultaneously. The interphalangeal joints of the fingers cannot be extended beyond a straight line. The distal joints flex about 80° and the proximal ones about 100°; the metacarpophalangeal joints have approximately 90° of flexion and 30° of extension beyond a straight line.

The interphalangeal joint of the thumb flexes by about 80° and extends beyond a straight line by about 20°. The metacarpophalangeal joint contributes about 75° of flexion and 10° of extension, and the carpometacarpal joint about 15° of flexion and 20° extension. Abduction of the thumb to about 60° is permitted, the movement occurring at the carpometacarpal and metacarpophalangeal joints.

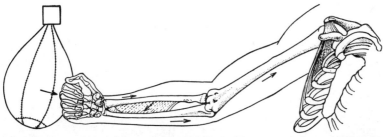

Fig. 144. Transmission of force in the upper limb. The arrows indicate the direction of the stresses set up.

Transmission of force in the upper limb

When a blow is struck with the fist, the force is taken by the heads of the metacarpals and is transmitted—in particular through the third metacarpal—to the carpal bones. Most of the force then passes through the line of the capitate and the scaphoid bones on to the lower end of the radius. The radius is not driven up against the humerus, because the fibres of the interosseous membrane transfer the force to the shaft of the ulna and so to the elbow joint and the humerus (Fig. 144). The humerus is driven upwards and the shock is taken by muscles which transfer the force to the scapula, and so to the muscles which tether the scapula down on the chest wall. Much of the shock also passes along the clavicle to the sternum, and a fall on the outstretched hand is the commonest cause of fracture of the clavicle.

When weight is taken in the hand the claw-like upper end of the ulna transfers the strain to the humerus and so to the muscles surrounding the shoulder joint, which prevent the humerus being dragged out of its socket. The force is also transferred to the clavicle by the muscles running from the clavicle

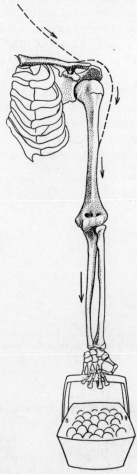

Fig. 145. Transmission of force in the upper limb. The arrows indicate the direction of the pull of gravity.

to the humerus and by the coraco-clavicular ligament. The clavicle is in turn suspended by muscles from the skull and vertebral column, and it is for this reason that carrying a heavy shopping basket may result in a pain at the back of the neck (Fig. 145).

Muscles

In the upper limb the origins of muscles are their proximal attachments, for their job is usually to move the hand about relative to the trunk. Nevertheless, the upper limb is often used to move the body about in such activities as climbing, crawling or swimming, and the muscles then act from their distal 'insertions' on their proximal 'origins'. Similarly, certain limb muscles can, when their distal attachments are fixed, act as accessory muscles of inspiration (p. 271) by helping to raise the ribs.

For obvious reasons the main arteries and nerves of a limb usually take the shortest and most sheltered course on their way to their destination. At the axilla (p. 341), elbow, and wrist the arrangement of the muscles is such as to produce the minimum disturbance of the great nerves and vessels when the joints are moved.

Muscles acting on the shoulder girdle

All muscles passing between the trunk and the clavicle, scapula, or humerus can produce movement at the sternoclavicular joint. Any movement of the scapula round the chest wall, whether in response to trunk-scapular or to trunk-humeral muscles, is necessarily accompanied by movement at the acromioclavicular joint (p. 296). Three main muscles have their primary action on these joints.

The **trapezius** has a long origin extending from the external occipital protuberance to the spine of the twelfth thoracic vertebra. In between these points it is attached to the ligamentum nuchae and the spines and supraspinous ligaments of the thoracic vertebrae. The fibres converge to be inserted into the crest of the spine of the scapula, the medial border of the acromion, and the posterior border of the lateral third of the clavicle (Fig. 146). The lower fibres depress the scapula; the upper fibres raise it, and take much of the weight of the limb. If the upper and lower fibres contract together, the scapula is twisted round like a wing-nut so that its inferior angle travels forwards and laterally (Fig. 147). The whole muscle acting at once retracts the scapula—i.e. pulls it back towards its fellow. The trapezius is supplied by the spinal part of the accessory nerve (p. 264).

The **serratus anterior** arises from the outer surfaces of the upper eight ribs on the lateral aspect of the chest and passes backwards, flat along the thoracic wall, to lie deep to the scapula and gain insertion to its medial border. It is supplied by the long thoracic nerve (p. 328), and it pulls the scapula forwards round the trunk (protraction). It is thus used in pushing movements or to prevent the scapula being forced backwards and upwards, as when the body is supported on the hands. In conjunction with the trapezius, the lower fibres of the muscle can draw the inferior angle of the scapula forwards and laterally in the movement of raising the arm above the head (p. 297). The serratus anterior also stabilizes the origins of other muscles from the scapula, and in this function it is assisted by a slip from the latissimus dorsi (p. 310) to the inferior angle of the scapula.

The **pectoralis minor,** a relatively small muscle, arises from the outer surfaces of the third, fourth and fifth costal cartilages and is inserted into the medial border of the coracoid process of the scapula (Fig. 148). It is supplied by fibres from the medial and lateral cords of the brachial plexus (p. 328), and it helps the lower fibres of the trapezius to depress the shoulder or to prevent it being driven upwards when the weight of the body is supported by the hands. If

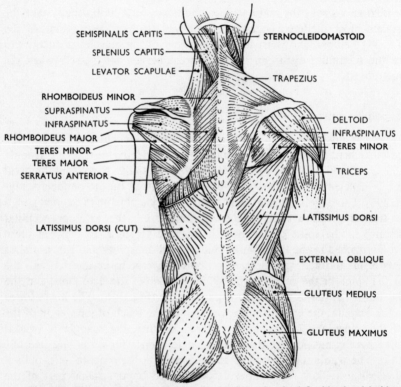

SEMISPINALIS CAPITIS
SPLENIUS CAPITIS
LEVATOR SCAPULAE
STERNOCLEIDOMASTOID
TRAPEZIUS
RHOMBOIDEUS MINOR
SUPRASPINATUS
INFRASPINATUS
RHOMBOIDEUS MAJOR
TERES MINOR
TERES MAJOR
SERRATUS ANTERIOR
DELTOID
INFRASPINATUS
TERES MINOR
TRICEPS
LATISSIMUS DORSI
LATISSIMUS DORSI (CUT)
EXTERNAL OBLIQUE
GLUTEUS MEDIUS
GLUTEUS MAXIMUS

Fig. 146. The muscles of the back and shoulder. On the left side the deltoid and trapezius have been removed and the latissimus dorsi cut short, in order to display the deeper muscles.

the scapula is fixed, the pectoralis minor can raise the ribs, and is thus an accessory muscle of inspiration (p. 271).

In addition to these main muscles three subsidiary ones lie deep to the trapezius and connect the medial border of the scapula to the vertebrae. The **levator scapulae** arises from the transverse processes of the first four cervical vertebrae and is inserted superior to the spine of the scapula. The **rhomboideus minor** and the **rhomboideus major** (Fig. 146) come from the lower part of the ligamentum nuchae and the spines of the upper thoracic vertebrae and are inserted opposite and inferior to the spine of the scapula respectively. All these muscles are supplied by the ventral ramus of the fifth cervical nerve, and the levator scapulae receives additional branches from the ventral rami of the third and fourth cervical nerves. They raise the scapula and retract it.

The pectoralis major and latissimus dorsi (p. 310) also affect the shoulder girdle because of their attachments to the trunk and to the humerous, but are dealt with in relation to the shoulder joint.

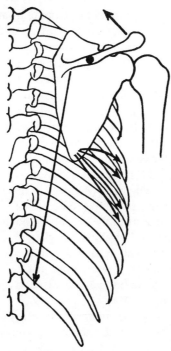

Fig. 147. The combined action of the trapezius and serratus anterior on the scapula; the arrows indicate the direction of pull of the muscles. The upper and lower fibres of the trapezius form a couple which rotates the scapula round the point indicated by the black dot. At the same time the lower fibres of the serratus anterior, massed on the inferior angle of the scapula, pull it laterally and forwards round the chest wall. The net result is that the glenoid socket turns to face upwards and forwards.

Summary of movements of shoulder girdle

Elevation:	**Trapezius,** levator scapulae, rhomboids.
Depression:	**Trapezius, pectoralis minor,** pectoralis major, latissimus dorsi
Retraction:	**Trapezius,** levator scapulae, rhomboids, latissimus dorsi
Protraction:	**Serratus anterior,** pectoralis major.
Rotation:	**Serratus anterior;** upper and lower fibres of **trapezius.**
Stabilization:	All the muscles passing between the shoulder girdle and the trunk; **serratus anterior** in particular.

Muscles acting on the shoulder joint

Because the shoulder joint has little ligamentous and no bony strength (p. 298), it is closely covered by a group of small powerful muscles which hold the humerus on to the scapula, and are called the **rotator cuff** (Fig. 138). The **subscapularis** takes origin from the costal surface of the scapula and runs upwards in front of the joint to be inserted into the lesser tubercle of the humerus. It is supplied by two nerves from the posterior cord of the brachial plexus (p. 328).

Above the joint lies the **supraspinatus**, coming from the supraspinous fossa of the scapula and inserting into the greater tubercle of the humerus, and behind is the **infraspinatus,** arising from the infraspinous fossa, and inserting posterior to the supraspinatus on the greater tubercle. Both are supplied by the suprascapular nerve, a branch of the upper trunk of the brachial plexus (p. 328).

Inferior to the infraspinatus lies the **teres minor** (Fig. 146), supplied by the axillary nerve (p. 336) and coming from the upper part of the lateral border of the scapula to be inserted into the lowest part of the greater tubercle posteriorly. All these muscles (Fig. 138) primarily protect the joint, but all can produce movement at it. Thus, the subscapularis medially rotates the humerus, the supraspinatus abducts it, the infraspinatus laterally rotates it, and the teres minor laterally rotates and adducts it. (All these actions may be readily deduced from the positions of the muscles relative to the joint.)

Helping to hold the humerus to the scapula is the long head of the biceps (p. 312), which takes origin from just above the glenoid socket; the tendon braces across the upper part of the joint. When the arm is abducted, the tendon of the long head of the triceps (p. 313), originating just below the socket, comes to underlie the head of the humerus, and so protects the otherwise unguarded inferior part of the capsule.

Four large muscles and one smaller one are primarily concerned with producing movement at the shoulder joint. The **pectoralis major** arises from the anterior surface of the sternum and the cartilages of the first six ribs, and also from the medial half of the clavicle on its anterior surface. The fibres converge to be inserted on the lateral lip of the intertubercular sulcus on the humerus in such a way that the upper fibres twist round the lower ones (Fig. 148), so forming the rounded anterior fold of the armpit, or **axilla**. The pectoralis major adducts the arm and medially rotates it. In addition, the upper (clavicular) fibres help to raise the arm in flexion, while the lower (sternocostal) fibres pull it back from the flexed position. The sternocostal fibres also protract and depress the shoulder girdle. Acting from its insertion, the pectoralis major is one of the main climbing muscles, for it can raise the trunk to the fixed arm; it can also act as an accessory muscle of inspiration (p. 271). It is supplied, like the pectoralis minor, by branches from the lateral and medial cords of the brachial plexus.

The **latissimus dorsi** has an extensive origin from the spines of the lower six thoracic vertebrae, and, by means of a sheet of the thoracolumbar fascia (Fig.

126), from the spines of the lumbar and sacral vertebrae and the posterior part of the iliac crest. The fibres run upwards and laterally (Fig. 146) over the inferior angle of the scapula, to which some of them become attached; the remainder insert by a flattened tendon into the floor of the intertubercular sulcus of the humerus. The muscle is supplied by the thoracodorsal nerve (p. 328). It is a powerful adductor of the arm and thus an important climbing muscle; it is also one of the chief extensors of the humerus, and comes into action in rowing and in the downstroke of the arm in swimming. It depresses and retracts the

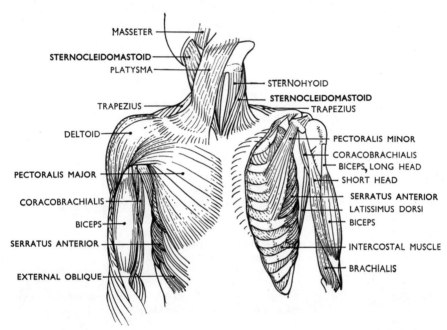

Fig. 148. The muscles of the front of the shoulder region. On the left side of the body the pectoralis major and deltoid have been cut away to expose the deeper muscles.

shoulder girdle. The fact that its nerve supply comes from the cervical region makes this muscle important in patients with paralysis of the lower limbs from a spinal cord injury. The latissimus dorsi, with its more proximal nerve supply still intact, can be used from its insertion to produce movement of the pelvis, and in this way the patient may achieve some independence of movement. In hospitals where spinal injuries are treated, efforts are made to strengthen the muscle by such exercises as archery and basket ball.

The **teres major** (Fig. 146) arises from the lower third of the lateral border of the scapula, inferior to the teres minor, and is inserted into the medial lip of the intertubercular sulcus. It is supplied from the posterior cord of the brachial plexus (p. 328), and it adducts and medially rotates the humerus.

The large fleshy **deltoid** muscle overlies the shoulder laterally (Figs. 146, 148). It comes from the crest of the spine of the scapula, the lateral border of the acromion, and the anterior border of the lateral third of the clavicle, and is inserted into the deltoid tuberosity of the humerus, half way down the outer aspect of the shaft. It is supplied by the axillary nerve (p. 336), and is the chief abductor of the arm. Its anterior fibres are also the chief flexors of the arm, and the posterior fibres help the latissimus dorsi to extend it. The anterior fibres are medial rotators, and the posterior fibres lateral rotators. In fact, some part of the deltoid contracts during almost any movement at the shoulder, probably to help to hold the head of the humerus into its socket (p. 298). Even in adduction, the posterior fibres of the muscle may be in action, perhaps to resist the medial rotation which would otherwise be produced by pectoralis major and latissimus dorsi.

The **coracobrachialis** is a much smaller muscle than the others, and arises from the tip of the coracoid process of the scapula. It is inserted into the medial aspect of the humerus opposite the deltoid tuberosity, and helps to adduct and flex the shoulder joint. It is supplied by the musculocutaneous nerve (p. 335).

Summary of movements at the shoulder joint

Flexion:	**Deltoid,** clavicular head of pectoralis major, coracobrachialis and short head of biceps.
Extension:	**Latissimus dorsi,** sternocostal head of pectoralis major, posterior fibres of deltoid.
Abduction:	**Deltoid,** supraspinatus.
Adduction:	**Pectoralis major, latissimus dorsi,** teres major, teres minor.
Lateral rotation:	**Infraspinatus,** teres minor, deltoid.
Medial rotation:	**Pectoralis major, teres major,** latissimus dorsi, deltoid, subscapularis.
Stabilization:	All muscles crossing the joint, particularly **subscapularis, supraspinatus, infraspinatus and teres minor.**

Muscles acting on the elbow joint

Only two movements are permitted at the elbow—flexion and extension. There are three main flexors. The **brachialis** arises from the anterior surface of the lower half of the humeral shaft and is inserted into the coronoid process and the tuberosity of the ulna. A few fibres are attached to the capsule of the joint. As the joint flexes, the anterior part of the capsule becomes slack, and might be nipped between the moving bones were it not that these fibres pull it up out of the way. The muscle is supplied by the musculocutaneous nerve (p. 335).

Like many an honest worker, the brachialis is obscured by a more showy partner, the **biceps brachii**, which lies in front of it. The long head of the biceps comes from the upper edge of the glenoid socket of the scapula, and the tendon runs through the upper part of the joint to emerge into the intertubercular sulcus. The short head arises, with the coracobrachialis, from the tip of the coracoid process, and the two heads unite into a fleshy belly which tapers again to be inserted by tendon into the posterior aspect of the radial tuberosity (Figs. 140, 148, 150). From the medial side of the tendon a strong sheet of fibrous tissue, the bicipital aponeurosis, passes medially to blend with the fascia over the flexor muscles of the forearm. The biceps is also supplied by the musculocutaneous nerve, but unlike the brachialis it has several actions. Not only does it flex the elbow, but it steadies the humerus at the shoulder joint and helps to flex it. It is one of the chief muscles of supination (p. 300), for it pulls the posterior part of the radial tuberosity forwards, causing the radius to rotate laterally. The biceps has been said to 'put in the corkscrew and pull out the cork'.

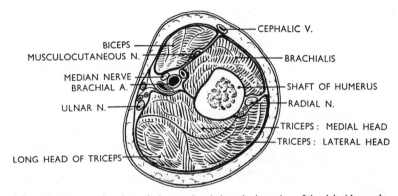

BICEPS
MUSCULOCUTANEOUS N.
MEDIAN NERVE
BRACHIAL A.
ULNAR N.
LONG HEAD OF TRICEPS
CEPHALIC V.
BRACHIALIS
SHAFT OF HUMERUS
RADIAL N.
TRICEPS : MEDIAL HEAD
TRICEPS : LATERAL HEAD

Fig. 149. Cross-section through the arm just below the insertion of the deltoid muscle.

The third main flexor of the elbow is the **brachioradialis** (Figs. 151, 153), which arises from the upper two-thirds of the lateral border of the humerus and from the strong lateral intermuscular septum which is attached to this border and separates the flexors from the extensors. The muscle crosses in front of the lateral part of the elbow joint and runs down to be inserted at the base of the styloid process of the radius. It is supplied by the radial nerve (p. 333). The brachioradialis is particularly active as a 'shunt' muscle during rapid flexion.

Several muscles flex the elbow as a subsidiary duty. The common origin of the forearm flexor muscles is on the front of the medial epicondyle of the humerus, and that of the extensors is from the front of the lateral epicondyle. Many of these muscles, whose primary action is on the wrist and hand, thus pass in front of the elbow joint, and so are able to flex it. Perhaps the most important in this respect are the pronator teres (p. 314) and the extensor carpi radialis longus (p. 315).

There is only one important extensor of the elbow. The **triceps** (Fig. 149), as its name implies, has three origins. The long head comes from the inferior part of the glenoid socket of the scapula. The large medial head comes from the posterior aspect of the shaft of the humerus below and medial to the radial groove; the smaller lateral head arises above and lateral to this groove. The triceps is inserted by a tendon into the olecranon process of the ulna, and supplies a few fibres to the posterior part of the capsule of the elbow joint, to pull it out of the way of the bones as the joint extends. In addition to extending the elbow, the triceps supports the abducted shoulder joint. It is supplied by the radial nerve (p. 333).

The other extensor of the elbow is the **anconeus**, a small triangular muscle coming from the back of the lateral epicondyle of the humerus and inserted into the upper part of the posterior surface of the ulna.

Summary of movements at the elbow joint

Flexion:	**Brachialis, biceps, brachioradialis,** flexor and extensor muscles of the wrist.
Extension:	**Triceps,** anconeus.

Muscles acting on the radioulnar joints

Of the two main supinators, the **biceps brachii** has already been described (p. 312). The other is the **supinator** muscle which lies deeply in the upper part of the forearm; it arises from the radial collateral ligament of the elbow joint and the lateral epicondyle of the humerus, and also from the lateral part of the shaft of the ulna adjacent to the coronoid process. The fibres pass laterally and downwards posterior to the shaft of the radius and are inserted all round the upper third of the shaft like a string wound round a top. When the muscle contracts the radius 'unwinds' by rotating laterally (Fig. 150). The supinator is supplied by the radial nerve (p. 333).

Assisting the biceps and the supinator is the extensor pollicis longus (p. 323), which is able to supinate because it comes from the shaft of the ulna and runs obliquely laterally across the forearm to be inserted into the thumb.

The first of the two main pronators is the **pronator teres** (Fig. 151), which comes from the 'common flexor origin' on the front of the medial epicondyle, and from the medial part of the coronoid process of the ulna. The whole muscle passes obliquely across to be inserted on a rough area half-way down the lateral surface of the shaft of the radius. This insertion is at the point of maximum curvature of the radial shaft, and the muscle gets the best possible purchase. It is supplied by the median nerve (p. 329), and can also flex the elbow.

The **pronator quadratus** is a thick quadrangular muscle which runs straight across the lower quarter of the front of the ulna to the lower quarter of the front of the radius (Fig. 152). It is supplied by the median nerve (p. 329).

Two flexor muscles, the flexor carpi radialis and the variable palmaris longus, pass obliquely across the forearm from medial to lateral side, and so can assist in pronation.

Summary of movements at radioulnar joints

Supination:	**Biceps brachii, supinator,** extensor pollicis longus.
Pronation:	**Pronator teres, pronator quadratus,** flexor carpi radialis, palmaris longus.

Muscles acting on the radiocarpal and midcarpal joints

Many muscles cross the wrist to enter the hand. Some of them stop short at the carpus, but others act primarily upon the digits. Whatever the primary action, every tendon crossing the wrist is capable of acting upon it.

The three primary flexors of the radiocarpal and midcarpal joints all arise from the common flexor origin in front of the medial epicondyle of the humerus (Fig. 151). The **flexor carpi radialis** is the most lateral of the three. It is inserted into the anterior aspect of the bases of the second and third metacarpals, and can abduct the wrist as well as flex it. The muscle also flexes and pronates the forearm; it is supplied by the median nerve (p. 329).

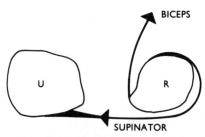

Fig. 150. Schematic cross-section through the bones of the forearm, showing the direction of pull of the biceps and supinator. Both muscles rotate the radius laterally, in the movement of supination. U, ulna; R, radius.

The **palmaris longus** is a thin muscle which is not always present. Its tendon crosses the wrist almost exactly in the midline, and can be made to stand out by flexing the wrist against resistance; it is just medial to the tendon of the radial flexor. It is inserted into the palmar aponeurosis (p. 325) and is a pure flexor of the wrist; it can also help in flexion and pronation of the forearm.

The **flexor carpi ulnaris** comes from the common origin but has an additional attachment to the medial side of the olecranon process and the upper part of the posterior border of the ulna. Like the other two, it becomes tendinous halfway down the forearm, and is inserted into the base of the fifth metacarpal and the hook of the hamate. The pisiform lies in the substance of the tendon like a sesamoid bone. It is supplied by the ulnar nerve (p. 331), and can also adduct the hand.

To balance the three flexor muscles there are three primary extensors, all of which are supplied by the radial nerve (p. 333).

The **extensor carpi radialis longus** (Fig. 153) comes from the lower part of the lateral border of the humerus, distal to the origin of the brachioradialis, and is inserted by a long tendon into the dorsum of the base of the second metacarpal. It helps to flex the elbow, and extends and abducts the wrist.

The **extensor carpi radialis brevis** arises from the common extensor origin in front of the lateral epicondyle, and is inserted by a long tendon into the dorsum of the base of the third metacarpal. Its actions resemble those of the long radial extensor.

The **extensor carpi ulnaris** comes from the common origin, but, like the ulnar flexor, it has an additional origin from the posterior border of the ulna. Again like the flexor, the extensor inserts into the medial side of the base of the fifth metacarpal. It extends the wrist and adducts it.

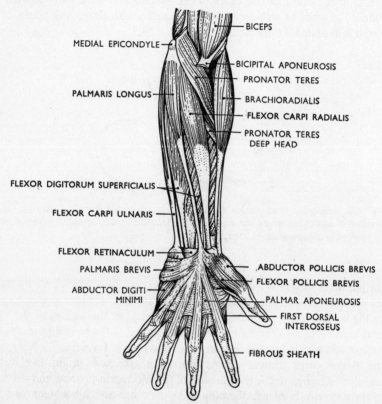

Fig. 151. Superficial muscles of the front of the forearm and hand.

Summary of movements at the wrist

Flexion:	**Flexor carpi radialis, flexor carpi ulnaris,** palmaris longus, flexor muscles of fingers and thumb.
Extension:	**Extensor carpi radialis longus** and **brevis, extensor carpi ulnaris,** extensors of fingers and thumb.
Abduction:	**Extensor carpi radialis longus, extensor carpi radialis brevis:** abductor pollicis longus, extensor pollicis brevis (p. 323); **flexor carpi radialis.**
Adduction:	**Flexor carpi ulnaris, extensor carpi ulnaris.**

Muscles acting on the fingers

There are two main flexors of the fingers. The **flexor digitorum superficialis** (Figs. 151, 155) arises from the common flexor origin, the coronoid process of the ulna, and the anterior border of the radius. It divides into four tendons which cross the wrist and fan out to the fingers. As they approach their insertion

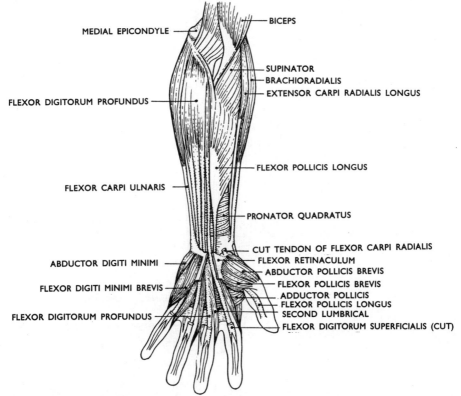

BICEPS

MEDIAL EPICONDYLE

SUPINATOR
BRACHIORADIALIS
EXTENSOR CARPI RADIALIS LONGUS

FLEXOR DIGITORUM PROFUNDUS

FLEXOR POLLICIS LONGUS

FLEXOR CARPI ULNARIS

PRONATOR QUADRATUS

CUT TENDON OF FLEXOR CARPI RADIALIS
FLEXOR RETINACULUM
ABDUCTOR DIGITI MINIMI
ABDUCTOR POLLICIS BREVIS
FLEXOR DIGITI MINIMI BREVIS
FLEXOR POLLICIS BREVIS
ADDUCTOR POLLICIS
FLEXOR POLLICIS LONGUS
SECOND LUMBRICAL
FLEXOR DIGITORUM PROFUNDUS
FLEXOR DIGITORUM SUPERFICIALIS (CUT)

Fig. 152. Deep muscles of the front of the forearm and hand. The brachioradialis has been pulled laterally to show the extensor carpi radialis longus which underlies it; the tendon of this muscle passes round on to the back of the wrist.

each divides into two slips, one going to each side of the middle phalanx. Through the tunnel thus formed the flexor digitorum profundus tendon runs to the terminal phalanx. The superficial flexor bends the proximal interphalangeal and metacarpophalangeal joints of all the fingers, and also helps to flex the midcarpal, radiocarpal and elbow joints. It is supplied by the median nerve (p. 329).

The **flexor digitorum profundus** comes from the anterior surface of the ulna proximal to the pronator quadratus, and also from the front of the interosseous membrane (Figs. 152, 155). It too divides into four tendons, each of which is inserted into the front of the base of the distal phalanx of a finger, reaching its destination by tunnelling behind the tendon of the superficialis. The muscle flexes all the joints of all the fingers, as well as the radiocarpal and midcarpal joints. The part of the muscle going to the index and middle fingers is supplied

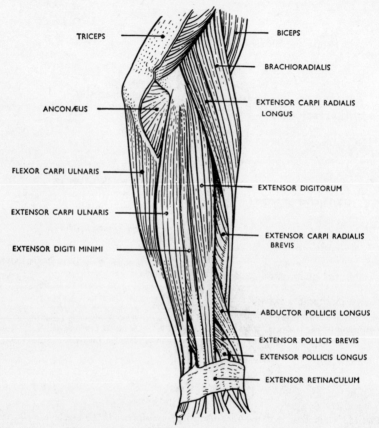

TRICEPS

BICEPS

BRACHIORADIALIS

EXTENSOR CARPI RADIALIS LONGUS

ANCONÆUS

FLEXOR CARPI ULNARIS

EXTENSOR DIGITORUM

EXTENSOR CARPI ULNARIS

EXTENSOR DIGITI MINIMI

EXTENSOR CARPI RADIALIS BREVIS

ABDUCTOR POLLICIS LONGUS

EXTENSOR POLLICIS BREVIS

EXTENSOR POLLICIS LONGUS

EXTENSOR RETINACULUM

Fig. 153. Superficial muscles of the back of the forearm.

by the median nerve (p. 329), and the part going to the ring and little fingers is supplied by the ulnar nerve (p. 331).

The little finger has a small flexor muscle of its own—the **flexor digiti minimi**—which comes from the hook of the hamate and inserts into the base of the proximal phalanx of the finger (Fig. 152). It is supplied by the ulnar nerve.

The extensor apparatus of the fingers is more complicated. The **extensor digitorum** (Fig. 153) comes from the common extensor origin in front of the lateral epicondyle of the humerus, and crosses over the wrist as a bundle of four tendons which fan out to the four fingers. Over the dorsum of the proximal phalanx of each finger the extensor tendon expands into a membranous hood which replaces the capsule of the metacarpophalangeal joint. This expansion receives the insertion of the lumbricals and the interossei and is attached to the sides of the middle phalanx and the base of the dorsal surface of the distal

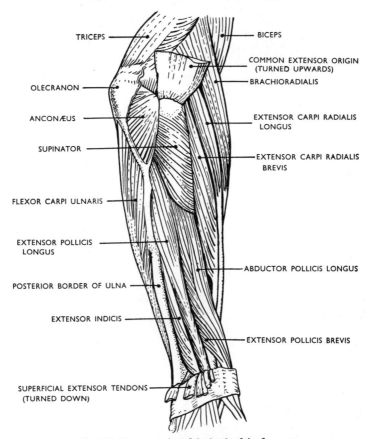

TRICEPS

BICEPS

COMMON EXTENSOR ORIGIN
(TURNED UPWARDS)

OLECRANON

BRACHIORADIALIS

ANCONÆUS

EXTENSOR CARPI RADIALIS
LONGUS

SUPINATOR

EXTENSOR CARPI RADIALIS
BREVIS

FLEXOR CARPI ULNARIS

EXTENSOR POLLICIS
LONGUS

POSTERIOR BORDER OF ULNA

ABDUCTOR POLLICIS LONGUS

EXTENSOR INDICIS

EXTENSOR POLLICIS BREVIS

SUPERFICIAL EXTENSOR TENDONS
(TURNED DOWN)

Fig. 154. Deep muscles of the back of the forearm.

phalanx (Fig. 156). The tendons of the extensor digitorum are connected with each other on the back of the hand, so that extension of the middle and ring fingers cannot be effected separately without difficulty. The index and little fingers, however, receive their own extensor muscles, and so can be extended independently.

The **extensor indicis** (Fig. 154) comes from the posterior surface of the ulna just below its middle, and is inserted into the extensor expansion of the index finger. The **extensor digiti minimi** is really only a partially separated portion of

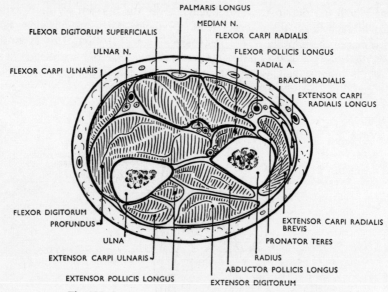

PALMARIS LONGUS

MEDIAN N.

FLEXOR DIGITORUM SUPERFICIALIS

FLEXOR CARPI RADIALIS

ULNAR N.

FLEXOR POLLICIS LONGUS

RADIAL A.

FLEXOR CARPI ULNARIS

BRACHIORADIALIS

EXTENSOR CARPI RADIALIS LONGUS

FLEXOR DIGITORUM PROFUNDUS

EXTENSOR CARPI RADIALIS BREVIS

ULNA

PRONATOR TERES

EXTENSOR CARPI ULNARIS

RADIUS

ABDUCTOR POLLICIS LONGUS

EXTENSOR POLLICIS LONGUS

EXTENSOR DIGITORUM

Fig. 155. Cross-section through the middle of the forearm.

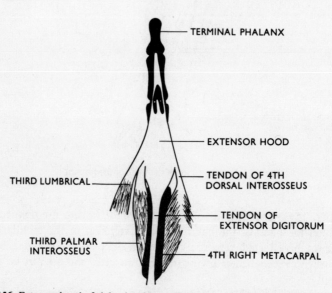

TERMINAL PHALANX

EXTENSOR HOOD

THIRD LUMBRICAL

TENDON OF 4TH DORSAL INTEROSSEUS

TENDON OF EXTENSOR DIGITORUM

THIRD PALMAR INTEROSSEUS

4TH RIGHT METACARPAL

Fig. 156. Extensor hood of right ring finger. The sides of the hood, which normally lie flat at the sides of the finger, at right-angles to the centre of the hood, have been pulled dorsally into the same plane in order to show the insertions of the muscles.

the extensor digitorum. The three extensor muscles extend all the joints of the fingers, and assist the extensors of the wrist in extending the radiocarpal and midcarpal joints. The extensor digitorum can spread (abduct) the fingers by hyper-extending the metacarpophalangeal joints. All three extensors are supplied by the radial nerve (p. 333).

The small cylindrical **lumbrical** muscles arise from the lateral sides of the four flexor digitorum profundus tendons in the palm, and, each swinging round the lateral side of its corresponding finger, are inserted into the extensor expansions. Their line of pull is in front of the metacarpophalangeal joint, which

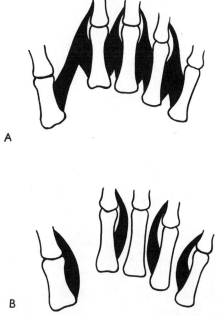

Fig. 157. Diagram of the attachments of the interossei. A: dorsal interossei. B: palmar interossei.

they therefore flex (Figs. 152, 158), but their insertion into the extensor expansion causes them to extend the two interphalangeal joints of each finger. The medial two muscles are supplied by the ulnar nerve, and the lateral two by the median nerve, though this arrangement is variable.

Also inserted into the extensor expansion are the tendons of the **interossei**, which are all supplied by the ulnar nerve. The interossei come from the shafts of the metacarpal bones, and their line of pull also passes in front of the metacarpophalangeal joints. Like the lumbricals, they flex these joints and extend the interphalangeal joints. The four **dorsal interossei** (Fig. 157) abduct the fingers from the imaginary line running through the middle finger (p. 44). They are larger than the palmar interossei, and each arises from the adjacent

surfaces of two metacarpals. The four **palmar interossei** (Fig. 157) adduct the
thumb and the fingers to the line of the middle finger, and each arises only from
the shaft of the metacarpal of its own digit. The lumbricals and interossei make
possible fine movements of individual fingers—they represent the fine
adjustment of the flexor-extensor mechanism, and if they are paralysed
movements like playing the piano, typing, or sewing become impossible. Without them the extensor digitorum cannot extend the phalanges against the tonic
pull of the flexors. The first dorsal interosseous muscle stabilizes the thumb and
adducts it to the index finger (p. 324).

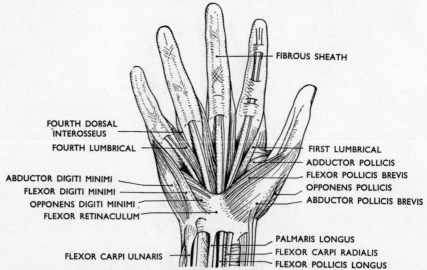

Fig. 158. The short muscles of the hand. The fibrous flexor sheath of the index finger
has been opened to expose the tendons it contains.

The little finger can be abducted away from the midline of the fingers by the
small but strong **abductor digiti minimi**, which arises from the pisiform and the
tendon of flexor carpi ulnaris and is inserted into the base of the proximal
phalanx of the little finger (Fig. 158). The **opponens digiti minimi** comes from
the hook of the hamate and inserts into the shaft of the fifth metacarpal. It can
roll the little finger partly round so that the palm is cupped. Both these small
muscles are supplied by the ulnar nerve.

Summary of movements at metacarpophalangeal joints of fingers

Flexion:	**Flexor digitorum profundus, flexor digitorum superficialis,** lumbricals, interossei. Flexor digiti minimi.
Extension:	**Extensor digitorum. Extensor indicis.** Extensor digiti minimi.
Abduction and Adduction:	**Interossei.** Lumbricals may help in abducting index and middle finger and in adducting ring and little finger. **Abductor digiti minimi.**
Rotation:	Lumbricals (specially index) and interossei.

Summary of movements at interphalangeal joints of fingers

Summary of movements at interphalangeal joints of fingers

Flexion: **Flexor digitorum profundus** (both joints); **flexor digitorum superficialis** (proximal joint only).

Extension: **Extensor digitorum; interossei; lumbricals.** Extensor indicis. Extensor digiti minimi.

Muscles acting on the thumb

There are two flexors, two extensors, two abductors and two adductors of the thumb, and a single muscle which opposes it to the fingers.

The **flexor pollicis longus** (Fig. 152, 155) is a deep muscle of the forearm, partner to the flexor digitorum profundus. It arises from the front of the shaft of the radius, proximal to the insertion of the pronator quadratus, and its tendon lies deep in the wrist on its way to be inserted into the front of the base of the distal phalanx of the thumb. It flexes all the joints of the thumb as well as the radiocarpal and midcarpal joints. (Remember that flexion of the thumb is by definition in a plane at right angles to that of flexion of the fingers.) The **flexor pollicis brevis** is a small muscle arising from the lateral side of the front of the carpus and inserting into the lateral side of the base of the proximal phalanx of the thumb. It flexes the carpometacarpal and metacarpophalangeal joints of the thumb. Both flexor muscles are supplied by the median nerve (but see p. 330).

The **extensor pollicis longus** comes from the dorsal surface of the ulna about its middle, just proximal to the origin of extensor indicis (Fig. 154), and also from the interosseous membrane. Its tendon runs straight down to the back of the lower end of the radius, where it hooks round the radial tubercle as it changes direction and runs obliquely laterally to insert into the dorsum of the base of the distal phalanx of the thumb, having on its way partly replaced the capsule of the metacarpophalangeal joint. The **extensor pollicis brevis** (Fig. 154) comes from a small area on the dorsal surface of the radius roughly corresponding to the origin of the extensor indicis from the ulna. The tendon of the muscle runs in close company with the tendon of the abductor pollicis longus, and with it forms the anterior boundary of the 'anatomical snuff-box—a depression on lateral side of the wrist bounded dorsally by the tendon of the extensor pollicis longus. It is conveniently placed for the taking of snuff. The extensor brevis also blends with the capsule of the metacarpophalangeal joint, and is inserted into the dorsum of the base of the proximal phalanx. Both extensors extend the wrist and the carpometacarpal and metacarpophalangeal joints of the thumb, but the extensor longus can also extend the interphalangeal joint and supinate the forearm. The extensor brevis can abduct the wrist. They are both supplied by the radial nerve.

The **abductor pollicis longus** (Fig. 154) comes from the dorsal surfaces of both radius and ulna, proximal to the two extensor muscles of the thumb, and also from the interosseus membrane between. Its tendon passes down on the lateral side of the wrist, accompanied by the tendon of the extensor pollicis brevis, to be inserted into the base of the first metacarpal. It can abduct the wrist and helps to flex it. The action on the thumb is a combination of abduction and extension; the thumb stands out at an angle of about 45° to the palm when the muscle is stimulated. The **abductor pollicis brevis** is a pure abductor, the thumb being pulled out at right angles to the palm. It arises from the tubercle of the scaphoid, and is inserted into the lateral side of the base of the first phalanx (Fig. 151). The long abductor is supplied by the radial nerve, the short abductor by the median nerve (but see p. 330).

The two adductors are the **adductor pollicis** and the first dorsal interosseus (p. 322). The adductor pollicis arises by two heads from the front of the carpus and the third metacarpal bone (Fig. 158); the insertion is to the medial side of the base of the first phalanx of the thumb. It pulls the thumb into the plane of the palm. Like the first dorsal interosseus, it is supplied by the ulnar nerve.

Perhaps the most important muscle of all, the **opponens pollicis**, arises from the radial side of the front of the carpus and is inserted along the radial half of the front of the shaft of the first metacarpal. It is supplied by the median nerve (see p. 330) and rolls the metacarpal round so that the pad of the thumb faces the pads of the other digits.

Summary of movements of thumb

Carpometacarpal joint

Flexion:	**Flexor pollicis longus; flexor pollicis brevis.**
Extension:	**Extensor pollicis longus; extensor pollicis brevis;** abductor pollis longus.
Abduction:	**Abductor pollicis longus; abductor pollicis brevis.**
Adduction:	**Adductor pollicis;** first dorsal interosseus.
Rotation:	**Opponens pollicis;** abductor pollicis brevis; flexor pollicis brevis.

Metacarpophalangeal joint

Flexion:	**Flexor pollicis longus: flexor pollicis brevis.**
Extension:	**Extensor pollicis longus; extensor pollicis brevis.**
Abduction:	**Abductor pollicis brevis.**
Adduction:	**Adductor pollicis.**
Rotation:	Abductor pollicis brevis: flexor pollicis brevis.

Interphalangeal joint

Flexion:	**Flexor pollicis longus.**
Extension:	**Extensor pollicis longus.**

Fascial arrangements in the wrist and hand

In front of the carpus the deep fascia of the forearm is thickened to form the **flexor retinaculum.** This dense fibrous sheet stretches from the pisiform and the hook of the hamate to the tubercles of the scaphoid and the trapezium, and so completes the bony and fascial **carpal tunnel** (Figs. 158, 159) through which the tendons of the long flexors of the fingers and the thumb travel on their way into the hand. This arrangement prevents the tendons springing away from the bones—a condition graphically called 'bowstringing'—and so losing control of the joints over which they pass as well as destroying the capacity of the hand to grip. Such a provision is even more necessary in the hand and fingers, and the flexor retinaculum is continuous distally with a dense sheet of fascia called the **palmar aponeurosis**, which lies in front of the tendons as they traverse the middle of the palm. On the medial side of the palmar aponeurosis lie the short muscles of the little finger, forming the **hypothenar eminence**, and on the lateral side the short muscles of the thumb give rise to the **thenar eminence**. In both these regions mobility is essential, and the fascia is thin and delicate.

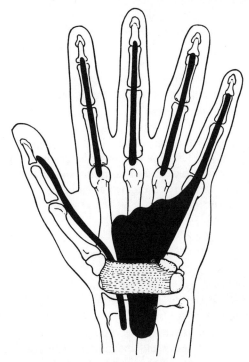

Fig. 159. The synovial sheaths of the flexor muscles. The extent and position of the sheaths are shown in black.

The palmar aponeurosis splits distally into four divisions, as the fingers of a glove stem from its palm. Each division is continuous with a dense fibrous sheath attached to the sides of the proximal and middle phalanges and to the base of the distal phalanx. This **fibrous digital sheath** prevents the flexor tendons bowstringing by providing them with a semi-rigid tunnel in which to operate (Fig. 158). The flexor tendons of the fingers thus play against a strong layer of deep fascia from the moment they enter the hand, and to reduce the friction they are provided with lubricant **synovial sheaths** (p. 43; Fig. 159). Each finger has its own digital synovial sheath, extending from the insertion of the profundus tendon to the neck of the metacarpal, and enveloping both the profundus and the superficialis tendons. At the wrist another large sheath wraps round the superficialis and profundus tendons, and the fifth digital synovial sheath communicates with this by a proximal extension (Fig. 167). The thumb has its own synovial sheath surrounding the tendon of the flexor pollicis longus. These sheaths are readily infected after injury to a finger or the thumb, and infection can spread up and down them, disintegrating the tendon they contain, and causing a permanently stiff finger.

On the back of the wrist a strong **extensor retinaculum** is attached to the radius and ulna, with separate compartments for the extensors as they pass into the hand. On the back of the hand the tendons communicate with each other (p. 319) forming a strong tendinous layer which partly replaces the deep fascia. The fingers cannot be extended much beyond a straight line, and thus there is no fibrous sheath on the dorsum of the finger. The attachment of the extensor expansion to the phalanges in itself effectively prevents any bowstringing in those people who have unusually mobile interphalangeal joints, and at the metacarpophalangeal point the restraining influence is the attachment of the expansion to the capsule of the joint (p. 319). The synovial sheaths of the extensor muscles are less important than those of the flexor muscles, because they are less often infected. Those of the finger extensors stop about the middle of the dorsum of the hand and do not continue into the fingers; those of the thumb muscles behave similarly, though the abductor longus often retains its sheath right up to its insertion.

The use of the hand

The number of muscles available gives some idea of the complexity of movements possible when different combinations of these muscles are employed. The **power grip** (Fig. 160) utilizes the long flexors of the fingers and thumb and enables a firm grasp to be taken. It is used for carrying heavy objects, and for controlling tools and machinery. The **precision grip** is not so powerful, and depends largely on the small muscles. It is used for all fine work with small objects, such as sewing and writing. The object gripped is held between the finger pads and the thumb pad, or in the cup formed by the fingers, thumb and palm. Since these grips involve different groups of muscles, they may be independently disturbed following a nerve injury or a muscle-tearing accident. Thus a patient who cannot button up his clothes (precision grip) may be able to carry and wield effectively a heavy pick and shovel. There appears to be little objective difference between the grips exerted by the 'preferred' hand and the other. In right handed people, for example, the power of the grip exerted by the left hand is usually within 5 per cent or so of the power of the right-handed grip.

The individual movements of the fingers are mediated by the small muscles (lumbricals and interossei). The index finger has to a considerable extent become liberated from the others because it is supplied with a separate extensor tendon, and thus is used for pointing and exploring. Loss of movement in the index is therefore a more serious matter than loss of movement of (say) the little finger.

All gripping movements are accompanied by extension of the wrist, to give the flexor tendons the best possible mechanical advantage. The most efficient extension of the fingers can only be achieved when the wrist is either straight or slightly flexed; if we try to point with the wrist extended, the flexor tendons are mechanically stretched, and resist further stretching. Passive flexion of the wrist

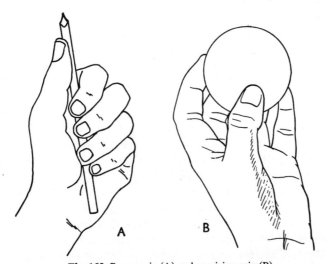

Fig. 160. Power grip (A) and precision grip (B).

automatically extends the fingers, and passive extension flexes them. When the metacarpophalangeal joints are fully flexed spreading of the fingers becomes impossible (p. 305), and rotation at these joints is abolished. Rotation is most marked in the index and little fingers, whose pads can be brought to lie at right angles to each other when the fingers are spread. This movement is used, together with cupping of the palm by the thenar and hypothenar muscles, to facilitate the grip of an object in the palm.

If a fibrous flexor sheath becomes scarred after injury, the affected finger is drawn down to the palm by the contraction of the resulting scar tissue. In the condition known as Dupuytren's contracture the palmar aponeurosis on the medial side of the palm spontaneously contracts, and pulls down the little and ring fingers. This may cause surprisingly little disability, but if the middle and index fingers become affected the hand is virtually useless.

The nerves

The brachial plexus

The upper limb is supplied by the ventral rami of the 5th, 6th, 7th and 8th cervical nerves and that of the 1st thoracic nerve, which together form the **brachial plexus**. The 5th cervical is reinforced by a twig from the 4th cervical, and the 1st thoracic receives a branch from the 2nd. The central nerve of the plexus is the 7th cervical (Fig. 161). These ventral rami are the **roots** of the plexus, and lie in the neck deep to the sternocleidomastoid muscle. As they travel obliquely downwards and laterally to enter the upper limb, the upper two roots (C5 and C6) unite to form the **upper trunk** of the plexus, and C8 and T1 unite to form the **lower trunk.** C7 goes on alone, but is now called, for the sake of symmetry, the **middle trunk** of the plexus. The trunks of the plexus can usually be felt in the supraclavicular fossa, though they are covered over by deep fascia. Behind the clavicle each of the three trunks divides into an anterior and a posterior **division**. The divisions are short-lived, for they unite with each other almost immediately to form **cords**. The three posterior divisions join to form the **posterior cord**, the anterior divisions of the upper and middle trunks unite to form the **lateral cord**, and the anterior division of the lower trunk goes on alone, changing its name to the **medial cord**. In the axilla each cord gives off branches, and continues on its way down the limb under another name. Thus, the posterior cord gives off two **subscapular** branches to the subscapularis and the teres major, the **thoracodorsal** nerve to the latissimus dorsi and the **axillary** nerve to the deltoid and the teres minor, and then pursues its course as the **radial** nerve. The medial cord gives off a branch to the two pectoral muscles (the **medial pectoral** nerve), the **medial cutaneous nerve of the arm** and the **medial cutaneous nerve of the forearm**, and is then continued as the **ulnar** nerve. The lateral cord also gives a branch to the pectoral muscles (the **lateral pectoral** nerve), and then becomes the **musculocutaneous** nerve. The other great nerve of the upper limb—the **median** nerve—is formed by the union of two branches, one from the lateral, and the other from the medial cord (Fig. 161).

It is not only the cords of the brachial plexus which give off branches. From the roots come several nerves, among them the **long thoracic nerve**, which supplies serratus anterior. The upper trunk sends the **suprascapular** nerve posteriorly to supply the supraspinatus and infraspinatus muscles.

The roots and trunks of the brachial plexus lie in the neck, above the clavicle, the divisions lie behind it, and the cords and their branches are inferior to it. All the branches of the plexus have come off by the time the lateral border of the pectoralis minor muscle has been reached, and the four main nerves of the limb (median, radial, ulnar and musculocutaneous) are said to begin at this point. In the axilla the plexus is covered over by the two pectoral muscles anteriorly, and posteriorly it lies on the outer surface of the first rib, the serratus anterior and the subscapularis.

The median nerve

The median nerve arises by two heads from the lateral and medial cords of the brachial plexus, and contains fibres derived from all the roots of the plexus. It runs down the medial side of the arm in the groove between the brachialis and biceps muscles, and eventually reaches the elbow medial to the biceps tendon. It is closely related to the ulnar nerve, which leaves it halfway down the arm to travel backwards and medially, and also to the brachial artery. The nerve is at

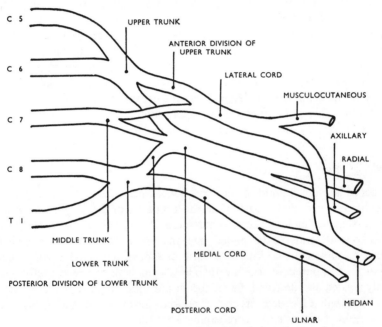

Fig. 161. Diagram of brachial plexus. Only the main branches are shown.

first on the lateral side of this artery, but crosses over it halfway down the arm to lie on its medial side, which is the position it holds at the elbow (Fig. 167). At the elbow the median nerve has gained the midline of the limb (hence its name), and it retains this position from this level downwards. It enters the forearm between the two heads of the pronator teres muscle (p. 314), and becomes closely adherent to the deep surface of the flexor digitorum superficialis. At the wrist it emerges from under that muscle and lies between the tendons of the palmaris longus and the flexor carpi radialis (Fig. 168). It passes into the hand deep to the flexor retinaculum and ends in the palm by dividing up into terminal branches which supply the skin of the fingers and the small muscles of the thumb.

Distribution

The median nerve supplies all the muscles on the front of the forearm except flexor carpi ulnaris and the medial half of flexor digitorum profundus. In the hand it gives branches to the lateral two lumbricals, the abductor pollicis brevis, the opponens pollicis and the flexor pollicis brevis. However, in something like 5–10 per cent of patients all the thenar muscles may be supplied by the ulnar nerve. The median nerve has no cutaneous branches until it reaches the hand, where it supplies the lateral half of the palm and the palmar surfaces of the thumb, index and middle fingers. The palmar surface of the ring finger is disputed territory, the most usual arrangement being that the median nerve supplies one half and the ulnar nerve supplies the other. Sometimes the ulnar nerve takes over the whole finger, and even part of the middle finger as well. Conversely, the median nerve may supply the whole of the ring finger. The terminal branches of the nerve twine round the fingers to supply the dorsal surface of the distal two phalanges (also the proximal phalanx in some people). Like all the other main nerves of the upper limb, the median gives branches to the joints which it crosses—the elbow, the wrist, and the digital joints.

Results of injury

The nerve is often injured at the wrist by such accidents as putting the hand through a glass window. Most of the short muscles of the thumb will usually be put out of action and the thumb, deprived of the tone of the short abductor and the opponens, tends to be pulled into the plane of the palm by the intact adductor and first dorsal interosseous. It cannot be abducted at right angles to the palm or opposed to the fingers. The thenar eminence eventually becomes grossly wasted and flattened. (If the thenar muscles are all supplied by the ulnar nerve they will of course escape). The long flexor of the thumb will still be working, for it receives its nerve supply from the median in the forearm above the injury. The lateral two lumbrical muscles will usually be paralysed, and this will affect the capacity of the index and middle fingers to make independent movements, but their power will not be much affected, since the long flexors will still be working, to say nothing of the interossei, which are supplied by the ulnar nerve. The sensory loss is of extreme importance. Superficial and deep sensation will be lost over an area of variable size which always includes the pads of the thumb and index fingers, and usually the pad of the middle finger also. The patient is thus deprived of his chief tactile exploratory organs. Further, his sense of joint movement is abolished in the interphalangeal joints of the thumb and the two most important functional fingers. If he smokes he has to hold the cigarette between the ring and little fingers or he may burn his fingers without being aware of it, and cooking becomes a hazardous occupation.

If the median nerve is cut above the elbow all the flexors of the wrist are out of action except the flexor carpi ulnaris and the medial two tendons of the flexor digitorum profundus. Flexion and abduction of the wrist are therefore very considerably weakened though not abolished, and pronation of the forearm is lost. The condition of the thumb is similar to that in a wrist injury of the nerve, but the long flexor is now paralysed, so that the thumb cannot be flexed at all. It is in fact almost completely useless, though it can act like the claw of a lobster to nip articles between its medial surface and the index finger (intact adductor and first dorsal interosseous). The long flexor tendons to the index and middle finger are paralysed, so that these fingers cannot be flexed at the interphalangeal joints, though flexion at the metacarpophalangeal joints is still possible because of the intact interossei; the movement is, however, grossly weakened. The superficialis tendons to the ring and little fingers are knocked out, but the profundus tendons to these two fingers (ulnar nerve) are still intact, so flexion of all joints of these fingers is possible. The sensory loss in this lesion is exactly the same as in a lesion at the wrist.

The grip of such a hand is grossly deficient, though small articles can be held between the two medial fingers and the palm. Nothing heavy can be carried, and the loss of the thumb makes any precision grip impossible. A patient in this condition could not use his hand to dress himself, or to perform any manual labour. He can write by holding the pen in an unnatural position between thumb and index or between two fingers and using his shoulder muscles.

The ulnar nerve

The ulnar nerve is the continuation of the medial cord of the brachial plexus. It travels down the medial side of the arm medial to the median nerve and the brachial artery, and halfway down the arm it swings posteriorly through the medial intermuscular septum which separates the flexor muscles from the extensors. It now lies anteromedial to the triceps muscle, and so reaches the groove between the medial epicondyle of the humerus and the trochlea (Fig. 162). Here it can be felt against the bone, and pressure in this situation may produce tingling in the distribution of the nerve.

The nerve enters the forearm between the two heads of the flexor carpi ulnaris, and becomes adherent to the deep surface of this muscle, which it follows right down to the wrist. It crosses into the hand superficial to the flexor retinaculum just lateral to the pisiform bone, under the shelter of which it divides into branches which supply muscles and skin in the hand.

Distribution

The ulnar nerve supplies nothing in the arm. In the forearm it supplies the flexor carpi ulnaris and the medial half of the flexor digitorum profundus. In the hand it sends branches to all the interossei, the medial two lumbricals, the short muscles of the little finger, and the adductor pollicis. The cutaneous distribution complements and corresponds to that of the median nerve; it is usually confined to one and a half fingers and the medial side of the palm. A dorsal branch sent round to the back of the hand while the nerve is still in the forearm supplies the dorsal surfaces of the proximal phalanges of the medial one and a half or two and a half fingers, as well as a corresponding area on the back of the hand.

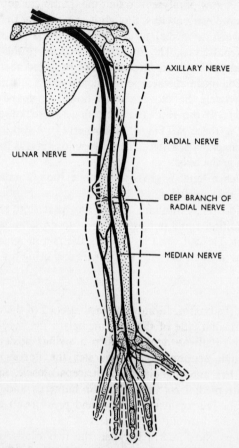

AXILLARY NERVE

RADIAL NERVE

ULNAR NERVE

DEEP BRANCH OF
RADIAL NERVE

MEDIAN NERVE

Fig. 162. The median, ulnar, and radial nerves. Most of their branches have been omitted for the sake of clarity.

Results of injury

If the nerve is injured at the elbow the hand takes up a characteristic posture called 'claw-hand' (Fig. 163). The fingers are hyperextended at the metacarpophalangeal joints and flexed at the interphalangeal joints because the paralysis of the interossei and the medial two lumbricals upsets the balance of power round these joints (pp. 322, 323). The two radial fingers are less flexed at the interphalangeal joints than the two ulnar ones because the two lateral lumbricals are usually intact, being supplied by the median nerve. On the other hand, the tendons of the flexor digitorum profundus to the two medial fingers are paralysed, so no flexion is possible at the distal interphalangeal joints of these fingers. The small muscles of the little finger are all paralysed, but the small muscles of the thumb are usually intact except for the adductor (unless they all happen to be supplied by the ulnar nerve). Ulnar deviation of the wrist is weakened because of the loss of the flexor carpi ulnaris. Tests for ulnar nerve paralysis are (a) to make the patient abduct the little finger against resistance (abductor digiti minimi) and (b) to make him grip a card between finger and thumb without bending the joints of the finger or the tip of the thumb (adductor pollicis and first dorsal interosseus).

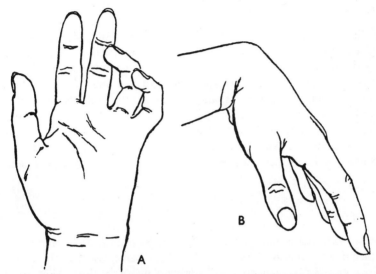

Fig. 163. Postural deformity following nerve injury. A: 'claw-hand' (ulnar nerve lesion. Notice wasting of hypothenar muscles). B: 'wrist-drop' (radial nerve lesion).

In an ulnar nerve lesion at the wrist the flexor carpi ulnaris and the medial part of the flexor digitorum profundus will escape paralysis.

The sensory loss in ulnar nerve paralysis involves the two least important fingers and the medial side of the hand. The motor loss chiefly affects the fine movements. The fingers can still be flexed, and the grip is a strong one except for the two medial fingers. Heavy tools can still be carried and wielded. But anything like sewing or knitting, playing the piano, writing or dealing cards is now extremely difficult. The eventual deformity is striking and characteristic. The hypothenar eminence is flattened, the interosseous spaces are empty of their muscles, and the metacarpals stick out under the skin.

The radial nerve

The radial nerve (Fig. 162) is the continuation of the posterior cord of the brachial plexus. It winds round the back of the humerus to enter the radial groove (p. 293), in which it is carried, between the lateral and medial heads of the triceps, round to the lateral side of the front of the arm, emerging between the brachialis and brachioradialis. In the forearm it sticks close to the deep surface of the brachioradialis, but leaves it about half-way down to swerve backwards on to the dorsum of the wrist and hand, where it ends in cutaneous branches. As the nerve enters the forearm it gives off a large **deep branch** which runs between the two heads of the supinator muscle (p. 314) to the back of the forearm.

Distribution

The radial nerve supplies all the heads of the triceps and the anconeus, and it also gives branches to the brachialis, the brachioradialis, and the extensor carpi radialis longus. All the other muscles on the back of the forearm are supplied by its deep branch. The skin of the posterior and the lateral surfaces of the arm, the back of the forearm, and a small area on the dorsum of the hand (Fig. 164) are all supplied by the radial nerve. There is a variable sharing out of skin between the radial and median nerves on the back of the digits, but usually the radial supplies the skin of the index and middle fingers as far as the proximal interphalangeal joints, and extends to the level of the interphalangeal joint of the thumb.

Results of injury

The most likely site of injury to the radial nerve is in the arm, where it is liable to be torn if the humerus is fractured. There is a paralysis of all the extensors of the wrist and digits, causing the characteristic 'wrist-drop' (Fig. 163). The lumbricals and interossei can straighten the fingers at the interphalangeal joints, and while the hand is in its 'dropped' position the metacarpophalangeal joints are not, as one might expect, flexed, but extended by the passive stretching of the extensor tendons (p. 327). Because of this also the grip is poor. When the wrist is passively dorsiflexed the metacarpophalangeal joints flex under the pull of the flexor muscles and the hand curls up, the thumb coming across the palm. In this position the flexor muscles can take firm hold of anything put into their grasp, and the hand becomes more useful; dorsiflexion is readily obtainable by wearing a small 'cock-up' splint. The hand is not, however, fully restored to function, as the fingers cannot open properly to lay hold of their objective, which has therefore to be inserted into the hand between the half-closed fingers. A partial remedy is to transplant one of the flexor tendons—say the palmaris longus—through the interosseous membrane and attach it to the extensor digitorum. The patient can then re-learn his movements sufficiently to be able to use this flexor muscle to extend the fingers a little, so improving the use of his grip.

Injury to the nerve in the radial groove does not normally affect the branches going to the triceps, since these come off more proximally. If they are severed, the whole of the triceps may be paralysed, so that the elbow cannot be extended against resistance. The forearm will, however, hang down straight by the side, because gravity is sufficient to overcome the tone of the flexors. Supination is weakened, but not abolished, because the biceps is still working (musculocutaneous nerve). The sensory loss on the back of the hand, forearm and arm in radial nerve paralysis is not a vital matter. Of the three major nerves—ulnar, median, and radial—paralysis of the radial is the least disabling, providing the correct measures are taken.

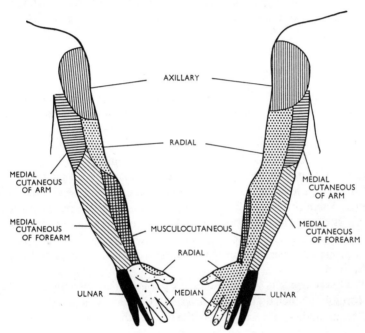

Fig. 164. Cutaneous nerve territories in the upper limb.

The musculocutaneous nerve

The musculocutaneous nerve is the continuation of the lateral cord of the brachial plexus. It supplies and then pierces the coracobrachialis muscle, and arrives in the plane between the biceps and the brachialis muscles, both of which it supplies. It continues to drift laterally, and emerges on the lateral side of the front of the arm close to the radial nerve. From here it is continued down as far as the ball of the thumb as the **lateral cutaneous nerve of the forearm**. Injury to this nerve interferes with flexion of the elbow, but it is not completely paralysed, for the brachioradialis and the flexor muscles of the forearm are still active. The branch given to the brachialis by the radial nerve (p. 334) is thought to be sensory, for the brachialis is usually completely paralysed if the musculocutaneous nerve is cut. Supination of the forearm is weakened by the loss of the biceps. The sensory loss of a musculocutaneous lesion is of little practical importance.

The axillary nerve

The axillary nerve is a branch of the posterior cord of the brachial plexus; it supplies the deltoid and teres minor, as well as a small area of skin over the insertion of the deltoid. Paralysis of the nerve leads to an inability to abduct the arm from the side more than about 30° (this much can be done by the supraspinatus) and considerable weakness of flexion and extension of the shoulder.

Injury to the brachial plexus

The commonest form of injury to the brachial plexus itself is a tear of the **upper trunk**. This may follow a blow on the shoulder forcing apart the head and the point of the shoulder, or it may occur as a birth injury. The paralysis produced is called Erb's paralysis, and is due to the loss of function of the myotomes and dermatomes of C5 and C6. The deltoid and supraspinatus are paralysed, so that the arm cannot be abducted; the flexors of the elbow are all paralysed, and the loss of the biceps and supinator means that supination of the forearm is greatly weakened. There is also weakness of the pectoralis major and latissimus dorsi, and the loss of the infraspinatus gives rise to a medial rotation of the shoulder. The arm thus hangs at the side, medially rotated at the shoulder and with the forearm pronated. There is anaesthesia over an area on the lateral side of the arm (Fig. 165).

If the **lower trunk** is torn (Klumpke's paralysis) by an injury in which the arm is forcibly abducted above the head, there is paralysis of the small muscles of the hand and weakness of the flexors of the fingers and thumb, with anaesthesia along the medial border of the forearm and in the ulnar distribution. The condition resembles an ulnar injury plus paralysis of the thenar muscles and weakness of the flexors of the digits.

Nerve territories of the upper limb

Figures 164 and 165 show the distribution of the cutaneous nerves and dermatomes in the upper limb. Because of the gross overlap between dermatomes no outlines have been drawn.

Motor points

Figure 166 shows the position of the motor points of the more important muscles of the upper limb. The motor point of a muscle is merely a convenient

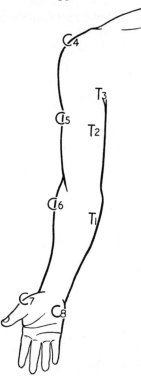

Fig. 165. Approximate scheme of the segmental innervation of the skin of the upper limb.

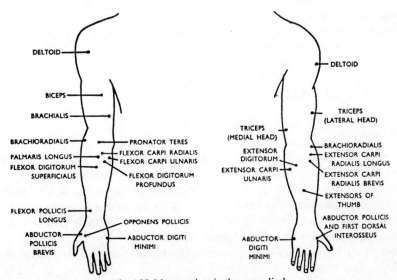

Fig. 166. Motor points in the upper limb.

place where it lies close to the skin, and where there are no nerve trunks in the vicinity; electrical stimulation at the motor point will cause the muscle to contract without disturbing its neighbours.

The vessels

The arteries

The subclavian artery (p. 266) is continued in the axilla as the **axillary** artery, which becomes the **brachial** artery at the outer border of the teres major. The course of this continuous vessel, the main arterial trunk of the limb, is given by a straight line drawn from the mid-point of the clavicle to the centre of the bend of the elbow when the arm is abducted to a right angle with the palm looking forwards. The medial, lateral and posterior cords of the brachial plexus are named from their relation to the middle part of the axillary artery, and the brachial artery continues to be closely associated with the median nerve in its course down the medial side of the arm, in the groove between the biceps and the brachialis (Fig. 167).

The **ulnar** artery is the larger of the two terminal branches of the brachial. From the centre of the bend of the elbow it runs down the medial side of the front of the forearm lateral to the ulnar nerve, crosses anterior to the flexor retinaculum, and ends in the palm by communicating with one of the terminal branches of the radial artery to form the **superficial palmar arch** (Fig. 168). In the forearm it gives off **interosseous** branches to the muscles of the front and back of the forearm.

The **radial** artery runs down the lateral side of the front of the forearm medial to the radial nerve (the arteries lie *within* the boundaries formed by the nerves) and winds backwards across the lateral ligament of the radiocarpal joint to reach the first dorsal interosseous muscle. Here it turns medially, pierces the muscle, and enters the palm to end by communicating with a deep branch of the ulnar artery to form the **deep palmar** arch. Branches from the palmar arches supply the hand and fingers.

The veins

The **deep** veins accompany the arteries, and the **superficial** veins lie outside the deep fascia for most of their course; they have no relation to arteries. These superficial veins are commonly used for giving or taking blood, or for giving intravenous injections. The **basilic** vein runs up the medial side of the limb and pierces the deep fascia halfway up the arm to drain eventually into the axillary vein. The **cephalic** vein, on the lateral side of the limb, pierces the deep fascia opposite the lower margin of the pectoralis major; it then runs in the groove between that muscle and the deltoid to enter the axillary vein just below the clavicle. At the elbow a communication passes from the cephalic to the basilic veins; the pattern of this communication is variable, but usually it is possible to recognize a **median cubital vein** (Fig. 169), which crosses just superficial to the median nerve and the brachial artery; a careless injection can injure either of them, sometimes irreparably.

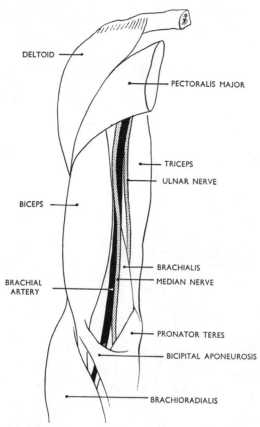

Fig. 167. The brachial artery and the median and ulnar nerves in the arm (branches of the artery have been omitted).

The lymphatics

The deep lymphatics accompany the veins which run with the main arteries; the superficial ones tend to run with the superficial veins. A few **cubital nodes** lies just above the elbow, but the main collection is the **axillary** group lying in the axilla. The various members of this group drain the upper limb, the skin of the trunk above the level of the umbilicus, and the breast. They make their presence felt after a vaccination or an inoculation into the deltoid muscle (a convenient, but sometimes rather painful, site for the injection of many kinds of unpleasant

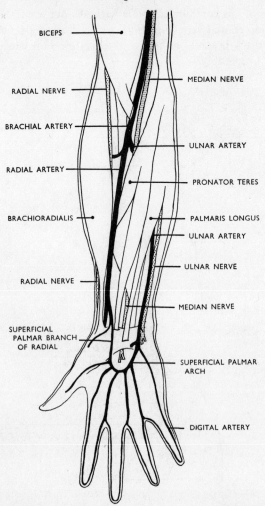

Fig. 168. The main arteries and nerves of the front of the forearm and hand (branches mostly omitted).

material). From the axillary nodes the lymph drains into the main lymph trunks which open into the venous system.

Surface anatomy

The position and relationships of the various bony landmarks have already been described. The muscles can be demonstrated by making them contract against resistance. Thus the biceps will stand out if an attempt is made to bend the elbow while someone else tries to prevent the movement. Details of muscle testing procedure for individual muscles will be found in larger books (p. 396).

The **axilla** is a pyramidal fossa with an apex formed by the middle third of the clavicle, the outer border of the first rib, and the upper border of the scapula; its base is formed by skin. The medial wall is formed by the serratus anterior, and the lateral wall by the coracobrachialis and the short head of the biceps. In its anterior wall are the pectoral muscles, and the pectoralis major forms the thick anterior axillary fold. The posterior wall contains the subscapularis and teres major, round which runs the flattened tendon of the

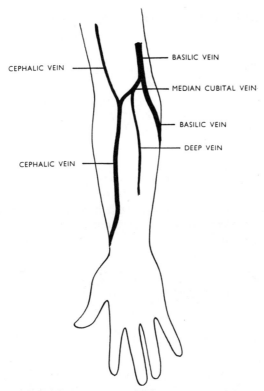

Fig. 169. The superficial veins in front of the elbow.

latissimus dorsi. Palpation will identify, through the thick fascia of the axillary base, the pulsations of the axillary artery and the vague resistance of the head of the humerus. The axilla contains the axillary vesels, the cords and main branches of the brachial plexus, and the axillary lymph nodes.

The **cubital fossa** is a triangular hollow in front of the elbow. Its base is an artificial line drawn between the two epicondyles, and its apex points distally down the forearm. The medial boundary is the lateral border of the pronator teres, and the lateral boundary is the medial border of the brachioradialis. The fossa contains the tendon of the biceps, and its anterior wall is formed by the bicipital aponeurosis blending with the fascia of the forearm. The biceps tendon may be picked up between finger and thumb, and a finger can also be pushed under the upper border of the aponeurosis. Just medial to the biceps tendon the pulsations of the brachial artery may be felt, and medial to this again the median nerve may sometimes be rolled under the finger.

At the wrist the tendons passing into the hand can all be readily identified. The 'anatomical snuff-box' on the radial side of the wrist (p. 323) contains in its floor the second part of the radial artery and the radial collateral ligament of the wrist.

The median nerve is medial to the tendon of the flexor carpi radialis at the wrist; it can sometimes be felt. The ulnar nerve is easily identified behind the medial epicondyle of the humerus, and its deep branch can be rolled against the hook of the hamate (p. 296).

Just above the wrist the radial artery is easily felt on the lower end of the radius; for this reason it is usually selected when it is desired to 'feel the pulse'. Nevertheless the pulse can also be felt in the ulnar artery above the wrist, and in the brachial artery on the medial side of the arm. The only satisfactory place where the arterial system of the upper limb can be occluded in order to stop bleeding is halfway down the arm, where the brachial artery can be compressed by forcing it laterally against the humerus. This is the site chosen for applying a tourniquet.

The superficial veins can be made to stand out by compressing the arm and so obstructing the venous return; if the limb is warm they are often visible without any such manoeuvre.

26 · The lower limb

Introduction

Unlike the hand, the foot has become highly adapted in one direction, and as a result has lost most of its more primitive functions. That these are still only dormant can be proved by people who are born without arms, and who by the use of their toes are able to cut up their food, write, paint, play cards, type, and even play the piano. In normal adults the foot is used only to support the body weight, and in civilized peoples it is cramped from birth into more or less unsuitable footwear. It is greatly to the credit of the foot that it continues to function so efficiently—even when it is tipped up on end by high-heeled shoes, forbidden to spread out to distribute the body weight, and denied the natural movements of its toes.

The lower limb is an instrument of locomotion and postural support, and the primary function of the majority of its muscles is to move the trunk relative to the ground, whether in walking or in preventing the body from falling over. The upright posture is an unstable equilibrium, and if the centre of gravity falls outside its base we tend to topple over; this is stopped by the pull of the muscles whose insertions are indirectly fixed by friction between the ground and the sole of the foot. This reversed function of the muscles is particularly important when the size of the 'base' of the body is reduced by standing on one leg, or taking a stride. Of course the lower limb is frequently moved on the trunk, but the reversed action of the muscles is of at least equal importance to the 'direct' action, and every 'insertion' is also an 'origin'.

The general plan of the lower limb is similar to that of the upper limb. There is a pelvic girdle corresponding to the shoulder girdle, a single bone in the thigh, two bones in the leg, a number of small bones in the ankle and foot. But all this framework has become grossly modified in response to the job it has to do. The pelvic girdle does not move freely on the trunk, as the shoulder girdle does; it is tightly fixed by some of the strongest ligaments in the body. It has also to support the pelvic and abdominal viscera, and to allow the process of childbirth. The hip joint has lost the mobility of the shoulder in exchange for great strength. No movements corresponding to pronation and supination can take place in the leg. The tarsus is more massive than the carpus. The big toe is in the same plane as the other toes and not at right angles to them. And so on. The lower limb is at once simpler and more specialized than the upper limb; in view of its more restricted functions, it need not be discussed in such great detail.

The skeleton

The pelvic girdle or **hip bone** consists of three bones—the **ilium**, the **ischium**, and the **pubis**—which are separate in the child, but fused together in the adult. The **ilium** is the largest component; it has a large flattened body which ends superiorly in a free border called the **iliac crest**. The whole length of this crest can easily be felt in the living body, as can the **anterior** and the **posterior superior iliac spines** at each end of it (Fig. 170). Below the two ends of the iliac crest lie the **anterior** and **posterior inferior iliac spines;** neither is easily felt. The lower end of the ilium is expanded to make the upper part of the socket for the hip joint, the **acetabulum.** This portion of the socket takes the greatest strain, and is thick and very strong. To the outer surface of the ilium several of the muscles of the buttock are attached, marking it with the **anterior, posterior,** and **inferior gluteal lines** (Fig. 165). The medial surface of the ilium faces the abdomen (p. 277), and posterior to this abdominal surface is the roughened region which articulates with the sacrum; this **auricular** surface is so-called because it is shaped something like an ear. Between the posterior inferior spine and the acetabulum is the wide smoothly curving border of the **greater sciatic notch**.

The **ischium** is the next largest component of the hip bone, and its most prominent feature is the **ischial tuberosity** on which we sit (Fig. 170). The ischium takes as big a share as the ilium in the formation of the acetabulum, but the ischial portion of the socket does not have to bear so much weight, and is not so strongly built. Just above the tuberosity the sharp **spine** of the ischium juts out medially, and between this and the tuberosity the margin of the ischium forms the **lesser sciatic notch**. From the anterior aspect of the tuberosity springs a strong bar of bone called the **ramus** of the ischium, and this unites with the **inferior ramus** of the pubis to form the **conjoined ramus**, which gives attachment to several of the muscles of the thigh.

The inferior ramus of the **pubis** leads up to its **body**, which articulates at the median plane with the body of its fellow in the **pubic symphysis** (p. 278). The body is flattened, with a superior **pelvic** and an inferior **femoral** surface. The pelvic surface is smooth, and helps to support the urinary bladder (p. 190); the femoral surface is rough, and its anterior border stands out as the **pubic crest**, at the lateral end of which is the **pubic tubercle** (Fig. 170). Running laterally from the body of the pubis is the **superior ramus,** which expands at its lateral end to make about one-fifth of the acetabular socket (the ilium and ischium make two-fifths each). At the point where the pubis joins the ischium the lower margin of the acetabulum is deficient, the gap being called the **acetabular notch**.

Running laterally from the pubic tubercle the **pectineal line** forms the sharp anterior border of the ramus. The conjoined ramus, the body of the pubis, the superior ramus and the body of the ischium together form the margins of the **obturator foramen** (Fig. 170), which is virtually filled by the **obturator membrane** (p. 277), and only transmits the relatively small obturator nerve and vessels (p. 376).

The relative broadening of the female pelvis (p. 278) pushes the hip joints further apart in the female than in the male. As the knees are similarly close together in both sexes, it follows that the femur is more oblique in a woman than in a man, and consequently that the angle between it and the bones of the leg is greater.

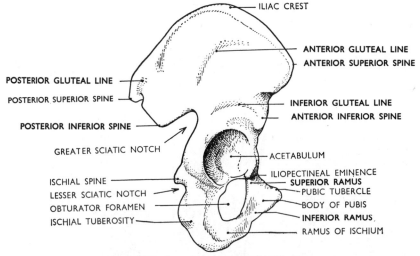

Fig. 170. The lateral aspect of the right hip bone.

The **femur** is the longest bone in the body. Its upper end carries the **head**—a rounded ball forming rather more than half a sphere—which fits into the acetabulum. It has a small pit on its smooth surface. Below the head is the **neck**, which leads to the two trochanters (Fig. 171). The **greater trochanter** is a large quadrangular chunk of bone which can be felt on the outer side of the thigh. In the male the upper border of the trochanter lies opposite the middle of the margin of the trouser pocket; in the female there is as a rule no such reliable guide. The widest part of the female thigh is not, as in the male, at the level of the greater trochanter, but rather further down, because a pad of fat accumulates in this situation. The lateral surface of the greater trochanter is roughened for the attachment of muscles, and the concave medial surface sticks up above the neck of the bone to form the **trochanteric fossa**. From the posterior border of the greater trochanter a marked **intertrochanteric crest** runs down to the **lesser trochanter**, which lies inferior to the greater trochanter on the

posteromedial aspect of the bone (Fig. 172). Joining the two trochanters in front
is a much fainter **intertrochanteric line.**

The shaft of the femur has three surfaces and three borders, only one of
which is well defined. This is the **linea aspera**, which lies posteriorly, in the
middle one-third of the shaft. Inferiorly it peters out into two diverging ridges
which lead down to the **condyles** of the femur (Fig. 172), and superiorly it leads
into two other ridges. The lateral one is the **gluteal tuberosity**, and the medial
one winds round to become continuous with the intertrochanteric line. The shaft

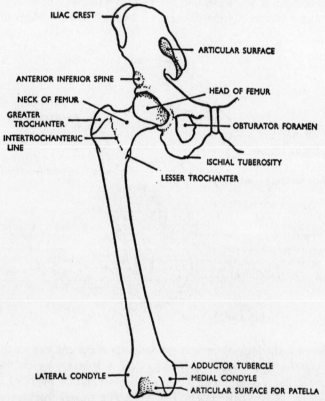

Fig. 171. Anterior aspect of the right hip bone and femur.

of the femur joins the neck at an angle of about 120°, and is twisted so that a
line through the centre of the head and the centre of the greater trochanter is at
an angle of some 15° with a line drawn between the centres of the two condyles.

The lower end of the femur bears massive **medial** and **lateral condyles** for
articulation with the tibia at the knee joint. From the medial condyle protrudes
the **adductor tubercle**—so called because the adductor magnus muscle is
inserted into it—and on the lateral surface of the lateral condyle is the less
obvious **lateral epicondyle** (Fig. 172). Below the lateral epicondyle is an

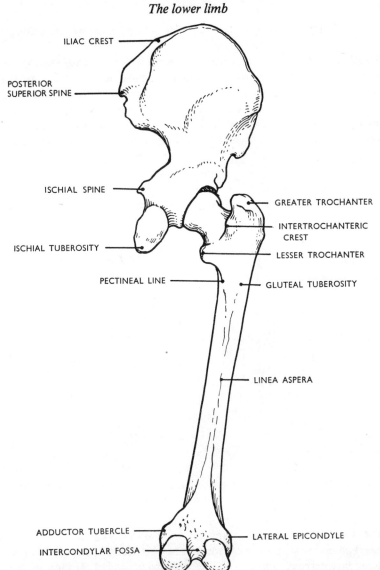

ILIAC CREST

POSTERIOR
SUPERIOR SPINE

ISCHIAL SPINE

ISCHIAL TUBEROSITY

PECTINEAL LINE

GREATER TROCHANTER

INTERTROCHANTERIC
CREST

LESSER TROCHANTER

GLUTEAL TUBEROSITY

LINEA ASPERA

ADDUCTOR TUBERCLE

INTERCONDYLAR FOSSA

LATERAL EPICONDYLE

Fig. 172. Posterior aspect of the right hip bone and femur.

irregular pit and groove for the accommodation of the popliteus muscle (p. 364). Between the two condyles at the back of the bone is the **intercondylar fossa**; in front the condyles are more or less continuous through the articular surface for the patella.

The **patella** is a small roughly triangular bone in the tendon of the quadriceps femoris, the muscle which extends the knee. It is the largest sesamoid bone, and is frequently broken. The patella is readily felt in the living subject, and if the quadriceps is relaxed, can be moved about between finger and thumb.

The **tibia**, the second largest bone in the body, is the medial bone of the leg. Its upper end is expanded to receive the femoral condyles, and has a **medial** and a **lateral condyle** of its own. Between these two lies the **intercondylar eminence** (Fig. 173). In front of the upper end of the tibia protrudes the large roughened **tuberosity** of the tibia (easily felt in front of the knee), to which is attached the tendon of the quadriceps muscle. Laterally lies a small facet for the head of the fibula.

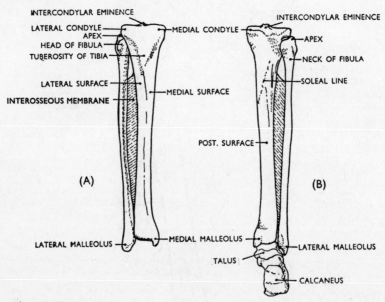

Fig. 173. The bones of the leg. A: anterior aspect of the right leg. B: posterior aspect, showing the relationship of the talus and calcaneus to the tibia and fibula.

The shaft of the tibia is triangular in cross-section, having a prominent **anterior border** which is subcutaneous in its whole length, a **medial** border, and a lateral **interosseous** border. The subcutaneous medial surface of the tibia forms the shin, and has no muscular attachments except at its upper end. The lateral and posterior surfaces are deeply clothed with muscles.

The lower end of the tibia spreads out into a quadrangular block which is the weight-bearing surface of the socket of the ankle joint. Medial to this articular surface there extends downwards the **medial malleolus**, a well defined bony prominence easily felt on the medial side of the ankle. Its lateral surface articulates with the talus. Laterally, the lower end of the tibia articulates with the lower end of the fibula, which completes the socket of the joint on the lateral side (Fig. 173).

The **fibula** is a long thin strut of bone which does not enter into the knee joint, and is so tightly bound to the tibia that its lower end can only twist slightly when the ankle joint is moved. The **head**, at the upper end, articulates with the lateral side of the upper end of the tibia; it bears a projection called the **apex**, and is used as a landmark on the lateral side of the knee. The shaft is triangular in cross-section, with an **interosseous** border facing the interosseous border of the tibia. The details of the other borders and surfaces are of little importance. The lower end of the fibula forms the **lateral malleolus**, which

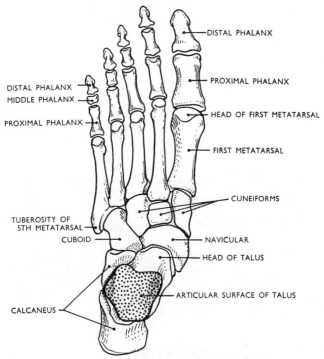

DISTAL PHALANX

PROXIMAL PHALANX

DISTAL PHALANX
MIDDLE PHALANX

HEAD OF FIRST METATARSAL

PROXIMAL PHALANX

FIRST METATARSAL

CUNEIFORMS

TUBEROSITY OF
5TH METATARSAL

CUBOID

NAVICULAR

HEAD OF TALUS

ARTICULAR SURFACE OF TALUS

CALCANEUS

Fig. 174. The bones of the left foot from above.

articulates with the lateral surface of the talus and is a well marked feature on the lateral side of the living ankle about a finger's breadth distal to the level of the medial malleolus. Just above the malleolus the lateral surface of the fibula becomes subcutaneous for about one fifth of its length.

The **tarsus** corresponds to the carpus in the hand (p. 295), and consists of seven separate tarsal bones. The **talus** is the keystone of the arch which they form (p. 372). It has a large protuberance on the superior surface of its **body**; this **trochlea** has upper, medial and lateral articular surfaces which fit into the socket of the ankle joint (Fig. 174). Anteriorly the trochlea is wider than posteriorly; this fact is of importance in relation to the stability of the ankle joint (p. 355). The **neck** of the talus juts out from the body at a slight angle medially; it carries the ovoid **head**, articulating with the navicular (Figs. 174, 175). The inferior surface of the talus has two main facets, which articulate with the calcaneus.

The **calcaneus** is the largest tarsal bone. Its posterior surface receives the tendo calcaneus (p. 365), and its anterior surface articulates with the cuboid (Figs. 174, 175). The lateral surface is smooth, and the medial surface is rather concave. Jutting out from the junction of the medial and superior surfaces is a

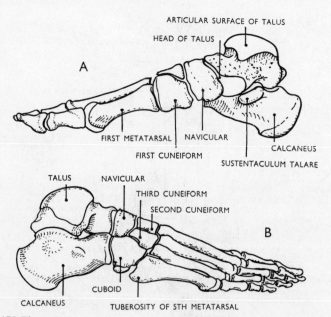

Fig. 175. The bones of the right foot. A: from the medial side. B: from the lateral side.

large projection called the **sustentaculum talare** (Fig. 175). Another large facet on the upper surface of the calcaneus bears the main weight of the talus, and between this and the sustentacular surface a groove forms one wall of the **sinus tarsi**, a cavity between the talus and the calcaneus.

The **cuboid** is much smaller. It articulates with the anterior surface of the calcaneus posteriorly, and with the fourth and fifth metatarsals anteriorly. On its inferior surface a groove accommodating the peroneus longus tendon distinguishes it from the other small tarsal bones.

The **navicular** has a concave posterior surface articulating with the head of the talus, and a convex anterior surface articulating with the three cuneiform bones. Its **tuberosity** protrudes medially, and can be felt by running the finger along the edge of the shoe (here again female clothing is an erratic guide). The navicular does, as its name implies, vaguely resemble a very unseaworthy boat.

The three **cuneiform** bones are wedge-shaped, and intervene between the navicular and the three medial metatarsal bones (Fig. 174).

The **metatarsals** resemble the metacarpals (p. 296), but they are flattened from side to side, and their heads are relatively narrow (except for that of the first metatarsal). The **first metatarsal** can be felt along all its length on the medial side of the foot. Its large head forms the ball of the great toe, which is one of the main weight-bearing points of the foot, and its shaft is thick and very strong. The **fifth metatarsal** has a large **tuberosity** which protrudes from its base to form a landmark on the lateral side of the foot.

The **phalanges** of the foot are similar to those of the hand (p. 296); the differences need not concern us.

Joints and movements

The **hip joint** is a ball and socket joint between the head of the femur and the acetabulum (p. 344). The ball is completely clasped by the socket—in striking contrast to the shoulder joint (p. 298), in which the socket is a very poor fit for the ball. The acetabulum is deepened by the attachment all round it of a fibrocartilaginous **labrum acetabulare** (cf. the labrum glenoidale of the shoulder joint), and the acetabular notch is filled in by the **transverse ligament of the acetabulum**. The capsule of the hip joint is attached above to the margins of the acetabulum and the labrum, and below to the intertrochanteric line in front and the neck of the femur behind. It is strongest in front of the joint, where the **iliofemoral ligament** is developed (Fig. 176). This thickening of the capsule is shaped like an inverted 'Y', and is attached below to the intertrochanteric line, and above to the iliac portion of the rim of the acetabulum. Medially in the capsule lies the much weaker **pubofemoral** ligament, and behind is the spirally arranged and less well-defined **ischiofemoral** ligament. Within the joint cavity a thin band runs from the pit of the head of the femur to the transverse ligament of the acetabulum. This **ligament of the head of the femur** transmits a small artery to supply the femoral head.

Flexion of the joint when the knee is extended is limited by the tension of the hamstring muscles (p. 363), but when these are relaxed the knee can be brought up against the abdominal wall, the amount of fat on the thigh and the abdomen being then the limiting factor. Extension from the anatomical position is limited to about 15° because the ilio-femoral ligament becomes tight, and the slightly ovoid head of the femur 'screws home' into the ovoid acetabular socket. Abduction and adduction are limited by the tension of the opposing muscles; about 60° of abduction and 30° of adduction are possible from the anatomical position (the opposite thigh must be abducted to allow adduction to take place). Rotation is limited by the opposing rotators. The toes can be turned outwards through an angle of 130° and inwards through an angle of about 90°; most of this movement takes place at the hip.

The hip is rarely dislocated; the femur usually breaks before the joint will give way. However, congenital dislocation of the hip is not uncommon; the baby is born with the head of the femur outside the acetabulum, which is usually imperfectly formed. It is often necessary to fashion a new acetabulum before the femur can be permanently replaced.

The **knee joint** is the largest and most complicated joint in the body. The main articulation is between the condyles of the femur and the tibia, but the front of the lower end of the femur also articulates with the posterior surface of the patella. This patellar surface gives the lateral condyle of the femur a greater total articular surface than the medial condyle, but more of the medial than of the lateral condyle comes in contact with the tibia. The femoral condyles are

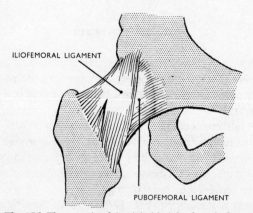

Fig. 176. The capsule of the right hip joint from in front.

curved, and the tibial condyles are almost flat, providing apparently most unstable platforms. Two semilunar intra-articular **menisci,** with built-up outer margins, lie on top of the tibial condyles, converting their platforms into shallow cups (Fig. 177). The menisci are joined together anteriorly by the **transverse ligament of the knee**, and their 'horns' are attached to the intercondylar eminence (Fig. 178). The medial meniscus is 'C'-shaped, and the lateral one is nearly circular, since its anterior horn is attached very close to its posterior horn, within the jaws formed by the 'C' of the medial cartilage. The medial meniscus is fixed to the medial ligament of the knee, but the lateral meniscus is more freely movable, and can curl up rather more round the femoral condyle.

Also attached to the intercondylar eminence are two important ligaments. The **anterior cruciate** ligament springs from the front of the eminence, and passes upwards, outwards, and backwards to be attached to the medial surface of the lateral condyle of the femur. Conversely, the **posterior cruciate** ligament, coming from the posterior part of the eminence, passes upwards, forwards, and inwards to gain attachment to the lateral surface of the medial condyle of the femur. Their chief function is to maintain the anteroposterior stability of the knee; if they are damaged the femur tends to slide off the tibia either forwards or backwards.

The **tibial collateral** ligament is a broad band in the capsule of the joint running from the medial surface of the medial condyle of the femur to the medial surface of the upper end of the tibia. Many of its fibres are firmly

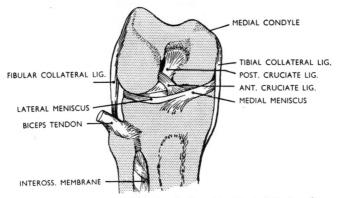

Fig. 177. The right knee-joint from in front (the joint is fully flexed).

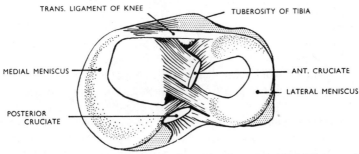

Fig. 178. The upper end of the right tibia from above, showing the attachments of the menisci and the cruciate ligaments.

attached to the medial meniscus. The **fibular collateral** ligament is an extrinsic ligament standing well away from the capsule on the lateral side. It is a thin cord attached to the lateral epicondyle of the femur and to the head of the fibula. The tibial and fibular collateral ligaments prevent dislocation of the joint medially or laterally.

In front, the capsule of the joint is replaced by the tendon of the quadriceps femoris (p. 361), which contains the patella as its sesamoid bone. The part of the tendon distal to the patella is called the **patellar ligament**, and is attached below to the tuberosity of the tibia. The capsule of the joint is reinforced by the tendons of other muscles which blend with it. Among these are the semimembranosus tendon (p. 363) which replaces the capsule behind, and the iliotibial tract of the fascia lata (p. 357) on the anterolateral aspect. The capsule itself is attached above to the femoral condyles and all round the margins of the upper end of the tibia, dipping down in front to include the tibial tuberosity.

The synovial membrane is very complicated. It wraps round the cruciate ligaments, excluding them from the joint cavity, and a pouch of the membrane is prolonged upwards above the patella to form a **suprapatellar bursa**. Many other bursae develop in relation to the muscles, and some communicate with the joint.

The knee is essentially a hinge joint permitting flexion and extension, but it also allows rotation (about 30° of lateral and 20° of medial rotation when the joint is flexed to a right angle). Flexion is usually limited by the thickness of the soft parts, and extension by the tension of the muscles and the ligaments. In the last stages of extension the femur rotates medially on the tibia, and this screws the ligaments up tight, so converting the joint into a rigid column. The movement is given as a medial rotation of the *femur*, because the tibia is usually fixed by the foot being planted on the ground. This medial rotation is one of the factors which screws home the head of the femur into the acetabulum (p. 351). The ligaments of the knee are not wound up tight unless the joint is slightly hyperextended (about 10° is possible), and in the normal standing position, with the knees not braced back, the ligaments are not as tight as they could be. The stability of the knee in the ordinary standing position thus probably depends on muscles rather than on ligaments.

The medial meniscus, by virtue of its attachment to the tibial collateral ligament, is liable to be torn when that ligament is suddenly stretched. This occurs when a sudden turn has to be made while running, and such an injury is common in professional footballers. It is therefore interesting that if a torn meniscus is removed, the patient is often able to continue to play first class football.

The **superior tibiofibular joint** is a small plane joint which has no connexion with the knee joint. It allows slight gliding movements of the head of the fibula on the tibia, particularly in relation to stress applied to the ankle joint. The interosseous borders of the tibia and fibula are united by a strong **interosseous membrane** in just the same way as the radius and the ulna are joined in the forearm (p. 300). The **inferior tibiofibular joint** is a very strong syndesmosis (p. 39). The roughened lower ends of the tibia and fibula are bound together by the **posterior tibiofibular** ligament, the lower margin of which forms part of the socket of the ankle joint. The joint is almost immovable, and has just sufficient 'give' to allow some twisting of the talus without this leading to a break of the fibula.

The **ankle joint** is a synovial hinge joint between the trochlea of the talus and the socket formed by the tibia, the fibula, and the posterior tibiofibular ligament. As in all hinge joints, there are strong medial and lateral ligaments. The **deltoid** ligament stretches from the medial malleolus to the sustentaculum talare, the medial surface of the talus, and the upper part of the navicular, as well as to the strong **plantar calcaneonavicular** ligament which runs forwards parallel to the ground (Fig. 179). On the lateral side of the joint the **anterior talofibular** and **posterior talofibular** ligaments form part of the capsule, and an extrinsic band, the **calcaneofibular** ligament, runs from the tip of the lateral

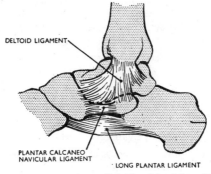

Fig. 179. Ligaments on the medial side of the right ankle joint.

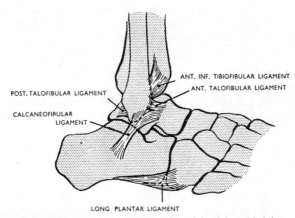

Fig. 180. Ligaments on the lateral side of the right ankle joint.

malleolus downwards to the lateral surface of the calcaneus about its middle (Fig. 180). In front and behind, as in all hinge joints, the capsule is relatively weak.

The movements of the ankle are **plantarflexion** and **dorsiflexion**; in plantarflexion the toes are pressed against the ground, and in dorsiflexion they point upwards. Measured from the position of rest (in which the foot is not quite at right angles to the leg) about 30° of plantarflexion is possible, and rather more dorsiflexion. In dorsiflexion the wider anterior part of the trochlea of the talus comes between the tibia and fibula; this brings the joint into the 'close-packed' condition (p. 46), and tightens several of the ligaments, rendering the joint more stable. Plantarflexion brings the narrower posterior end of the trochlear surface between the bones; the joint is then least stable, and permits a certain amount of twisting and rocking from side to side so that the foot can be abducted or adducted.

The talus articulates with the calcaneus by two ellipsoid joints. The posterior, **subtalar** joint is between the calcaneus and the talus alone; the anterior, **talocalcaneonavicular** joint is between the ovoid head of the talus and a socket formed by the calcaneus, the navicular, and the plantar calcaneonavicular ligament. Activity in these joints, taken along with the sum of activity in the other intertarsal joints (which are mostly plane or sellar), permits the sole of the foot to be turned to face medially. This movement of **inversion** is a compound of plantar flexion, adduction, and what might be termed supination; it is due largely to the calcaneus and the navicular twisting on the talus. Conversely, the movement of **eversion** causes the sole to face laterally (or as much laterally as possible). Eversion is compounded of dorsiflexion, abduction, and pronation, and owing to some confusion in terminology is sometimes simply called pronation of the foot. The common condition in which the foot is habitually held and used in the pronated posture must be distinguished from the less usual true flat foot, in which there is a descent of the medial arch of the foot (p. 373).

When the foot is off the ground, eversion and inversion are little more than parlour tricks; their real function is seen when walking over uneven ground, for the foot is able to adapt itself to the inequalities of the surface. (When the foot is fixed to the ground it is the talus which moves on the other bones, carrying with it the tibia and fibula.)

The total amount of eversion, measured in terms of alteration of the plane of the pads of the toes, is about 25°; inversion is more extensive, allowing about 40° of movement.

The movements of the foot are very complicated to describe, since none of the axes round which movements occur lie in any of the three descriptive planes of the anatomical position; they are all oblique. Most movements also involve conjunct rotation (p. 45), and it is not surprising that there should be confusion between anatomists and orthopaedic surgeons, physiotherapists and physical educationists.

The **tarsometatarsal** joints are not of great importance; the first of these joints, between the medial cuneiform and the first metatarsal, is much the largest and also the most mobile. The **metatarsophalangeal** and **interphalangeal** joints are similar to the metacarpophalangeal and interphalangeal joints in the hand (p. 304). The first metatarsophalangeal joint is especially important, for it takes a great part of the weight of the body during the thrust forward of the trunk on the limb during walking. The joint is extremely robust, and it varies considerably in the degree to which it can be dorsiflexed.

The muscles

Muscles acting on the hip joint

The main muscle at the back of the hip joint is the **gluteus maximus** (Fig. 181), which arises from the area behind the posterior gluteal line on the lateral surface of the ilium (Fig. 170), from the dorsum of the sacrum and coccyx, and from the sacrotuberous ligament. The fibres cross the fold of the buttock obliquely. One-quarter of them are inserted into the gluteal tuberosity of the femur, and the remaining three-quarters are attached to the lateral side of the upper end of the tibia and the capsule of the knee joint by means of the **iliotibial tract of the fascia lata**. This is a tough thickened band of the deep fascia (fascia lata), and it receives also the insertion of the **tensor fasciae latae**, which comes from the outer lip of the anterior part of the iliac crest (Figs. 181, 186). The iliotibial tract can be felt as the most anterior of the ridges at the lateral side of the knee when the joint is partly flexed.

The gluteus maximus is supplied by the inferior gluteal nerve (p. 378), and the tensor fasciae latae by the superior gluteal nerve (p. 377), both of which come from the sacral plexus. The tensor fasciae latae abducts the thigh, but its main function is to steady and strengthen the knee joint. The gluteus maximus is mainly an extensor of the thigh on the trunk, as in climbing, or of the trunk on the thigh, as in getting up from a seated position. The lower fibres of the muscle adduct the thigh and rotate it laterally and those inserted into the iliotibial tract brace the lateral side of the knee joint and may help to rotate the tibia laterally.

Deep to the gluteus maximus lie several other muscles. The **gluteus medius** (Fig. 181) comes from the lateral surface of the ilium between the posterior and anterior gluteal lines, and the **gluteus minimus** (Fig. 182) comes from the area between the anterior and the inferior gluteal lines. Both muscles are inserted into the lateral surface of the greater trochanter of the femur, the minimus in front of the medius. Both abduct the thigh on the trunk or the trunk on the thigh. This action is used in walking, where the glutei of one side tilt the trunk over laterally while the opposite foot is raised off the ground, so keeping the balance as the weight is transferred (p. 374). They are both supplied by the superior gluteal nerve.

The **piriformis** (Fig. 182) runs from the front of the sacrum out through the greater sciatic notch to be inserted on the medial side of the greater trochanter. The tendon of the **obturator internus** (p. 282) emerges through the lesser sciatic notch to reach a similar insertion. The **quadratus femoris** comes from the lateral margin of the ischial tuberosity and inserts into the region of the intertrochanteric crest. All three muscles act as lateral rotators of the thigh on the trunk, or vice versa, and in addition they probably have an important protective action during movement of the hip, like the rotator cuff of the shoulder joint (p. 310). In this they are assisted by the **obturator externus**, which comes from the outer aspect of the obturator membrane and winds round under the hip joint to reach the greater trochanter. This muscle is supplied by the

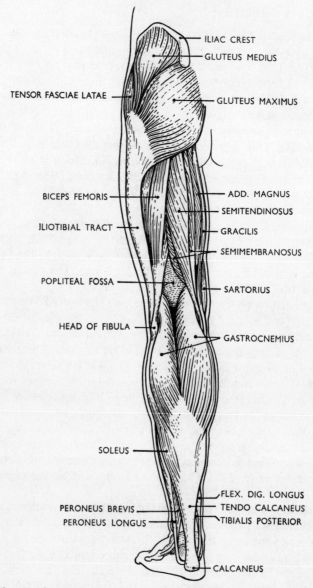

Fig. 181. The superficial muscles of the left lower limb seen from behind.

obturator nerve (p. 376); the other three are supplied by branches of the sacral plexus (p. 377).

The chief flexor of the hip is the **iliopsoas**, which has already been described (p. 382). It is assisted by the **rectus femoris** (Fig. 184), a component of the

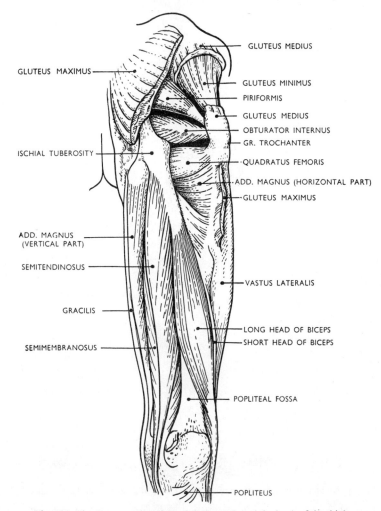

GLUTEUS MEDIUS

GLUTEUS MAXIMUS

GLUTEUS MINIMUS

PIRIFORMIS

GLUTEUS MEDIUS

OBTURATOR INTERNUS

GR. TROCHANTER

ISCHIAL TUBEROSITY

QUADRATUS FEMORIS

ADD. MAGNUS (HORIZONTAL PART)

GLUTEUS MAXIMUS

ADD. MAGNUS (VERTICAL PART)

SEMITENDINOSUS

VASTUS LATERALIS

GRACILIS

LONG HEAD OF BICEPS

SHORT HEAD OF BICEPS

SEMIMEMBRANOSUS

POPLITEAL FOSSA

POPLITEUS

Fig. 182. The deep muscles of the right buttock and the back of the thigh.

quadriceps (p. 361), which arises from the anterior inferior spine by one head, and from the bone above the acetabulum by another.

On the medial side of the hip joint lie the adductor muscles (Figs. 183, 185). The uppermost member of the group is the **pectineus**—a small muscle coming from the pectineal line of the pubis and inserted into the back of the shaft of the femur above the linea aspera. Inferior to this the **adductor longus** runs from the

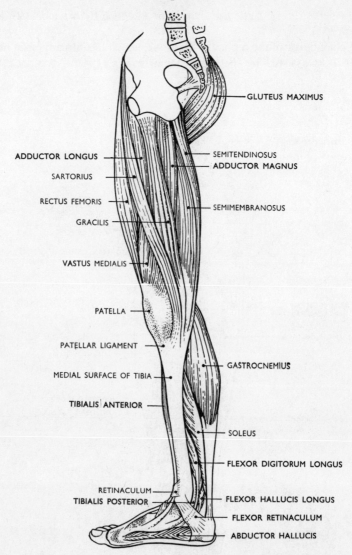

GLUTEUS MAXIMUS

ADDUCTOR LONGUS

SEMITENDINOSUS
ADDUCTOR MAGNUS

SARTORIUS

RECTUS FEMORIS

SEMIMEMBRANOSUS

GRACILIS

VASTUS MEDIALIS

PATELLA

PATELLAR LIGAMENT

GASTROCNEMIUS

MEDIAL SURFACE OF TIBIA

TIBIALIS ANTERIOR

SOLEUS

FLEXOR DIGITORUM LONGUS

RETINACULUM
TIBIALIS POSTERIOR

FLEXOR HALLUCIS LONGUS

FLEXOR RETINACULUM

ABDUCTOR HALLUCIS

Fig. 183. The superficial muscles of the right lower limb seen from the medial side.

femoral surface of the body of the pubis to the linea aspera. Inferior to this again is the **adductor brevis**, which arises from the body and inferior ramus of the pubis and is inserted into the linea aspera and part of the shaft of the femur above it. The **gracilis** is attached to the medial side of the inferior ramus of the pubis and runs straight down the medial side of the thigh to the upper part of the medial surface of the tibia (Figs. 181, 183). The great **adductor magnus** (Fig. 182) comes from the ischial tuberosity and the ramus of the ischium. Some of its fibres pass obliquely across to the linea aspera, others incline downwards to the ridge leading to the medial condyle, and the most medial ones run straight down to the adductor tubercle and the medial side of the knee joint. All the adductors adduct the thigh, but the pectineus also flexes it, and the brevis and longus also have a slight flexor action. The vertical fibres of the adductor magnus, on the other hand, can help in extension of the hip joint. The main function of the adductor group is to prevent or rectify lateral overbalancing; by acting from their insertions they pull the centre of gravity back over the supporting limb (p. 372). The pectineus is supplied by the femoral nerve (p. 375), but all the others are supplied by the obturator nerve (p. 376), and the adductor magnus also receives a branch from the sciatic nerve (p. 378).

Summary of movements at the hip joint

Flexion:	**Iliopsoas; rectus femoris;** pectineus; adductor longus; adductor brevis: sartorius (p. 364).
Extension:	**Gluteus maximus:** adductor magnus and the hamstring group (p. 363).
Abduction:	**Gluteus medius; gluteus minimus;** tensor fasciae latae.
Adduction:	**Adductor group** (adductors brevis, longus and magnus; pectineus; gracilis); **gluteus maximus** (lower fibres); quadratus femoris.
Lateral rotation:	**Gluteus maximus;** obturator internus; obturator externus; piriformis; quadratus femoris; adductors longus, brevis and magnus.
Medial rotation:	**Iliopsoas;** tensor fasciae latae; gluteus minimus (anterior fibres).

Medial rotation is weak by comparison with the strong movement of lateral rotation; similarly adduction at the hip is much stronger than abduction, and extension than flexion. These differences can all be related to the mechanics of walking (p. 373).

Muscles acting on the knee joint

The only extensor of the knee is the complex **quadriceps femoris**, which consists of four parts: the **rectus femoris**, the **vastus intermedius**, the **vastus lateralis**, and the **vastus medialis** (Figs. 184, 185). The origin of the rectus femoris has already been dealt with (p. 359), and all the other components of the muscle lie in relation to the shaft of the femur in the position indicated by their names. The vastus intermedius is concealed anteriorly by the rectus femoris; it arises from the upper two-thirds of the anterior and lateral surfaces of the shaft. The vastus lateralis comes mainly from the upper half of the linea aspera and the gluteal tuberosity, but some of its fibres reach up as far as the greater trochanter and the intertrochanteric line, and many take origin from the fascia lata which covers the muscle. The larger vastus medialis arises from the lower part of the intertrochanteric line, the linea aspera, and the medial intermuscular septum (Figs. 6, 185). All four components combine to form a tendon which is inserted into the tuberosity of the tibia, the patella being developed in it as a sesamoid bone. Fibres of the vastus medialis and the vastus lateralis blend with the capsule of the knee joint on either side of the patella and a few fibres of the

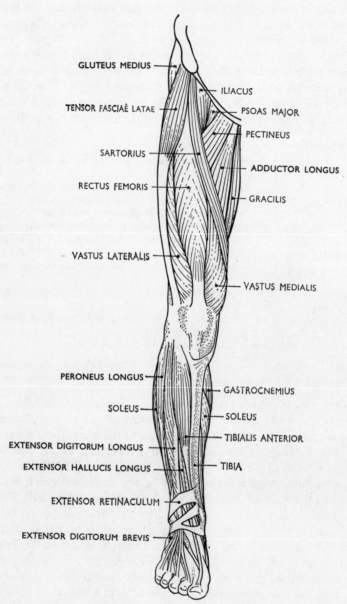

Fig. 184. The superficial muscles of the right lower limb seen from in front.

vastus intermedius are inserted into the capsule covering the suprapatellar bursa, and 'take up the slack' as the joint extends. The whole muscle is supplied by the femoral nerve (p. 375).

The flexors of the knee are more numerous. The **hamstring** group (Figs. 181, 182) comprises the **biceps femoris**, the **semitendinosus** and the **semimembranosus**. All three have a common origin from the ischial tuberosity, but the **biceps** receives a second head from the middle of the linea aspera. The two heads join and run down to be inserted into the head of the fibula, the tendon being split into two by the fibular collateral ligament. The **semitendinosus** runs from the common origin down the back of the thigh as a

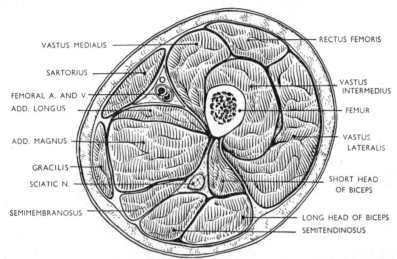

Fig. 185. Cross-section through the middle of the thigh: note the position of the sciatic nerve and the femoral blood vessels.

thin tendon (hence the name) to insert into the upper end of the medial surface of the tibia, close by the insertion of the gracilis. The **semimembranosus** is attached to the back of the medial condyle of the tibia, where many of its fibres spread out to reinforce the capsule of the knee joint. Its name describes the flat membranous sheet which is a feature of the muscle. All the hamstrings are supplied by the sciatic nerve, and all flex the knee. All except the short head of the biceps assist the gluteus maximus in extension of the hip, and their tension opposes flexion of the hip when the knee is straight.

As the biceps swings laterally away from the semitendinosus and semimembranosus in the distal third of the thigh, a depression is left in the midline called the **popliteal fossa** (Fig. 182). This fossa is a feature of the back of the knee when it is slightly flexed, but when the joint is extended the fossa disappears, and is converted into a bulge, because the semimembranosus moves laterally to fill up the space on the back of the joint.

The **sartorius** is a feeble muscle which springs from the anterior superior spine of the ilium and runs obliquely across the front and medial side of the thigh to be inserted into the upper end of the medial surface of the tibia, near the insertions of the gracilis and the semitendinosus. It puts the limb into the cross-legged position formerly adopted by tailors, and this accounts for its name. It is supplied by the femoral nerve (p. 375).

The **popliteus** (Fig. 182) is a small muscle which arises from the lateral condyle of the femur just below the lateral epicondyle by a thin tendon which rapidly gives place to fleshy fibres running downwards and medially to the posterior surface of the tibia in its upper third. It flexes the knee and laterally rotates the femur on the tibia, thus undoing the movement of medial rotation of the femur which occurs at the end of extension. It is supplied by the tibial nerve (p. 379).

The other important flexor of the knee, the **gastrocnemius,** is described below, for its main action is on the ankle.

Summary of movements at the knee

Flexion:	**Hamstring group** (biceps, semitendinosus, semimembranosus) **gastrocnemius;** popliteus; sartorius.
Extension:	**Quadriceps femoris.**
Lateral rotation of femur on tibia:	(This movement is equivalent to medial rotation of tibia on femur) **popliteus;** semitendinosus; semimembranosus; sartorius.
Medial rotation of femur on tibia:	(This movement is equivalent to lateral rotation of tibia on femur) **biceps;** iliotibial tract.

Muscles acting on the ankle joint

The only muscle whose primary action is to dorsiflex the ankle is the powerful **tibialis anterior**, which comes from the upper two-thirds of the lateral surface of the shaft of the tibia and from the interosseous membrane. It is inserted into the medial sides of the base of the first metatarsal and the medial cuneiform bone (Fig. 183, 184), and is supplied by the deep peroneal nerve (p. 379). The muscle is also an invertor (p. 356) of the foot. The movement of dorsiflexion at the ankle is notably assisted by the extensor hallucis longus and the extensor digitorum longus, which act primarily on the toes (p. 367).

The chief plantarflexor is a composite muscle often called the **calf muscle** or **triceps surae.** The **gastrocnemius** (Figs. 181, 186) arises by two heads, one from the posterior surface of the femur above the medial condyle, and the other from the posterior part of the lateral surface of the lateral condyle. The **plantaris,** a small and unimportant component of the muscle, arises from the posterior aspect of the femur above the lateral condyle, in a position corresponding to the medial head of the gastrocnemius. It is sometimes absent. The **soleus** (Fig. 181, 183), which lies deep to the gastrocnemius, arises from the posterior aspect of the head and the upper third of the shaft of the fibula, from the **soleal line** which runs obliquely downwards and medially across the posterior surface of the

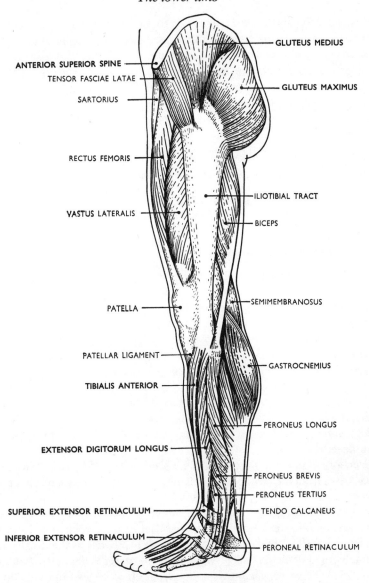

Fig. 186. The superficial muscles of the left lower limb seen from the lateral side.

tibia to its medial border, and from part of the medial border below this line. All three components of the calf muscle are inserted by a common tendon, the **tendo calcaneus** (Achilles tendon), into the middle of the posterior surface of the calcaneus, with a small bursa between it and the bone. The calf muscle is supplied by the tibial nerve (p. 379). All three components plantarflex the ankle, but the gastrocnemius (and the plantaris to an insignificant extent) can also flex the knee because of their attachments to the back of the femur. The soleus is an important postural muscle in the standing position (p. 371).

Other plantar flexors are the tibialis posterior, the flexor digitorum longus, the flexor hallucis longus and the peroneus longus, all of which have their primary actions on other joints. All the dorsiflexors and plantarflexors of the ankle contract from their insertions to stop the body overbalancing backwards or forwards respectively.

Summary of movements at the ankle joint

Dorsiflexion:	**Tibialis anterior; extensor digitorum logus;** extensor hallucis longus; peroneus tertius (p. 367).
Plantarflexion:	**Calf muscle** (gastrocnemius, soleus, plantaris); **tibialis posterior;** flexor digitorum longus; flexor hallucis longus; peroneus longus.

Muscles acting on the tarsal joints

All the muscles crossing the tarsal joints on their way to the toes have an action on these joints. However, inversion and eversion take place mostly at the talocalcanean and talocalcaneonavicular joints (p. 356), and are mainly subserved by four powerful muscles. Inversion is carried out by the **tibialis anterior** (p. 364) and the **tibialis posterior**, which is a strong muscle arising from the posterior surface of the shaft of the fibula and the lateral part of the posterior surface of the tibia (Fig. 187). Its tendon grooves the back of the medial malleolus and is inserted mainly into the tuberosity of the navicular, though it has membranous attachments to all the bones of the tarsus except the talus, and also to the middle metatarsals. It is the deepest muscle on the back of the leg, and is supplied by the tibial nerve. Besides inverting the foot, it plantarflexes the ankle.

Eversion is due to the peroneal muscles (Figs. 181, 186). The **peroneus longus** arises from the head of the fibula and the lateral surface of the upper two-thirds of its shaft. Its long tendon runs behind the lateral malleolus and enters the groove on the under surface of the cuboid, which conducts it across the sole to its insertion into the lateral sides of the base of the medial cuneiform and the first metatarsal. It is thus attached to the same two bones as the tibialis anterior, so completing a tendinous sling which helps to preserve the anterior arch of the foot (p. 372). The **peroneus brevis** comes from the lower two-thirds of the lateral surface of the fibula (overlapping the peroneus longus) and is inserted into the base of the fifth metatarsal. The **peroneus tertius**, not always present, comes from the lower part of the anterior surface of the fibula and is inserted with the peroneus brevis. The peroneus longus and brevis are supplied by the superficial peroneal nerve, and the peroneus tertius is supplied by the deep peroneal nerve. All the peronei evert the foot: the peroneus tertius in addition dorsiflexes the ankle.

The invertors and evertors play an important part in preventing lateral or medial overbalancing when the weight is carried on one foot. Acting from their insertions, they pull the centre of gravity of the body back over the narrow supporting pillar.

Summary of movements at tarsal joints

Inversion:	**Tibialis anterior; tibialis posterior.**
Eversion:	**Peroneus longus: peroneus brevis:** peroneus tertius.
Other movements:	Gliding movements summing to allow dorsiflexion, plantar flexion, and also some abduction, are produced by the muscles acting on the toes.

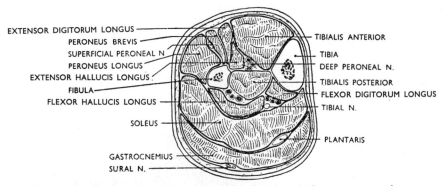

Fig. 187. Horizontal section through the middle of the leg: the interosseous membrane separates the tibialis posterior from the tibialis anterior and extensor hallucis longus.

Muscles acting on the toes

The **extensor digitorum longus** (Figs. 184, 187) comes from the upper two-thirds of the anterior surface of the shaft of the fibula, and splits into four tendons which run to the four lateral toes. These behave exactly as do the extensor tendons of the fingers (p. 319), and develop expansions on the dorsum of the proximal phalanges. To these are attached the tendons of the lumbricals, but not (p. 321) the interossei.

The **extensor digitorum brevis** arises from the anterior part of the upper surface of the calcaneus, and sends three slender tendons to be inserted into the extensor expansions of the middle three toes; a fourth tendon of the muscle is inserted into the base of the proximal phalanx of the great toe, and there is no tendon running to the little toe.

The **extensor hallucis longus** comes from the middle portion of the anterior surface of the fibula and the interosseous membrane. It is inserted into the base of the dorsum of the terminal phalanx of the great toe (Figs. 184, 187).

All the extensors of the digits are supplied by the deep peroneal nerve, and all can extend the metatarsophalangeal and interphalangeal joints of the toes, as well as dorsiflexing the ankle joint and the intertarsal joints. As in the hand, extension of the interphalangeal joints is greatly assisted by the lumbricals.

The **flexor digitorum longus** (Fig. 187) comes from the posterior surface of the tibia, medial to the tibialis posterior. Its tendon crosses the posteromedial aspect of the ankle joint and is inserted by four slips into the proximal ends of the plantar surfaces of the terminal phalanges of the lateral four toes, just as the

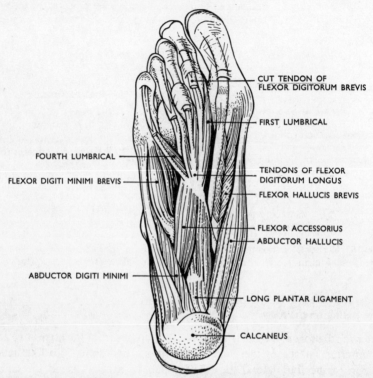

Fig. 188. The muscles on the sole of the right foot. The plantar aponeurosis and the flexor digitorum brevis have been removed to display the deeper muscles.

flexor digitorum profundus is inserted in the hand (p. 318). The **lumbricals** (Fig. 188) which spring from the sides of these tendons also behave similarly to the lumbricals in the hand, winding round the toes on their *medial* side to gain insertion to the extensor expansion on the dorsum. An additional muscle, not found in the hand, is the **flexor accessorius**, which runs from the plantar surface of the calcaneus to be inserted into the tendon of the flexor digitorum longus before it splits up. It changes the direction of pull of the tendon, so that the toes are pulled straight back to the heel and not towards the medial side of the ankle (Fig. 188).

The **flexor hallucis longus** (Figs. 187, 189) comes from the lower two-thirds of the posterior surface of the fibula, overlapping the tibialis posterior. It passes behind the ankle joint, and is inserted by a long tendon into the base of the distal phalanx of the great toe. Both the long flexors of the toes are supplied by the tibial nerve, and flex all the joints of their own digits, as well as the intertarsal joints. Both plantarflex the ankle joint and help to maintain the medial arch of the foot (p. 372).

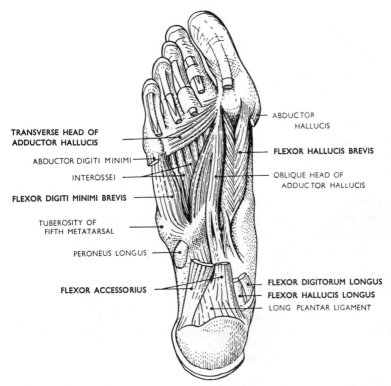

ABDUCTOR
HALLUCIS

TRANSVERSE HEAD OF
ADDUCTOR HALLUCIS

ABDUCTOR DIGITI MINIMI

INTEROSSEI

FLEXOR DIGITI MINIMI BREVIS

TUBEROSITY OF
FIFTH METATARSAL

PERONEUS LONGUS

FLEXOR ACCESSORIUS

FLEXOR HALLUCIS BREVIS

OBLIQUE HEAD OF
ADDUCTOR HALLUCIS

FLEXOR DIGITORUM LONGUS

FLEXOR HALLUCIS LONGUS

LONG PLANTAR LIGAMENT

Fig. 189. The deep muscles on the sole of the right foot. The flexor digitorum longus, the abductor hallucis, the abductor digiti minimi, and the flexor accessorius have been cut away to give a better view.

The **flexor digitorum brevis** (Fig. 188) is a poor affair when compared with the flexor digitorum superficialis of the fingers. It has an exactly comparable insertion, to the middle phalanges of the toes, and is perforated by the long flexor tendons in exactly the same way (p. 317). The muscle arises from the plantar surface of the calcaneus, and is confined to the sole of the foot.

The great toe has a number of short muscles attached to it. The **abductor hallucis** (Fig 188) runs along the medial side of the foot from the medial surface of the calcaneus to the base of the proximal phalanx of the toe; it abducts the great toe (i.e. pulls it *medially*) and supports the medial arch of the foot during the final phase of the 'push-off' in walking (p. 374). The **flexor hallucis brevis** (Fig. 189) comes from the plantar surfaces of the cuboid and lateral cuneiform bones, and is inserted into the two sides of the base of the proximal phalanx of the toe. The two sesamoid bones in its tendons of insertion are said to act as roller bearings in this important weight-bearing area of the foot (p. 372). The **adductor hallucis** (Fig. 189) is a strong muscle with two heads. The transverse head comes from the metatarsophalangeal joints of the lateral four toes, and the oblique head comes from the plantar surfaces of the bases of the middle metatarsals: both are inserted into the lateral side of the base of the proximal phalanx in common with the lateral head of the flexor brevis. All the muscles of the thumb except the opponens and the abductor longus (p. 323) are thus represented in the foot.

The little toe has an **abductor** (which deviates it *laterally*) and a **flexor brevis** (Fig. 188, 189), but these are of little importance.

The **interossei** (Fig. 189) correspond exactly to those of the hand (p. 321) except that they are inserted into the phalanges and not into the extensor hoods and that the dorsal group are arranged so as to abduct the toes away from an imaginary line drawn through the second toe (not the third); the plantar muscles adduct towards the same line. They flex the metatarsophalangeal joints and help to maintain the anterior arch of the foot (p. 372).

All the small muscles of the sole are supplied by branches of the tibial nerve (p. 379). In addition to their primary actions, they all have some effect in maintaining the arches of the foot.

Summary of movements of the toes: metatarsophalangeal joints

Flexion:	**Flexor digitorum longus** (flexor accessorius): **flexor hallucis longus;** flexor digitorum brevis; lumbricals; interossei. Flexor hallucis brevis; flexor digiti minimi brevis.
Extension:	**Extensor digitorum longus; extensor hallucis longus;** extensor digitorum brevis.
Abduction and adduction	**Interossei; abductor hallucis;** lumbricals; abductor digiti minimi.

Summary of movements of the toes: interphalangeal joints

Flexion:	**Flexor digitorum longus; flexor hallucis longus;** flexor digitorum brevis; (proximal joints only).
Extension:	**Extensor digitorum longus; extensor hallucis longus; lumbricals;** extensor digitorum brevis.

Fascial arrangements at the ankle and in the sole

Bowstringing (p. 325) of the dorsiflexors at the ankle is prevented by two **extensor retinacula** developed as thickenings of the deep fascia in front of the ankle (Fig. 186). Similarly, there is a strong **flexor retinaculum** (Fig. 185) on the medial side of the ankle, and the flexor tendons pass deep to this on their way into the sole, accompanied by the plantar vessels and nerves. Smaller bands restrain the peroneal muscles on the lateral side of the calcaneus.

An arrangement of this sort means that all the tendons must develop synovial sheaths to prevent friction. These are not exposed to such risks as those in the hand, and in consequence they become infected much less frequently. Their exact anatomy is thus less significant.

In the sole, a dense **plantar aponeurosis** corresponds to the palmar aponeurosis of the hand (p. 325). It is attached posteriorly to the under surface of the calcaneus, and anteriorly it splits up into digital fibrous sheaths just as in the palm. The long flexor tendons play in these sheaths, and in consequence develop synovial sheaths. The plantar aponeurosis is important in maintaining the arches of the foot. As in the palm, its central portion is much stronger than the portions which underlie the short muscles of the great and little toes.

Mechanism of standing

In the anatomical position the skull tends to fall forwards because its centre of gravity is in front of the atlanto-occipital joint. The muscles at the back of the neck, chiefly the semispinalis capitis, act as postural muscles to prevent this, and the erector spinae and the other long muscles of the vertebral column are also in constant action to maintain its secondary curvatures.

The centre of gravity of the body lies near the second sacral vertebra, and a perpendicular dropped through it passes behind the hip joint and in front of the knee and ankle joints. The muscles in front of the hip joint and behind the knee and ankle joints are thus put on the stretch and theoretically ought to serve as postural antigravity muscles. However, electromyography (p. 59) in subjects standing 'at ease' can detect no consistent electrical activity except in the calf muscles. It may be that the bony and ligamentous arrangements at the hip and knee joints suffice to maintain a stable posture, while those of the ankle joint do not. In the hip, the immensely strong iliofemoral ligament tightens in extension and the joint is screwed home by medial rotation of the femur so that the bony surfaces become 'close packed' (p. 46). The corresponding screwing-home mechanism of the knee does not operate until the joint is hyperextended (p. 354). In the ankle, the tendency to topple forwards brings the broad part of the talus between the tibia and fibula and tightens up the ligaments (p. 355).

The weight of the body is transferred through the talus to the other tarsal and metatarsal bones. The feet have been compared to a dome sawn in half down the middle; when the two feet are side by side the dome is reconstituted. The base of each half dome is composed of the posterior end of the calcaneus behind, the cuboid and fifth metatarsal laterally, and the heads of the metatarsals anteriorly (Figs. 174, 175). This analogy is useful in describing the configuration of the foot and the structure of its bones, but it is misleading because the foot is not rigid and static, but a mobile springy apparatus which continuously adapts itself to the varying stresses put upon it.

In the standing position about half the weight is taken by the pad of the heel under the calcaneus, and the remainder by the pads under the heads of the metatarsals. About one-third of the metatarsal share falls on the great toe and the remainder is distributed among the other toes.

The medial side of the foot is normally free from the ground, and this **medial arch** is composed of the calcaneus behind, the talus as keystone, the navicular, the three cuneiforms, and the medial three metatarsals (Fig. 175). It is held up by the shape of the bones, the tension of the strong plantar fascia (p. 371), the muscles and ligaments which act as tie-beams to the arch, and the pull of the muscles which sling under it on their way to the tarsus or toes. The **lateral arch** consists of the calcaneus (which it shares with the medial arch), the cuboid and the lateral two metatarsals. It is practically flat on the ground, and its posterior and anterior ends are important bearing points. The **anterior arch**, between the shafts and heads of the metatarsals, is in the coronal plane, transverse to the other arches. It is kept up by the tension of the plantar fascia and of the ligaments and muscles of the intermetatarsal spaces, by the adductor hallucis, and by the slinging action of the tibialis posterior and the peroneus longus (p. 366).

In standing, there is a phase lasting about half a minute in which all the muscles directly associated with the foot, except for the calf muscle, are inactive, and the support for the arches is presumably provided mainly by the ligaments. This inactive phase alternates with an active phase resulting from the unstable equilibrium. As postural sway (p. 226) occurs, the centre of gravity is pulled back into its original position relative to the feet by contraction of the muscles which first of all balance the leg on the foot, then the thigh on the leg, and lastly the trunk on the thigh. For example, if the body begins to fall to the left side, the left tibialis anterior and posterior pull the left leg medially on the foot and the adductor group of the thigh pull the pelvis over towards the right. The right peroneus longus and brevis pull the right leg laterally on the foot, and the right gluteus medius and minimus tilt the pelvis over towards the right side. At the same time the right erector spinae pulls the centre of gravity towards the right.

Anteroposterior sway is coped with by the dorsiflexors and plantarflexors of the ankle and the extensors and flexors of the knee and hip joints. The balancing action of gluteus maximus and tensor fasciae latae acting on the posterior and anterior ends of the iliac crest is also an important factor.

The posture of the body depends to a considerable extent on the **pelvic tilt**, which is the angle between the horizontal and a line drawn through the promontory of the sacrum and the upper border of the symphysis pubis. This angle is normally about 60°, and is largely determined by the pull of the hamstrings, the hip flexors, the erector spinae, and the abdominal muscles. The tensions in these muscle groups are in turn influenced by the way the individual habitually stands. If a 'tense' habit of standing has been acquired, the pelvic tilt increases, so that the pelvis rotates forward on the thighs. This carries the lumbar spine forwards, and with it the centre of gravity. To compensate for this, the upper part of the body is thrust backwards, increasing the lumbar curvature. The neck flexes to keep the gaze horizontal, and is held stiffly, with the chin tucked in; the knees are usually hyperextended. Continued standing in this manner may lead to a shortening of the hip flexors and the erector spinae, while the hamstrings tend to lengthen; this naturally perpetuates the condition.

If the opposite habit of standing slackly has been developed, the pelvic tilt decreases, with consequent shortening of the hamstrings and strain on the iliofemoral ligament. The lumbar curvature decreases, bringing the centre of gravity backwards, and the head and thorax are thrust forwards to compensate. The thoracic curvature increases, the neck is extended, and the chin is poked forwards to bring the head level. The knees are usually slightly bent and thus mechanically unstable; this flexion in turn interferes with the stability of the ankle and the arches of the feet. The slack standing position is more common than the tense one, and it gives the familiar impression that its practitioner is giving away at nearly every joint and is about to fold up completely.

If a high heel is worn, the weight is shifted forwards, and a disproportionate amount of strain is taken by the anterior arch of the foot, which may flatten. The heads of the metatarsals are forced on to the sole, stretching the nerves going to the toes. The toes themselves are rammed by the weight of the body into the narrow constricting part of the shoe, and the consequences are evident in the numbers of smart shoes which are kicked off under restaurant tables.

A collapse of the medial arch gives rise to the commonest type of flat foot. Since the arch gives the foot its springiness and its ability to absorb shocks administered to it by running and jumping, the flat foot suffers under the strain and becomes painful. The pain is usually only marked when the arch is actually flattening and the nerves entering the sole are being stretched; when destruction of the arch is complete the foot, though relatively painless, is clumsy and the patient has a characteristic gait. The causes of flat foot range from the sudden strain imposed on the foot by the rapid increase in weight during adolescence to paralysis of the muscles maintaining the arch. It is particularly common in nurses and waiters, who are compelled by their calling to remain on their feet for long periods.

Mechanism of walking

In walking the centre of gravity of the body is pushed upwards (by about 5 cm) and forward by the contraction of the muscle groups which extend the hip and the knee and plantarflex the ankle. As the thrusting limb leaves the ground it is brought forwards in front of the body by contraction of the flexors of the hip, and this now helps the displaced centre of gravity to make the body fall forwards.

The dorsiflexors of the ankle and toes contract to prevent the toes from dragging on the ground, and after the limb passes in front of the centre of gravity, the hip and knee once more extend, so that the limb returns to the ground fully extended, with the ankle partly dorsiflexed. As the foot strikes the ground, the knee flexes, and the dorsiflexors of the ankle contract again, preventing the forepart of the foot from slapping down. The weight is first taken by the posterior end of the calcaneus, transferred along the lateral arch to the head of the fifth metatarsal, and thence medially along the anterior arch to the ball of the great toe, from which the final thrust as the step is completed transfers the force back along the medial arch. The lateral arch is thus chiefly put under strain when the foot is coming down, and the medial arch when it pushes off again. The great toe rocks on the two sesamoid bones in the flexor hallucis brevis, and as the heel is raised the metatarsophalangeal joint of the great toe dorsiflexes, so that the last part of the lower limb to retain contact with the ground is the phalangeal pad of the great toe. This dorsiflexion 'winds up' the plantar aponeurosis by pulling it forwards round the heads of the metatarsals like a cable being wound on a windlass, and helps to support the medial arch as the main strain is taken. Once the heel has been raised the long flexors of the toes supply the final impetus for the next thrust, and this would cause the toes to curl up underneath themselves were it not for the action of the lumbricals and the extensor tendons, which keep the interphalangeal joints straight.

As the thrusting limb leaves the ground, the trunk begins to fall over to the same side, away from the supporting limb. To prevent this happening the gluteus medius and minimus of the supporting limb tilt the body laterally and at the same time the erector spinae and the abdominal muscles on the side of the free limb contract, apparently to keep the pelvis from tilting. They are not entirely successful in this, so that there is a sideways lurch of about 5 cm with each step. If the hip abductors are paralysed, this movement is greatly exaggerated.

Finally, rotation takes place at both hips during a step, in order to maintain a steady forward progress in a relatively straight line. The trunk rotates laterally on the thigh of the limb that is on the ground, and the thigh that is free rotates laterally on the pelvis.

As walking proceeds, the body is kept steady on each foot in turn by the muscles acting from their insertions just as they do to control postural sway during standing.

A further balancing factor is the ballistic swing of the arms; the swing of the right upper limb counterbalances that of the left lower limb and vice versa.

When running, the heel does not touch the ground, and both feet are off the ground at once; otherwise the mechanism is essentially similar to that of walking.

The nerves

The lumbar plexus

The lumbar plexus lies within the substance of the psoas major muscle in the posterior wall of the abdomen (Fig. 191). It is a simple loop plexus between the ventral rami of the first four lumbar nerves, which increase in size from above

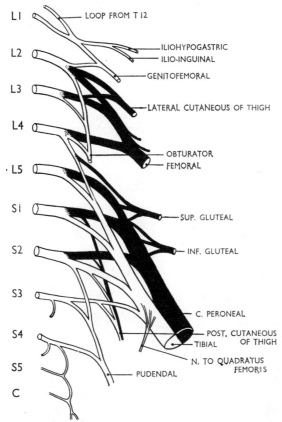

Fig. 190. Diagram of the lumbar and sacral plexuses. A number of branches have been omitted. The nerves represented in white spring from the anterior divisions of the ventral rami; those in black from the posterior divisions. The small unlabelled branches of S3 and S4 contain parasympathetic fibres which run to the pelvic plexuses of the autonomic system.

downwards. Its main branches are the femoral and obturator nerves, both of which contain fibres from the 2nd, 3rd and 4th lumbar nerves (Fig. 190).

The **femoral nerve** runs its course inside the abdomen, and enters the thigh just lateral to the midpoint of the inguinal ligament, where it at once breaks up into a number of branches which supply most of the skin on the front of the thigh as well as three muscles—the pectineus, sartorius and quadriceps. Its longest branch is the **saphenous nerve**, which supplies skin on the medial side of the leg and foot (Fig. 192). The femoral nerve is not often injured, because it lies deeply among the muscles of the abdomen, and its course in the thigh is very brief.

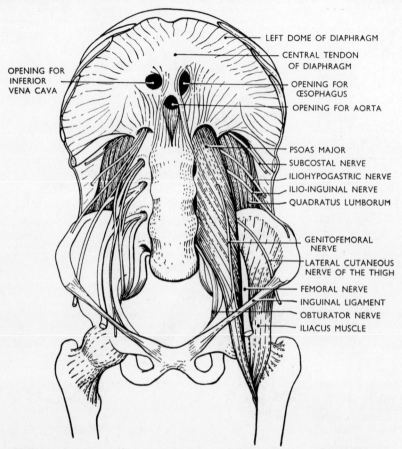

LEFT DOME OF DIAPHRAGM

CENTRAL TENDON OF DIAPHRAGM

OPENING FOR INFERIOR VENA CAVA

OPENING FOR ŒSOPHAGUS

OPENING FOR AORTA

PSOAS MAJOR

SUBCOSTAL NERVE

ILIOHYPOGASTRIC NERVE

ILIO-INGUINAL NERVE

QUADRATUS LUMBORUM

GENITOFEMORAL NERVE

LATERAL CUTANEOUS NERVE OF THE THIGH

FEMORAL NERVE

INGUINAL LIGAMENT

OBTURATOR NERVE

ILIACUS MUSCLE

Fig. 191. The lumbar plexus and the muscles of the posterior abdominal wall. On one side the psoas major has been removed to expose the manner in which the branches of the plexus are formed within its substance.

The **obturator nerve** also lies deeply. It runs between the psoas major and the brim of the pelvis (Fig. 191) to enter the thigh through the obturator foramen. It supplies the adductor muscles (including occasionally the pectineus, which then has a double supply) and the skin over the medial aspect of the thigh. Paralysis of this nerve is also rare.

The sacral plexus

The sacral plexus (Fig. 190) is formed from the ventral rami of the fifth lumbar and the upper three sacral nerves, with a large communication from the fourth lumbar above and a smaller one from the fourth sacral below. It lies on the anterior surface of the sacrum, separated from it by the fibres of the piriformis

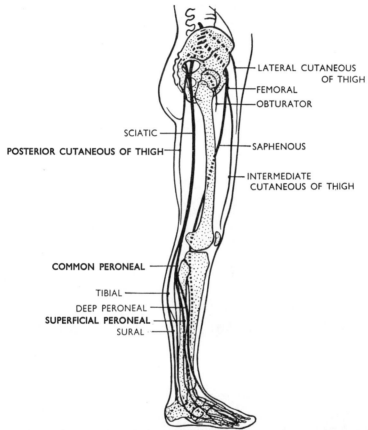

Fig. 192. The femoral, obturator, and sciatic nerves. Most of their branches have been removed for the sake of clarity.

muscle. The biggest branch of the plexus, the sciatic nerve, is formed at the margin of the greater sciatic foramen (Fig. 192).

The **superior gluteal nerve** emerges through the greater sciatic foramen with the sciatic nerve, and runs to supply the gluteus medius and minimus and the tensor fasciae latae. Section of this nerve paralyses abduction at the hip and produces a characteristic lurching limp (p. 374).

The **inferior gluteal nerve** emerges through the greater sciatic notch below the sciatic nerve, and confines itself to supplying the gluteus maximus. Injury to this nerve paralyses the chief extensor of the hip. In consequence, stability at the hip is imperilled during walking, and the patient cannot get out of a chair without using his hands to push him up. He also has difficulty in bending down without overbalancing, for the gluteus maximus is largely in control of this movement, 'paying out' as the trunk moves forwards. Lastly, the loss of the gluteus maximus leads to a loss of the protection afforded by the padding action of its thick bulk, and a bedridden patient may develop bedsores (p. 63).

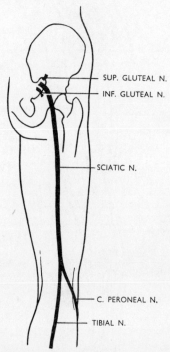

SUP. GLUTEAL N.

INF. GLUTEAL N.

SCIATIC N.

C. PERONEAL N.

TIBIAL N.

Fig. 193. The position of the sciatic nerve in the buttock.

The main branch of the sacral plexus, the chief nerve of the lower limb, and the largest nerve in the body, is the **sciatic nerve**. In consists of two divisions, the **tibial** and **common peroneal nerves**, which may leave the pelvis separately, but are usually wrapped up in a common sheath. The sciatic nerve is derived from all the roots of the sacral plexus except the fourth sacral, and leaves the pelvis through the greater sciatic notch, one-third of the way down a line joining the posterior superior spine to the ischial tuberosity. It leaves the buttock opposite the middle of a line joining the ischial tuberosity to the tip of the greater trochanter of the femur, and between these two points it describes a curve convex laterally. Intramuscular injections are very frequently given in this region, and may do disastrous damage if injected into the sciatic nerve. Accordingly such injections are usually made in the upper outer quadrant of the buttock, where they are furthest away from the danger zone (Fig. 193). In the buttock the nerve is covered posteriorly by the gluteus maximus and piriformis, and anterior to it lie the obturator internus and the quadratus femoris. It gives no branches in this part of its course.

The sciatic nerve runs straight down the middle of the back of the thigh on the posterior surface of the adductor magnus, being covered posteriorly by the long head of the biceps femoris. It usually ends about two-thirds of the way down the thigh, at the upper end of the popliteal fossa, by dividing into the tibial and common peroneal nerves. In the thigh it sends branches to the biceps, semimembranosus, semitendinosus, and the vertical fibres of adductor magnus.

The **common peroneal nerve** (Fig. 192) runs laterally along the margin of the biceps femoris to pass over the lateral head of the gastrocnemius. It crosses the neck of the fibula, where it can easily be felt by rolling it against the bone. In this position it is in danger of being cut with a scythe or billhook; injury to the nerve is an occupational danger for amateur hedgers and ditchers. After rounding this dangerous traverse the nerve breaks up into two branches, the **deep peroneal**, which supplies all the muscles in the front of the leg, and the **superficial peroneal**, which supplies the peroneal muscles, as well as giving cutaneous branches to the lateral side of the leg and the dorsum of the foot.

The **tibial nerve** (Fig. 192) continues the line of the sciatic nerve, and enters the leg behind the middle of the back of the knee joint. It supplies all the muscles and some of the skin on the back of the leg. It lies deeply among the muscles of the leg, and under the flexor retinaculum it divides into its two terminal branches, the **medial** and **lateral plantar nerves**, which supply the muscles and skin of the sole of the foot.

Cutting the sciatic nerve in the buttock or thigh thus paralyses all the muscles and skin of the leg and foot except for the skin on the medial side, which is supplied by the saphenous nerve (Fig. 194). In addition, the hamstring muscles will be paralysed if the cut is above the nerves supplying them. The result is a limb which will bear no weight and cannot be used for walking. If no recovery takes place, the patient is better off with an artificial limb.

Section of the common peroneal nerve is followed by paralysis of all dorsiflexors and evertors of the foot, and in consequence the foot tends to plantarflex every time the weight is taken off it—a condition known as 'foot drop' (compare wrist drop, p. 334). In addition, the invertors pull the sole of the dropped foot medially, so that the patient faces two hazards. When his foot is off the ground the dragging toes may catch in some obstacle and trip him up, and when he puts it down again he may well go over on the lateral side of his foot, spraining or breaking the ankle. This can be remedied by attaching to the toe of his shoe, on the lateral side of its tip, a small steel spring attached to an anklet. The strength of the spring is just such as to overcome the tendency of the plantarflexors and evertors to pull the foot down and out, and his foot will then

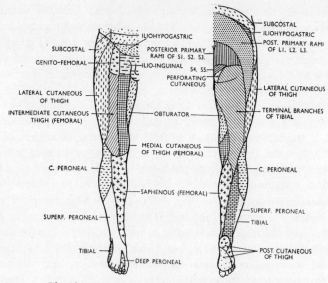

Fig. 194. Cutaneous nerve territories in the lower limb.

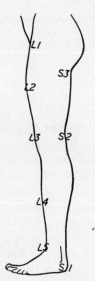

Fig. 195. Approximate scheme of the segmental innervation of the skin of the lower limb.

come down flat for the next step. His walk may be quite normal in appearance with such an appliance fitted, and he will be able to get along without any other assistance. The sensory loss is of little moment.

If, on the other hand, the tibial nerve is paralysed, there is a very different story. The plantarflexors are now paralysed, and the patient cannot 'shove off' with his foot when taking a step. The muscles of the sole are gone, and the arches of the foot collapse; the toes cannot be moved to add momentum to the thrust. The result is a rigid inelastic foot which is of little use as a support or for walking. But perhaps the most serious feature of the disability is the loss of sensation on the sole. If a small stone gets into the patient's shoe it will not be perceived, and an abrasion so produced may end as a large septic ulcer penetrating and destroying the bones of the foot. Such penetrating ulcers may result merely from bearing weight on the normal pressure points, and to prevent them the patient must take the greatest care of the skin of his soles, bathing them twice a day or oftener, changing his socks two or three times a day, and constantly inspecting the feet for signs of injury. It is perhaps not surprising that patients often prefer the certainties of an artificial limb to the doubts of a tibial paralysis.

Nerve territories in the lower limb

Figures 194 and 195 show the cutaneous distributions of the lower limb and the approximate positions of the dermatomes. The fourth and fifth lumbar dermatomes are of particular interest for they are the ones usually involved when an intervertebral disc prolapses backwards in the lumbar region. This is the commonest cause of 'sciatica', a name which merely means pain in the distribution of the sciatic nerve.

Motor points

Figure 196 shows the position of the main motor points on the lower limb (See p. 336).

The blood vessels and lymphatics

The main artery of the lower limb is the **femoral artery**, which enters the thigh at a point half way along the inguinal ligament as the continuation of the external iliac artery (p. 285). The femoral artery is medial to the femoral nerve and lateral to its own large vein. Its course in the thigh may be indicated on the surface by the upper two-thirds of a line joining the midpoint of the inguinal ligament to the adductor tubercle when the heel is placed on the patella of the opposite limb. Close to its origin the artery crosses in front of the hip joint, and can be compressed against the head of the femur by leaning on it with the closed fist. A tourniquet should be applied half way down the thigh so as to compress the artery against the shaft of the femur.

The femoral artery lies at first anterior to the adductor group of muscles, being crossed anteriorly by the sartorius and overlapped by the medial border of the vastus medialis. Two-thirds of the way down the thigh it passes through a hole in the adductor magnus, between it and the bone, and escapes on to the back of the thigh at the upper angle of the popliteal fossa. In the front of the thigh it is comparatively superficial, and may be injured by butchers using knives or choppers. This fact was also well known to Italian secret societies,

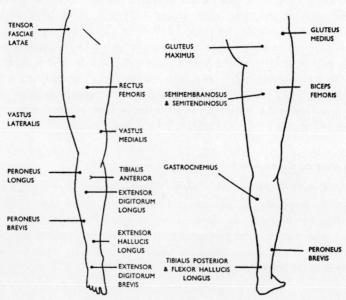

Fig. 196. Motor points in the lower limb.

whose members used a stab in the thigh as a favourite means of disposing of their victims. Such a stab can at one blow sever four large vessels—the femoral artery and vein, and their main offshoots, the **profunda** artery and vein. The profunda artery (Fig. 197) is distributed to the deeper tissues of the back of the thigh by branches which pass through holes in the adductor magnus and the other muscles of the adductor group.

In the popliteal fossa the femoral artery becomes the **popliteal** artery, which is covered posteriorly by its own vein and by the tibial nerve. It passes straight down behind the knee joint, and ends at the distal border of the popliteus muscle by dividing into the **anterior** and **posterior tibial** arteries.

The **posterior tibial** artery runs down alongside the tibial nerve, which lies on its lateral side, between the deeper muscles of the back of the leg, and enters the foot by passing deep to the flexor retinaculum. In the leg it gives off the

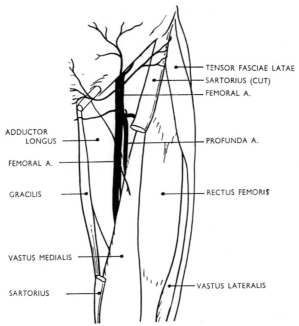

Fig. 197. The position of the femoral artery in the thigh.

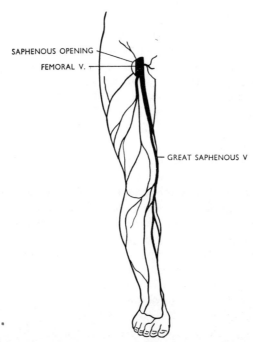

Fig. 198. The great saphenous vein (most of its tributaries have been omitted).

peroneal artery, which runs down along the medial border of the fibula among the muscles to end in branches which supply the lateral side of the ankle. In the foot the posterior tibial artery divides into a **medial** and a **lateral plantar** artery, which accompany the corresponding nerves to supply the structures in the sole and the plantar surfaces of the toes.

The **anterior tibial** artery pierces the upper part of the interosseous membrane to reach the front of the leg. It accompanies the deep peroneal nerve down to the ankle joint, where it changes its name and ends by making a wide sweep across the dorsum of the foot as the **dorsalis pedis** artery.

The pulse may be felt in the anterior tibial artery midway between the tips of the two malleoli, in front of the ankle joint. The pulse of the posterior tibial artery is midway between the tip of the medial malleolus and the medial margin of the heel.

The buttock is supplied by the **superior** and **inferior gluteal** arteries, which arise from the internal iliac artery and escape through the greater sciatic notch. The lower part of the back of the thigh is fed by branches of the profunda.

The **deep veins** of the lower limb accompany their corresponding arteries, and eventually drain into the femoral vein (gluteal veins in the case of the buttock). The superficial veins do not correspond to any artery, and run most of their course in the superficial fascia. The pressure of the long column of blood in the superficial veins may lead to them becoming varicose (p. 135). The **great saphenous** vein (Fig. 198) is the most important of the superficial group, and it extends from the medial side of the ankle up to the groin, where it pierces the deep fascia and ends in the femoral vein. On its way up it passes a hand's breadth posterior to the medial border of the patella. The **saphenous opening** in the deep fascia lies about three fingers' breadth distal and lateral to the pubic tubercle. The **small saphenous** vein, which comes up the back of the leg from the lateral side of the ankle, drains into the popliteal vein at the back of the knee. Together with their branches, the two saphenous veins drain nearly all the blood from the superficial tissues of the lower limb. They are often used for giving intravenous fluids or blood.

The **lymphatics** of the lower limb, like those of the upper limb, are divided into superficial and deep, and tend to accompany the veins. In the popliteal fossa a few nodes form the **popliteal group**, but the main filters are deployed just inferior to the inguinal ligament, where there are **deep** and **superficial inguinal** groups. The deep group receives afferent vessels from the limb, but the superficial group receives in addition lymphatics from the skin of the trunk below the umbilicus, from the external genitalia, and from the anus.

Surface anatomy

The bony landmarks in the lower limb have already been discussed. The demonstration of individual muscles by action against resistance follows the same lines as those laid down for the upper limb, and as in the upper limb, the muscles outline certain regions of anatomical importance. The only one of these which gives rise to clearly marked surface features is the popliteal fossa (p. 363) behind the knee. This is a diamond-shaped space containing the popliteal vessels and the tibial and common peroneal nerves, together with the popliteal lymph nodes and a quantity of fat. Medially the space is bounded by the semimembranosus and semitendinosus above and the medial head of gastrocnemius below (Fig. 181); the lateral boundaries are the biceps femoris and the lateral head of gastrocnemius. The fossa is filled up by the semimembranosus when the knee is extended (p. 363), but when it is flexed the biceps tendon forms a prominent feature of the lateral aspect of the knee and can be picked up between finger and thumb, as can the less immediately noticeable tendons forming the medial boundary of the upper part of the fossa. A little anterior to the biceps tendon in the flexed position the posterior margin of the iliotibial tract of the fascia lata may form a visible ridge.

The tendons in front of the ankle are easily identified, as are those of the peronei. But the tendons passing under the flexor retinaculum cannot be identified by palpation, with the exception of the tibialis posterior, which can be felt passing to its insertion distal to the lower edge of the retinaculum.

The surfaces features of the main nerves and vessels, and the position of the pulses, have already been discussed.

Glossary

This glossary is not intended to be complete, but merely to illustrate some of the ways in which anatomical terminology has been built up. For fuller information consult a medical dictionary. Most of the words are derived from the classical languages: L = Latin: G = Greek.

An approximate guide to pronunciation is:

ă as in lad	ĕ as in met	ĭ as in pill	ŏ as in rob	ŭ as in shut
ā as in late	ē as in meet	ī as in pile	ō as in robe	ū as in shute

Prefixes

A-, an-	Without, not
Ab-	From, away from
Ad-	Towards
Ante-	Before
Apo-	From (implies separation or derivation)
Bi-	Twice, two
Circum-	Around
Co-	Together
Contra-	Against, opposed
De-	Down, from, away
Dis-	Apart
Dys-	Bad, difficult
Endo-	Within
Epi-	Upon
Exo-	Outside
Extra-	Outside
Hemi-	Half
Hetero-	Different
Homo-	Same
Hyper-	Above, beyond, excessive
Hypo-	Below, under, deficient
Infra-	Beneath
Inter-	Between
Intra-	Within
Meta-	Next after, beyond
Para-	Beside
Peri-	Around
Post-	Behind, after
Pre-	Before
Pro-	Before
Sub-	Beneath
Supra-	Above
Syn-	Together

Suffixes

-blast	(G. *blastos*, germ) Applied to formative cells
-cyte	(G. *kutos*, cell) Applied to adult cells
-itis	Inflammation of
-oid	Like
-ology	Study of

Abdŏ′men (probably from L. *abdere*, to conceal) The belly

Abduct (L. *abducere*, to draw away) To draw away from the mid-line. The *abdūcent* nerve (6th cranial) is so-called because it innervates the muscle which abducts the eye

Accommodation (L. *accommodare*, to fit) The adjustment of the eye for close vision

Acetă′bulum (L.) A cup for holding vinegar; hence the cup-shaped socket of the hip joint

Acrŏ′mion (G. *akros*, extreme: *ōmos*, shoulder) The point of the shoulder

Adduct (L. *adducere*, to lead to) To draw towards the midline

Adenoid (G. *adēn*, gland; *eidos*, likeness) Gland-like; applied to the mass of lymphoid tissue at the back of the nasopharynx

387

Albū′min (L. *albus*, white) An important protein

Aliment′ary (L. *alimentum*, nourishment) Relating to food

Al′vĕŏlus (L.) A small cavity

Anǎ′bolism (G.) A building up

Anae′mia (G. *an-; haima*, blood) Deficiency in the number and/or quality of the red blood corpuscles

Anaesthē′sia (G. *an-; aisthēsis*, feeling) Absence of sensation

Anastomō′sis (G. *anastomoein*, to provide with a mouth) A communication between two blood vessels

Anǎ′tomy (G. *anatōme*, dissection)

An′ulus (L. *ānŭlus*, a ring) A circular opening; a ring-shaped area

Ānus (L.) A ring

Āor′ta (G. *aortē*) The main arterial trunk of the body

Aphā′sia (G. *a-; phăsis*, a saying) Term applied to conceptual disorders of production or reception of speech

Aponeurō′sis (G. *apo-; neuron*, tendon) A broad thin tendinous sheet

Appen′dix (L.) An appendage

Arach′noid (G. *arakhnē*, spider; *eidos*, form) Hence applied to the thin cobweb-like membrane, the second of the coverings of the central nervous system

Arĕ′olar (L. *areola*, dim. of *area*, space) Full of small spaces

Ar′tery (G. *artēria*, from *ăēr*, air, and *tērein*, to keep) The arteries are empty after death, for their blood is forced into the venous side of the circulation. They were thus regarded as air-tubes

Artē′riole (G. *artēria*) A small artery

Artí′cular (L. *articulus*, dim. of *artus*, a joint) Pertaining to a joint

Arytē′noid (G. *arutaina*, a ladle) Resembling a ladle; applied to two small cartilages in the larynx, which have no resemblance to ladles

Asper (L.) Rough; hence linea aspera, the rough line on the back of the femur

Āt′rium (L.) A hall or antechamber

At′rophy (G. *atrophia*, from *a-; trophē*, nourishment) Wasting

Au′ricle (L. *auricula*, dim. of *auris*, ear) Applied to the appendages of the atria because of their fancied resemblance to ears

Auscultā′tion (L. *auscultare*, to listen to) Method of examining by listening, often with a stethoscope (q.v.)

Autonom′ic (G. *autos*, self; *nomos*, law) Independent, self-controlling

Axil′la (L.) The armpit

Axon (G. *axon*, axis) The process of a nerve cell which conducts impulses away from the cell

Basil′ic (Arabic *al-basilic*; inner) The inner vein of the forearm

Bă′sophil (G. *basis*, base; *philein*, to love) Applied to cells which readily take up basic dyes

Bī′ceps (L. *bis*, twice; *caput*, head) A muscle with two heads

Bīfid (L. *bis*, twice; *findere*, to split) Split in two

Bī′pennate (L. *bis*, two; *penna*, feather) The arrangement of the fibres in a bipennate muscle resembles a feather

Brā′chial (G. *brakhiōn*, arm) Pertaining to the arm (*antebrachial*; pertaining to the forearm)

Brevis (L.) Short

Bronchus (G. *brogkhos*, windpipe) Originally referred to the trachea (q.v.) and subsequently to its branches

Buccinator (L.) A trumpeter; hence the muscle of the cheek used by trumpeters

Bursa (G.) A purse; hence a synovial pouch

Cae′cum (L. *caecus*, blind) The cul-de-sac at the beginning of the large intestine

Calcā′neus (L. *calx*, a heel) The bone of the heel

Cal′carine (L. *calcar*, a spur) Spur-shaped

Calcificā′tion (L. *calx*, lime; *facere*, to make) Deposition of calcium salts

Cal′lus (L.) Hard; applied to the formation of new bone round a fracture

Calorĭ′meter (L. *calor*, heat; G. *metron*, measure) An instrument for measuring the amount of heat produced by chemical substances when burned, or by the living body

Canal (L. *canalis*) A water-pipe; hence a tubular passage

Capil′lary (L. *capillus*, a hair) Hair-like; hence a vessel with a fine bore

Cǎ′pitate (L. *caput*, head) Head-shaped

Capi′tulum (L. *caput*, head) A little head

Cap′sule (L. *capsula*, dim. of *capsa*, a box) Hence the enveloping covering of a joint

Car′diac (G. *kardia*, heart) Pertaining to the heart

Carŏ′tid (G. *karoein*, to stupefy) Sudden compression of both common carotid arteries produces unconsciousness by reducing the blood supply to the brain

Car′pus (G. *karpos*, the wrist) The eight wrist bones

Cǎ′talyst (G. *kata*, down; *luein*, dissolve) A substance which brings about or speeds

Catalyst (*cont.*)
up a chemical reaction without itself being altered

Că'taract (G. *katarrhaktēs*, a waterfall) Opacity of the lens of the eye; the name derives from the appearance in the early stages

Cauda (L.) Tail: *Cauda equina* = horse's tail

Caudate (L. *cauda*, tail) The caudate nucleus has a long tail

Cava (L.) Hollow

Cĕpha'lic (G. *kephalikos*, pertaining to the head) The lateral vein of the forearm: so-called because letting blood from it was thought to cure headache

Cĕrebel'lum (L. Dim. of *cerebrum*, brain)

Cĕ'rebrum (L.) Brain (but the term now excludes the mid-brain and hind-brain)

Cerŭ'men (L. *cera*, wax) The secretion of the glands in the external auditory meatus

Cer'vix (L.) Neck

Chias'ma (G. *khiasmos*, a cross) An arrangement like the capital Greek letter Chi; hence optic chiasma, the place where some fibres of the optic nerves cross over to the opposite side

Chon'droblast (G. *khondros*, cartilage; *blastos*, germ) A cell which forms cartilage

Chŏ'rioid (G. *khorion*, membrane; *eidos*, form) The middle coat of the eye. Also applied to the filtration mechanisms producing cerebrospinal fluid

Chrŏ'mosome (G. *khrōma*, colour; *sōma*, body). The chromosomes stain deeply when prepared for histological examination

Ci'lia (L.) Eyelashes; hence also applied to the fine hair-like processes possessed by some cells

Clă'vicle (L. *clavicula*, dim. of *clavis*, key) The latin word was used of the sticks which boys used to bowl hoops; these were similar to the clavicle in shape

Clŏ'nus (G. *klōnos*, turmoil) Spasm of muscle with rapidly succeeding contractions and relaxations

Coc'cyx (G. A cuckoo) The term is said to be derived from a resemblance to the bill of the bird

Cŏch'lea (G. *kokhlos*, a snail) The shape of the organ closely resembles that of a snail shell

Coe'liac (G. *koilos*, hollow) Applied to the belly or body cavity; later confined to a restricted region of its upper portion

Col'lagen (G. *kolla*, glue; *gennan*, to produce) Glue and gelatin are made by boiling connective tissue

Colla'teral (L. *cum*, with; *lateralis*, from

latus, side) Hence running by the side of; accessory

Colli'culus (L. *collis*, hill) Hence a little hill; applied to the small mounds in the roof of the mid-brain

Cŏ'lon (G. *kolon*) The large intestine; possibly from G. *koilos*, hollow

Cŏma (G. *kōma*) A state of complete unconsciousness with no response apparent to any form of stimulation

Com'missure (L. *commissura*, a joining together) Hence applied to bundles of fibres connecting the two cerebral hemispheres

Con'cha (G. *kogkhē*, a shell) Applied to the three shell-like bones on the lateral wall of the nose

Con'dyle (G. *kondulos*, knuckle) Hence a rounded articular surface

Conjuncti'va (L. *conjungere*, to connect) Connects the eyeball with the lids

Cŏ'nus (G. *kōnos*) A cone

Cor'acoid (G. *korax*, a crow; *eidos*, form) Like a crow's beak

Cord (G. *khordē*, a string)

Cor'nea (L. *cornu*, a horn) The cornea is composed of collagen fibres which gave rise to the appellation 'horny'

Corŏ'na (L.) A crown. Corona radiata, the radiating crown of fibres entering and leaving the internal capsule; coronary, applied to the arteries which encircle the heart like a crown

Cŏ'ronoid (G. *korōnē*, crow; *eidos*, form) Applied to bony processes curved like a crow's beak

Cor'pus callŏ'sum (L. *corpus*, body; *callosus*, a beam or rafter) The corpus callosum could be said to resemble a beam passing from one cerebral hemisphere to the other

Cor'pus striā'tum (L. *corpus*, body; *stria*, a stripe or furrow) The caudate and lentiform nuclei have a striped appearance from the admixture of grey and white matter

Corpus'cle (L. *corpusculum*, dim. of *corpus*, body) A little body; used of the cells of the blood and also of some sensory endings of nerves

Cor'tex (L.) Rind

Cos'ta (L.) A rib; hence the adjective costal

Cri'coid (G. *krikos*, a ring: *eidos*, form) The cricoid cartilage is shaped like a signet ring

Crŭ'ciate (L. *crux*, a cross) In the form of a cross; the two cruciate ligaments of the knee joint form a St. Andrews cross whether looked at from in front or from the side

Crus (L.) The leg; anything having a process like a leg—e.g. the diaphragmatic *crura*

Cū′neiform (L. *cuneus*, wedge; *forma*, form) Wedge-shaped

Cusp (L. *cuspis*, a point) Used of the teeth and also of the flaps of the vascular valves

Cutā′neous (L. *cutis*, skin) Pertaining to the skin

Cȳ′toplasm (G. *kutos*, cell; *plassein*, to form) The protoplasm (q.v.) of the cell excluding the nucleus

Dē′cussate (L. *decussare*, to cross over) Hence decussation, a place where fibres cross over to the opposite side

Dē′ferens (L.) Carrying away

Del′toid (G. *delta*, the triangular fourth letter of the Greek alphabet) Triangular in shape

Den′drite (G. *dendron*, a tree) A branching process

Denti′culate (L. *dens*, tooth) Toothed

Diabē′tes (G. *dia*, through, *bainein*, to go) A condition in which much water is excreted. D. inspidus: due to disease of the hypophysis; the urine is not sweet (insipid). D. mellitus: due to disease of the Islets of Langerhans; sugar is excreted in the urine, giving it a sweet taste (L. *mel*, honey)

Diagnō′sis (G. *dia*, through; *gnōsis*, knowledge) The recognition of a disease

Di′aphragm (G. *diaphragma*, a partition wall)

Diă′stolĕ (G.) A drawing apart; hence the relaxation of the walls of the cavities of the heart

Digas′tric (G. *dis*, twice; *gastēr*, belly) Having two bellies

Dis′locate (L. *dis*, apart; *locare*, from *locus*, place) Displacement of a bone or organ from its normal position

Dis′tal (L. *distare*, to be distant) Peripheral, further away

Divertī′culum (L.) A turning aside, a blind ending

Duodē′num (L. *duodenarius*, containing twelve) The duodenum was originally described as being twelve fingers' breadths in length

Dura mater (L. *dura*, hard; *mater*, mother) The outermost and toughest of the three membranes enveloping the central nervous system

Ef′ferent (L. *effere*, to bear away) Carrying away from

Ejă′culate (L. *ex*, out; *jaculum*, a dart) Eject rapidly, especially of the seminal fluid

Elec′trocar′diogram (G. *ēlektron*, amber; *kardia*, heart; *gramma*, a mark) A record of the changes in electrical potential produced by the contracting heart muscle

Elec′tro-enceph′alogram (G. *ēlektron*, amber; *egkephalos*, brain; *gramma*, mark) A record of the changes in electrical potential produced by the brain

Elec′tromy′ogram (G. *ēlektron*, amber; *mus*, muscle: *gramma*, mark) A record of the changes in electrical potential produced by a contracting muscle

Em′bryo (G. *embruon*, from *en*, within; *bruein*, to swell) That which grows or swells within the womb

Endocar′dium (G. *endon*, within; *kardia*, heart) The lining membrane of the heart

En′docrine (G. *endon*, within; *krinein*, to separate) Endocrine secretion is the process of liberating substances into the blood and not on to an exterior surface

Endomȳ′sium (G. *endon*, within; *mus*, muscle) Delicate connective tissue covering each muscle fibre

Endoneu′rium (G. *endon*, within; *neuron*, sinew) Delicate connective tissue covering each nerve fibre

Endothē′lium (G. *endon*, within; *thēlē*, nipple) The lining membrane of blood and lymph channels, also of ducts, such as those which open at the nipple

En′zȳme (G. *en*, in; *zumē*, leaven) Any catalyst (q.v.) found in living cells and accelerating their chemical processes

Ēōsi′nophil (G. *ēos*, dawn; *philein*, to love) Used of cells staining with eosin, a red dye; possibly derived from the red colour of the dawn

Epider′mis (G. *epi*, upon; *dermis*, skin) The outer part of the skin

Epidi′dymis (G. *epi*, upon; *didumoi*, twins) Lying on the twins (the testicles)

Epiglot′tis (G. *epi*, upon; *glottis*, the larynx)

Epi′phȳsis (G. *epi*, upon; *physis*, growth) Name given to the separate portions of bone in which secondary centres of ossification appear, and between which and the shaft growth takes place at epiphyseal plates

Epi′thēlium (G. *epi*, upon; *thele*, nipple) A surface layer of cells

Ery′throcyte (G. *eruthros*, red; *kutos*, cell)

Excrete (L. *excernere*, to separate out) To expel waste matter from the body

Ex′ocrine (G. *ex*, out of; *krinein*, to separate) Externally secreting

Expirā′tion (L. *ex*, out; *spirare*, to breathe)

Exten′sion (L. *extendere*, to stretch out)

Ex′teroceptive (L. *exterus*, on the outside; *capere*, to take) Receiving information from outside the body

Extrapyra′midal (L. *extra*, outside; G. *puramis*, pyramid) Outside the pyramidal system of motor fibres

Extrin′sic (L. *extrinsecus*, from without) External—e.g. the ocular muscles moving the eyeball; ligaments at a distance from the joint capsule

Fa′cet (French *facette*, from L. *facies*, a face) A small articular surface on a bone

Falx (L.) A sickle; the *falx cerebri* could conceivably be called sickle-shaped

Fa′scia (L. *fascia*, band) The fascia may be looked on as a bandage enclosing the muscles

Fasci′culus (L.) A bundle

Fau′cēs (L.) The throat; the opening into the throat

Fē′mur (L.) Thigh

Fenes′tra (L.) Window

Fī′bril (L. *fibrilla*, dim. of *fibra*, fibre)

Fībroblast (L. *fibra*, fibre: G. *blastos*, germ) A fibre-producing cell

Fi′bula (L.) A pin or skewer; the reference is to the thinness of the bone

Fīlum (L.) A thread

Flac′cid (L. *flaccidus*, flabby)

Flā′vus (L.) Yellow; hence *ligamenta flava*, composed of yellow elastic tissue

Fontanelle (French, dim. of *fontaine*, fountain) The derivation is obscure

Forā′men (L.) A hole

Fossa (L.) A ditch or a grave

Fō′vea (L.) A pit

Fron′tal (L. *frons*, forehead) Pertaining to the forehead

Fūni′culus (L.) A little cord; hence a slender bundle

Fū′siform (L. *fusus*, spindle; *forma*, shape) Spindle-shaped

Gang′lion (G. *gagglion*, a tumour) First used in relation to abnormal swellings, later applied to the sympathetic ganglia, and subsequently applied to the swellings on sensory nerves and the masses of grey matter in the base of the cerebral hemispheres

Gastric (G. *gaster*, stomach) Pertaining to the stomach

Gastrocnē′mius (G. *gaster*, stomach, *kneme*, leg) Hence the muscle belly of the leg

Gĕni′culate (L. *genu*, knee) Bent like a knee

Glabel′la (L. *glaber*, bald) The space between the eyebrows devoid of hair

Gland (L. *glans*, an acorn)

Glē′noid (G. *glene*, socket; *eidos*, form) Applied to the flat socket of the shoulder joint

Glomĕr′ulus (L. *glomus*, a ball of thread)

Glos′sopharyngē′al (G. *glossa*, tongue; *pharugx*, pharynx) Pertaining to the tongue and pharynx

Glū′teus (G. *gloutos*, buttock) A buttock muscle: G. maximus, the largest buttock muscle, etc.

Goitre (French, from L. *guttur*, throat) Any enlargement of the thyroid gland

Go′nad (G. *gŏnē*, semen) A sexual gland, either testicle or ovary

Gracilis (L.) Thin or slender

Gut (O. English) The intestinal canal

Gy′rus (G. *guros*, round) A ring; applied to the convolutions of the brain between the sulci (q.v.)

Hāē′moglō′bin (G. *haima*, blood; L.*globus*, ball) The red protein which colours the erythrocytes of the blood and transports oxygen

Hăe′morrhage (G. *haima*, blood; *rhagia*, bursting forth)

Hal′lux (L. *allex*) The great toe

Hamate (L. *hamus*, a hook) Hooked

Hemiplē′gia (G. *hemi*, half; *plēgē*, a blow or stroke) Paralysis of one half of the body

Hĕpar (G.) The liver; hence hepatic

Hernia (L.) A protrusion

Hīlum (L.) The point of attachment of a seed; hence the point of entry of blood vessels, etc. into an organ

Hi′stiocyte (G. *histos*, web; *kutos*, cell) A connective tissue cell; the 'web' refers to the meshwork of connective tissue fibres

Histŏ′logy (G. *histos*, web; *logos*, discourse) Microscopic anatomy

Hor′mone (G. *horman*, to set in motion) A chemical excitant

Hū′merus (L.) The bone of the arm; from G. *omos*, shoulder

Hȳ′aline (G. *hualos*, glass) Like glass

Hȳdrocĕ′phalus (G. *hudor*, water; *kephalē*, head)

Hȳ′oid (G. *huoeides*) Shaped like the Greek letter upsilon; U-shaped

Hȳ′perextension (G. *huper*, over; L. *extendere*, to extend)

Hȳper′trophy (G. *huper*, over; *trophe*, nourishment) enlargement of cells or fibres

Hȳpoglos′sal (G. *hupo*, under; *glossa*, tongue)

Hȳpo′physis (G.*hupo*, under; *physis*, growth) The hypophysis grows under the brain

Hўpotha′lamus (G. *hypo*, under; *thalamos*, an inner chamber) Below the thalamus (q.v.)

Ĭ′lĕŭm (G. *eilein*, to twist) The lower part of the small intestine

Ĭ′liăc (L. *ilium*, the flank) Relating to the ilium, the bone of the flank or groin

Im′par (L.) Unpaired

Incus (L.) Anvil

Infundĭ′bulum (L.) Funnel; applied to the stalk of the hypophysis

In′guinal (L. *inguen*, groin)

Inspira′tion (L. *inspirare*, breathing in)

Intercală′ted (L. *intercalere*, to insert) Inserted, placed between

Interos′seous (L. *inter*, between; *os*, bone)

In′tima (L. *intimus*, innermost) The innermost coat of a blood-vessel

Intrin′sic (L. *intrinsecus*, inside, within) The intrinsic muscles of the tongue lie within the tongue itself; the intrinsic ligaments of a joint lie actually in its capsule

Inver′sion (L. *invertere*, to turn upside down) The movement by which the sole of the foot turns inwards

I′ris (G.) Rainbow; refers to the pigmentation of the iris

Is′chium (G. *iskhion*, hip)

Jejŭ′num (L. *jejunus*, empty) The upper part of the small intestine is found empty after death

Jŭ′gular (L. *jugulum*, neck) (Also jŭgular)

Kată′bolism (G. *kata*, down; *ballein*, to throw) The breakdown of complex molecules into simpler ones

Kĕ′ratin (G. *keras*, horn) A protein found in horny tissues—skin, hair, nails, etc.

Kўphŏ′sis (G. *kyphos*, crooked) An increased forward curvature of the vertebral column

Lă′crimal (L. *lacrima*, tear) Relating to tears or the apparatus for producing and disposing of them

Lac′teal (L. *lacteus*, milky)

Lamb′doidal (G. *lambda*, the letter of the alphabet) Shaped like a lambda or an inverted Y

Lamel′la (L. *lamella*, dim. of *lamina*, a plate)

Lă′mina (L.) A plate or leaf

La′teral (L. *latus*, side)

Latis′simus (L. superl. of *lātus*, wide) Broadest or widest

Len′tiform (L. *lens*, lentil; *forma*, shape) Shaped like a lens or lentil

Lĕū′cocyte (G. *leukos*, white; *kutos*, cell)

Levă′tor (L. *levare*, to lift)

Li′gament (L. *ligare*, to bind)

Li′nea (L.) A line

Li′ngua (L.) The tongue; hence *lingual*

Lobe (G. *lobos*, a pod) A part of an organ separated from the neighbouring parts by a fissure or a septum

Locomŏ′tor (L. *locus*, place; *movere*, to move) Producing movement in space

Lordŏ′sis (G. *lordos*, bent backwards) Increase in the backward curvature of the lumbar part of the vertebral column

Lum′bar (L. *lumbus*, loin)

Lum′brical (L. *lumbricus*, earthworm) The muscles are thin and more or less cylindrical, like a worm

Lŭ′men (L.) Light; hence the clear space within a tube

Lŭ′nate (L. *luna*, moon) Moon-shaped; the lunate bone looks like a crescent moon when viewed from the side. Hence also *semilunar*

Lymph (L. *lympha*, water) Hence *lymphatic*; a tube conveying lymph

Lymph′ocyte (L. *lympha*, water; *kutos*, cell) A lymph cell

Mă′crophage (G. *makros*, large; *phagein*, to eat) Hence a large cell capable of engulfing foreign particles or bacteria

Macroscop′ic (G. *makros*, large; *skopein*, to view) Visible to the naked eye

Mă′cula (L.) Spot

Mag′nus (L.) Large

Major (L.) Larger

Mal′lĕŏlus (L.) A small hammer

Mal′leus (L.) A hammer. The human malleus resembles the hammer used by the Romans to stun oxen before sacrificing them

Man′dible (L. *mandibula*, from *mandere*, to chew)

Manŭ′brium (L.) A handle

Massē′ter (G. *masētēr*, a chewer)

Mas′toid (G. *mastos*, breast; *eidos*, form) Resembling a breast

Mă′trix (L.) A breeding female animal; later used of formative issues such as the tissue which gives origin to the nails, and subsequently transferred to mean the groundwork of a tissue which shelters the cells—e.g. the matrix of cartilage

Maxil′la (L.) The upper jaw

Mĕā′tus (L.) A passage

Mĕ′dia (L.) Middle; from L. *medius*, as are the words medial and median

Mĕ′diăstĭ′num (L.) The origin is variously attributed; perhaps it comes from *medius*,

Mĕ′dĭastĭ′num (*cont.*)
middle, and *stans*, standing; thus 'standing in the middle' of the chest

Medul′la (L.) Literally means marrow, as bone-marrow. Hence the internal part of an organ, as suprarenal medulla, and the part of the nervous system enclosed by the vertebral canal—medulla spinalis and medulla oblongata (oblongata = oblong)

Megakă′ryocyte (G. *megas*, large; *karuon*, nut; *kutos*, cell) The cells are very large, with a large nucleus

Mellĭ′tus (L.) Sweet (v. diabetes)

Menin′gēs (G. *mēniggĕs*, plural of *mēnigx*, membrane)

Menis′cus (G. *mēniskos*, crescent) The intra-articular cartilages of the knee joint are crescentic

Mĕ′sentery (G. *mesos*, middle; *enteron*, intestine) The double layer of peritoneum which suspends the intestine from the posterior abdominal wall

Meta′bolism (G. *metabole*, change)

Metacar′pus (G. *meta*, beyond; *karpos*, wrist)

Metatar′sus (G. *meta*, beyond; *tarsos*, ankle)

Mi′crometre (G. *mikros*, small) One thousandth part of a millimetre

Minimus (L.) Smallest

Minor (L.) Smaller

Mitochon′dria (G. *mitos*, thread, *khondrion*, granule) The mitochondria are small threadlike or rod-shaped bodies in the cytoplasm

Mi′tral (G. *mitra*, head-dress) Hence the mitre worn by bishops, which is not unlike the cusps of the mitral valve of the heart

Modĭ′olus (L.) A hub of a wheel

Mŏ′lar (L. *mola*, a mill) A grinding tooth

Mŏ′nocyte (G. *monos*, single; *kutos*, cell) Monocytes are few and far between in the circulating blood

Mū′cus (L.) The viscid secretion of certain glands. Adjective mucous

Mȳ′elin (G. *muelos*, marrow)

Mȳocar′dium (G. *mus*, muscle; *kardia*, heart) The muscular layer of the heart wall

Mȳ′otome (G. *mus*, muscle; *tome*, section)

Myxoede′ma (G. *muxa*, slime; *oidein*, to swell)

Navĭ′cular (L. *navicula*, a small ship) Boat-shaped

Nerve (L. *nervus*, a sinew or nerve) The double meaning is understandable, for small nerves look very like bundles of connective tissue, and vice versa

Neural′gĭa (G. *neuron*, nerve; *algos*, pain) Intermittent pain in the distribution of a nerve or nerves

Neurilem′ma (G. *neuron*, nerve; *lemma*, husk)

Neuro′glĭa (G. *neuron*, nerve; **glia**, glue) Hence the connective tissue of the central nervous system

Neu′trophil (L. *neuter*, neither; G. *philein*, to love) Staining with dyes which are neither acid nor alkaline

Nĭgrum (L. *niger*, black) Substantia nigra = the black substance

Node (L. *nodus*, a knot)

Nū′cha (Arabic) At first meant the spinal cord, but later applied to the nape of the neck and Latinized to provide it with a genitive; Ligamentum nuchae = the ligament of the nape of the neck

Nū′cleus (L.) Kernel

Nȳstag′mus (G. *nustazein*, to nod, as in sleep) Involuntary rhythmic movements of the eyeballs

Obturātor (L. *obturare*, to close) The obturator membrane blocks up the obturator foramen

Occĭ′pital (L. *ob*, at; *caput*, head)

Oculomo′tor (L. *oculus*, eye: *movere*, to move)

Oedē′ma (G. *oidema*, a swelling) Accumulation of fluid in the tissues

Oesŏ′phagus (G. *oisophagos*, gullet)

Olĕ′cranon (G. *olēnē*, elbow; *kranos*, helmet) Hence point of the elbow

Omen′tum (L.) May be derived from *omen*, since the practice of divining the future from the inspection of the entrails of animals was common in Rome

Ophthal′moscope (G. *ophthalmos*, eye; *skopein* to view) An instrument for examining the interior of the eye

Orbĭ′cularis (L.) Circular muscles, as orbicularis oculi; orbicularis oris

Or′gan (G. *organon*, implement) A part of the body having a specific function or functions

Or′gasm (G. *orgān*, to swell) The culmination of the sexual act

Os, oris (L.) A mouth

Os, ossis (L.) A bone

Ossifica′tion (L. *os*, bone; *facere*, to make)

Os′teoblast (G. *osteon*, bone; *blastos*, germ) A bone-forming cell

Os′teoclast (G. *osteon*, bone; *klastos*, broken) A bone-destroying cell

Os′teocyte (G. *osteon*, bone; *kutos*, cell) A bone cell

Pan'creas (G. *pan*, all; *kreas*, flesh) The abdominal sweetbread

Parathy'roid (G. *para*, beside; *thureos*, a shield; *eidos*, form) The small glands behind the thyroid gland (which has little resemblance to a shield; the 'shield' is probably the thyroid cartilage close by)

Parŏ'tid (G. *para*, beside; *ous*, ear)

Patel'la (L.) A little pan

Pectorā'lis (L. *pectus*, breast) A breast muscle

Pĕdun'cle (L. *pedunculus*, a stalk or stem)

Pen'nate (L. *penna*, feather) Resembling a feather in the arrangement of its fibres; hence bipennate, unipénnate, etc.

Pericar'dium (G. *peri*, around; *kardia*, heart) The membrane surrounding the heart

Perimȳ'sium (G. *peri*, around; *mus*, muscle)

Periŏs'teum (G. *peri*, around; *osteon*, bone)

Peristal'sis (G. *peri*, around; *stalsis*, compression) The wavelike contractions of the wall of the alimentary canal

Peritonē'um (G. *peri*, around; *teinein*, to stretch) The lining membrane of the abdominal cavity, which is also stretched over the viscera

Peronē'al (G. *perone*, a pin or brooch) Relating to the fibula (q.v.)

Pĕt'rous (G. *petra*, stone) Stony hard

Pha'gocyte (G. *phagein*, to eat; *kutos*, cell) A cell which devours foreign material

Phă'lanx (G. *phalagx*) The small bones of the fingers and toes are arranged like soldiers drawn up for battle in ranks as a phalanx

Phren'ic (G. *phrēn*, midriff) Relating to the diaphragm—e.g. phrenic nerve

Pia (L. *pius*, faithful, devout) The pia mater faithfully follows the contours of the brain and spinal cord

Pi'riformis (L. *pirum*, pear; *forma*, form)

Pi'siform (L. *pisum*, pea; *forma*, form)

Pitu'itary (L. *pituita*, phlegm) At one time it was supposed that the hypophysis or pituitary gland secreted the mucus found in the pharynx

Pleura (G. *pleura*, rib) The membrane lining the ribs and covering the lungs

Plexus (L.) A plait

Poliomyeli'tis (G. *polios*, grey; *muelos*, marrow; *-itis*, denotes inflammation) Inflammation of the grey matter of the spinal cord

Pol'lex (L.) The thumb

Po'lymorph (L. *polus*, many; *morphē*, form) A white cell having nucleus which may have many different shapes

Pons (L.) A bridge

Poplitē'al (L. *poples*, the ham) Relating to the back of the knee

Presbyō'pia (G. *presbus*, an old man; *ōps*, eye) A defect of vision due to loss of elasticity of the crystalline lens with age

Prŏ'prioceptive (L. *proprius*, one's own; *capere*, to receive) Receiving stimulation from within the body (contrast exteroceptive, q.v.)

Prŏ'toplasm (G. *protos*, first; *plassein*, to form) Hence the basic material of life

Ptĕ'rygoid (G. *pterux*, a wing; *eidos*, form) Wing-shaped

Ptȳ'alin (G. *ptualon*, spittle) The digestive ferment in saliva

Pul'monary (L. *pulmo*, lung) Relating to the lungs

Pylŏr'us (G. *pyle*, a gate; *ouros*, a guard) Hence the 'sentry' at the exit from the stomach

Quadrā'tus (L.) Squared, or four-sided

Quad'riceps (L. *quadri-* four; *caput*, head) Four-headed

Radio'logy (L. *radius*, ray; G. *logos*, discourse) The science dealing with radiant energy, including X-rays

Rāmus (L.) A branch

Rectus (L.) Straight

Rĕ'nal (L. *ren*, kidney) Pertaining to the kidney

Retină'culum (L. *retinēre*, to retain) A restraining band

Retraction (L. *retrahere*, to draw back)

Sac'cule (L. *sacculus*) A little sack

Sā'crum (L. *sacer*, sacred) The 'sacred bone' was believed to be the immortal part of the body, and to form a basis for its resurrection

Sa'gittal (L. *sagitta*, arrow) The sagittal suture resembles an arrow because of the 'flight' provided by the lambdoidal suture at its posterior end

Săphē'nous (G. *saphēnēs*, manifest) The saphenous venous system is easily seen

Sarcolem'ma (G. *sarx*, flesh; *lemma*, husk) The sheath of an individual muscle fibre

Sartŏr'ius (L. *sartor*, a tailor) The tailor's muscle

Scala (L.) A flight of steps; but there are no 'steps' in the cochlea

Scā'lene (G. *skalēnos*, uneven) Applied to the scalene muscles, which were supposed to have three unequal sides

Scă'phoid (G. *skaphē*, boat; *eidos*, form) Boat-shaped

Sca'pula (L.) The shoulder blade

Sciă′tic (G. *iskhion*, ischium)

Sclē′ra (G. *sklēros*, hard) The outer coat of the eye

Se′ptum (L.) A wall

Sĕ′samoid (G. *sēsamē*) Resembling a sesame seed

Si′nus (L.) A hollow or cavity e.g. the air sinuses of the skull

Si′nusoid (L. *sinus*, a cavity)

Sŏ′leus (L. *solea*, a sole) The term is derived, not from the sole of the foot, but from a resemblance of the belly of the muscle to the body of the flatfish

Sphē′noid (G. *sphēn*, wedge; *eidos*, form) Wedge-shaped

Sphinc′ter (G. *sphigkter*, a binder) Any muscle surrounding and capable of narrowing or closing an orifice

Splanch′nic (G. *splagkhna*, the viscera)

Stā′pēs (L.) A stirrup

Ster′num (G. *sternon*, the breast)

Stĕ′thoscope (G. *stethos*, the chest; *skopein*, to view) The stethoscope is an instrument for listening and not viewing

Stȳ′loid (L. *stilus*, a writing instrument)

Subclā′vian (L. *sub*, below; *clavis*, key) Under the clavicle (q.v.), as in subclavian artery, which passes behind the clavicle on its way into the arm

Suprarē′nal (L. *supra*, above; *ren*, kidney)

Sustentā′culum (L.) A support

Sympathet′ic (G. *sumpatheia*, sympathy) The sympathetic system was thought to connect up different parts of the body without the mediation of the central nervous system, and to be the means whereby one part of the body 'sympathized' with another. We now call the phenomenon 'referred pain'.

Sȳm′physis (G. *sun*, together; *phusis*, growth)

Sȳnapse (G. *sun*, together; *haptein*, to touch)

Sȳndesmō′sis (G. *syndesmos*, a band or ligament)

Sȳs′tolĕ (G. *sustole*, contraction)

Tar′sus (G. *tarsos*, the ankle)

Tec′tum (L.) Roof

Tem′poral (L. *tempus*, time) The temples are where the white hairs indicating the passing of time first appear

Ten′don (L. *tendo*, from *tendere*, to stretch)

Tentōr′ium (L.) A tent

Tĕ′rĕs (L.) Round

Tes′tis (L. *testis*, a witness) In Roman law no man could be a witness unless his testicles were present

Tha′lamus (G. *thalamos*, an inner chamber) Yet the thalamus is solid

Thĕ′nar (G. *thĕnar*, hand). Later restricted to the lateral part of the hand, when the word hypothenar was applied to the medial side

Thō′rax (G.) The chest

Thȳ′roid (G. *thureos*, shield; *eidos*, form) Shield-shaped. The term was first applied to the thyroid cartilage and later, by association, to the thyroid gland which lies on it

Topogra′phical (G. *topos*, place; *graphein*, to write) Applied to descriptive anatomy and maps

Trache′a (G. *trakheia*, rough) The 'roughness' is due to the cartilaginous hoops. The windpipe was originally called the arteria trachea, the rough airpipe, in contrast to the arteria leia, the smooth airpipe, a term which was applied to the blood vessels leading from the heart (v. artery)

Tri′ceps (L. *tres*, three; *caput*, head)

Trigĕm′inal (L. *trigeminus*, triple) An allusion to the three large divisions of the nerve

Trochan′ter (G. *trokhanter*, runner) Possibly derived from the movement of the greater trochanter of the femur in running

Troch′lea (G. *trokhilia*, a pulley)

Trŏ′phic (G. *trophē*, nourishment)

Tū′ber (L.) A swelling. Hence tuberosity and tubercle

Un′cus (L.) A hook

U′reter (G. *ourētēr*, from *ouron*, urine)

U′tricle (L. *utriculus*, a little bag)

U′vula (L.) A little grape

Vagi′na (L.) A sheath

Vā′gus (L.) Wandering

Vas (L.) A vessel

Vas′tus (L.) Enormous

Ven′tricle (L. *ventriculus*) A little cavity

Ve′stibule (L. *vestibulum*, a fore-court)

Vil′lus (L.) A tuft of hair; hence hair-like processes from the intestinal wall

Vis′cus (L.) Any internal organ

Vit′reous (L. *vitreus*) Glassy

Xi′phoid (G. *xiphos*, sword; *eidos*, form) Resembling a sword

Zȳgomăt′ic (G. *zugōma*, a bolt or yoke) Hence applied to the bar of bone which stretches back from the cheek to the ear

Books to read or consult

CLARK W.E. LE GROS (1971) *The Tissues of the Body*, 6th edition. Oxford: Clarendon Press.

Cunningham's Textbook of Anatomy, 11th edition (1972) Edited by G.J. ROMANES. London: Oxford University Press.

GARDNER E., GRAY, D.J. & O'RAHILLY, R. (1969) *Anatomy*, 3rd edition. Philadelphia: Saunders.

GRANT J.C.B. (1972) *An Atlas of Anatomy*, 6th edition. Baltimore: Williams & Wilkins.

Gray's Anatomy, 35th edition (1973) Edited by R. WARWICK and P.L. WILLIAMS. London: Longmans.

Hewer's Textbook of Histology for Medical Students, 9th edition (1969) Edited by S. BRADBURY. London: Heinemann.

LOCKHART R.D. (1962) *Living Anatomy*, 6th edition. London: Faber.

MOREHOUSE L.E. & MILLER, A.T. (1971) *Physiology of Exercise*, 6th edition. St. Louis: Mosby.

Samson Wright's Applied Physiology, 12th edition (1971) Edited by C.A. KEELE & E. NEIL. London: Oxford University Press.

SINCLAIR D.C. (1973) *Human Growth After Birth*, 2nd edition. London: Oxford University Press.

WOOD JONES, F. (1942) *Principles of Anatomy as Seen in the Hand*, 2nd edition. London: Baillière Tindall & Cox.

Index

In most cases where several references are given, the main reference is printed in bold type.